Unified Theory and Practice:

Polymer Adhesion, X-Ray Diffraction & X-Ray Florescence

by Frank H. Chung, PhD

The contents of this work, including, but not limited to, the accuracy of events, people, and places depicted; opinions expressed; permission to use previously published materials included; and any advice given or actions advocated are solely the responsibility of the author, who assumes all liability for said work and indemnifies the publisher against any claims stemming from publication of the work.

All Rights Reserved
Copyright © 2020 by Frank H. Chung, PhD

No part of this book may be reproduced or transmitted, downloaded, distributed, reverse engineered, or stored in or introduced into any information storage and retrieval system, in any form or by any means, including photocopying and recording, whether electronic or mechanical, now known or hereinafter invented without permission in writing from the publisher.

Dorrance Publishing Co
585 Alpha Drive
Suite 103
Pittsburgh, PA 15238
Visit our website at *www.dorrancebookstore.com*

ISBN: 978-1-4809-5756-5
eISBN: 978-1-4809-5779

FOREWORD

Two unified theories and their practices are presented in this volume: The first unifies and clarifies the confusing status of polymer adhesion; the second unifies and simplifies the tedious work of quantitative X-ray Diffraction (XRD) and X-ray Fluorescence (XRF). These theories are integrated and summarized in the first two chapters, their implications and subtle details can be found in the remaining chapters. Each chapter has been peer-reviewed and published in well-known international journals or books. A list of "Chapter Titles vs. Published Articles" is appended for cross reference.

1. Unified Theory and Guidelines of Polymer Adhesion

There are eight adhesion theories in the literature. Seven older theories explain adhesion strength in terms of words and figures. Only one theory gives mathematical treatment based on quantum mechanics. The Unified Adhesion Theory derives a mathematical equation linking bond length, bond energy, and bond force (adhesion strength). It consolidates the seven older theories into one coherent concept, which reveals that strong adhesion between materials is governed by two criteria: Intimate molecular contact of closer than 9 Å (necessary condition) and maximum attractive force with minimum potential energy (sufficient condition). Under these conditions, the two sets of molecules experience the strongest bonding force yet remain the most stable state. A set of guidelines is compiled for ways to promote adhesion strength. The Unified Adhesion Theory integrates all prior adhesion theories. it is the current and the only adhesion theory in mathematical form.

2. Unified Theory and Practice of Quantitative XRD & XRF
Due to the complex matrix effect, the relationships between X-ray intensity and concentration are variable curves depending on the composition of unknown mixtures. The current textbooks on XRD & XRF teach the Internal Standard and the Spiking techniques. Both techniques require the use of calibration lines from certified standards, which is tedious and time-consuming, especially for multiple component analyses of non-routine and totally unknown mixtures. The Unified Theory of Quantitative XRD & XRF eliminates the matrix effects, the calibration lines, and the certified standards. Based on new insights, it derives a mathematical formula to decode both XRD & XRF signals. A single XRD or XRF scan quantifies the chemical compounds or chemical elements in any mixture. It reduces some 80% of the lab work performed in current practice. It is the first significant enrichment of the techniques since the discovery of XRD & XRF over 100 years ago by the father and son team of Henry & Lawrence Bragg.

Frank H. Chung

CONTENTS

1. Unified Theory and Guidelines of Polymer Adhesion 1
2. Three Laws of Quantitative XRD & XRF Analyses 35
3. Unified and Simplified Practice of Quantitative XRD & XRF 61
4. Matrix Flushing Technique for Quantitative X-ray Diffraction of Mixtures 85
5. Adiabatic Principle of X-ray Diffraction of Mixtures 107
6. Simultaneous Determination of Reference Intensities 125
7. Matrix Flushing Techniques for Quantitative X-ray Fluorescence Analysis 133
8. Thin-Film Technique of Quantitative X-ray Fluorescence Analysis 157
9. Quantitative X-ray Diffraction of Powder Samples 171
10. X-ray Diffraction in U.S. Industry 185
11. X-ray Diffraction Techniques and Instrumentation 207
12. Progress and Potential of X-ray Diffraction 245
13. The Principle of Diffraction Analysis 287
14. The Practice of Diffraction Analysis 307
15. Polymers and Pigments in Paint Industry 359
16. Determination of Crystallinity of Polymers by XRD 389
17. Vacuum Sublimation and Crystallography of Quinacridones . . 405
18. Crystallography of Toluidine Red Pigment 421
19. Synthesis and Analysis of Quartz, Cristobalite, and Tridymite . . 425
20. Imaging and Analysis of Airborne Dust for Silica 437
21. Characterization of Airborne Particulates 447

Unified Theory and Practice:

Polymer Adhesion, X-Ray Diffraction & X-Ray Florescence

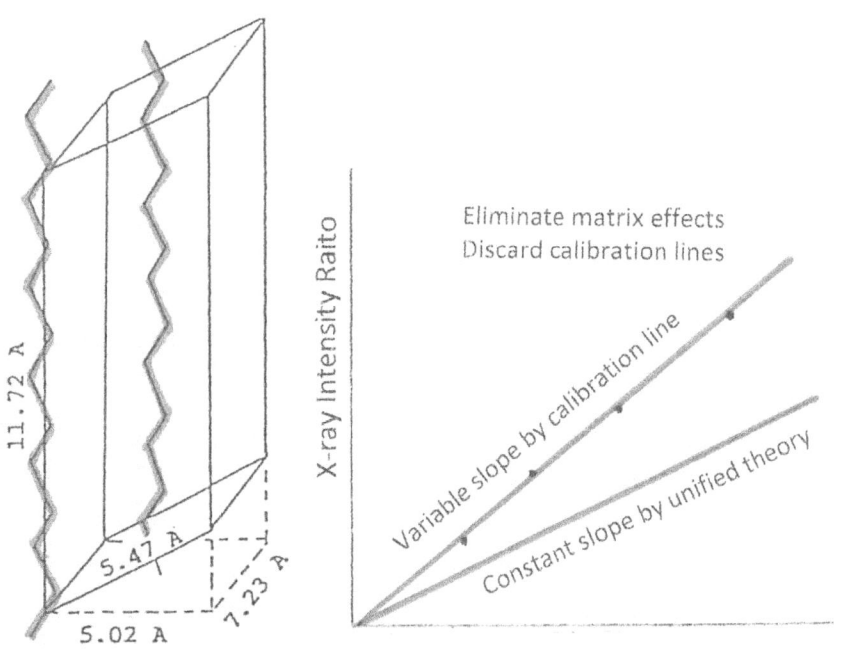

1.

Unified Theory and Guidelines of Polymer Adhesion

> **An educated mind is able to entertain a thought without accepting it**
>
> — Aristotle

1. ABSTRACT

A new approach was proposed to consolidate the many adhesion theories into one coherent concept. The maximum attractive force between two sets of molecules (adhesion strength) is derived from the Lennard-Jones potential function and calculated with measured bond length and bond energy. It leads to two criteria for strong adhesion: intimate molecular contact of closer than 9Å (necessary condition), and maximum attractive force with minimum potential energy (sufficient condition). The criteria conform to the key elements of most prior adhesion theories. Seven prior adhesion theories and their relevance to these two criteria were briefly reviewed and discussed. In order to draw up a set of guidelines on adhesion and supplement the missing pieces of information, 21 model polymers of varied functionality were synthesized to study (1) the effect of polar groups on adhesion and (2) the effect of polymer conformations on adhesion. The results indicate that polar groups are more effective in polymer backbone than in side chains for promoting adhesion. The presence of both hydrogen donors and acceptors in the same backbone maximizes adhesion. True (active) solvents enhance adhesion, co-solvents (latent solvents) boost adhesion by inducing favorable conformation of polymers in solution, but thinners (diluents) reduce adhesion. The set of guidelines covered the effects of functional groups, solvent blends, pigment loadings, adhesion promotion, and adhesion loss.

2. INTRODUCTION

A great deal of research has been done and many volumes have been written on adhesion.[1-8] However, the present status of adhesion theory is confusing. At least seven adhesion theories are currently in use. Each theory has its merits, but none is universally applicable. Three factors contributed to this situation: First, each theory deals with only part of the problem; second, similar ideas are presented in different languages depending on the expertise of the author; and, finally, each theory emphasizes its own insight and tends to expel others. With regard to recent advances in research tools for surface analysis, and demands for high performance coatings/adhesives in the automotive, aerospace, and micro-electronics industries,[7,8] probably the time is ripe to consider a consolidated theory of adhesion. Ideally, a consolidated theory of adhesion should be able to: (1) explain the origin of attractive forces, i.e., why things adhere, (2) suit all cases of adhesion such as polymer to metal/oxide (protective coatings, adhesives), polymer to polymer (basecoat/clearcoat, coating plastics), metal to metal (soldering), cement to steel (concrete), etc., (3) encompass all existing adhesion theories, (4) accommodate experimental facts, (5) reveal ways leading to strong adhesion, and (6) interpret the loci of adhesion failures. A consolidated theory of adhesion is proposed here to partially fill these demands. This is possible mainly due to the large number of prior studies. A matching set of guidelines on adhesion is compiled from prior work and supplemented by this study. Certainly, a lot more work is necessary to substantiate or modify the consolidated concept of adhesion.

3. UNIFIED ADHESION THEORY

The objectives of this study have been to clarify the status of adhesion theory and to shorten the gap between theory and practice by tying together all prior insights in a coherent concept, and by generating a set of guidelines on adhesion for field applications. The electronic attraction between molecules is the thread of logic used to approach these objectives. The details are discussed below.

3.1 Maximum Attractive Force (Theory)

Theoreticians attempt to solve the Schrodinger equation using perturbation methods, and then calculate the potential energy between molecules which predicts bond strength. But the calculations are extremely tedious even for small molecules and infeasible for polymers.[9]

To date, the bond lengths have been accurately measured by X-ray diffraction analysis and the bond energies by calorimetry. Instead of starting from quantum theory to predict bond strength, why not reverse the order by starting from known bond length and bond energy to compute the realistic bond strength, i.e., maximum attractive force? A scheme to pursue this new approach is presented here.

When two molecules approach each other, their interactions can be expressed by a potential energy function in terms of their distance of separation. Quite a few potential energy functions were adopted in quantum mechanics. The most commonly used one is the Lennard-Jones function [eq. (1)], which is valid for neutral molecules such as polymers, metals, and oxides[10,11]:

$$E = \frac{A}{X^{12}} - \frac{B}{X^6} \tag{1}$$

where E = potential energy, X = distance of separation between molecules, and A, B = constants.

The positive term is short range repulsive energy; the negative term is long range attractive energy. The repulsive inverse power may range from 9 to 15,[12] which slightly changes the steepness of the slope. The exponential repulsive functions (Buckingham and Morse) may represent the repulsive term better, but the improvement over the inverse 12th power function is less than 2%.[10] The attractive inverse power could be 1, 2, 3, or 7 in special cases such as Coulomb or retarded attractions.[12] The inverse sixth power attractive function is adequate for the discussion of attraction/bonding between neutral molecules.

According to eq. (1), the overall attractive force (F) between molecules is given by

$$F = -\frac{dE}{dX} = \frac{12A}{X^{13}} - \frac{6B}{X^7} \tag{2}$$

$$\frac{d^2E}{dX^2} = \frac{156A}{X^{14}} - \frac{42B}{X^8} \tag{3}$$

The minimum potential energy (E_0) occurs at X_0 when $dE/dX=F=0$, i.e., the attractive force equals the repulsive force, the net force is zero. The maximum attractive force (F_m) occurs at X_m when $dE^2/dX^2=F=0$. The zero potential energy, $E = 0$, occurs at X_e, which is the off limit of further penetration of electron clouds between molecules. We have

$$X_0 = (2A/B)^{1/6} = \text{bond length} \quad (4)$$

$$E_0 = -B^2/4A = \text{bond energy} \quad (5)$$

$$A = -E_0 X_0^{12} = \text{repulsive constant} \quad (6)$$

$$B = -2E_0 X_0^6 = \text{attractive constant} \quad (7)$$

$$X_m = 1.11 X_0 = \text{apex point} \quad (8)$$

$$F_m = -2.69 E_0/X_0$$
$$= \text{maximum attractive force} \quad (9)$$

$$X_e = 0.89 X_0 = \text{off limit} \quad (10)$$

$$E_e = 0 = \text{energy equilibrium} \quad (11)$$

$$E = E_0 \left[2\left(\frac{X_0}{X}\right)^6 - \left(\frac{X_0}{X}\right)^{12} \right] \quad (12)$$

Equation (8) indicates that the maximum attractive force and the minimum potential energy occur in close proximity (11% shift between X_0 and X_m). The fact that the maximum attractive force and the minimum potential energy do not fall on the same point sets off harmonic vibrations, which are the origins of infrared spectra. Equation (9) can be used to calculate the adhesion strength (F_m) from known bond length (X_0) and bond energy (E_0) The adhesion strength of common bonds are uniformly listed in Table i, which reveals that the O—H single bond is stronger than the C=C double bond.

From eq. (12), $E = 0.008E_0$ (less than 1% E_0) at $x = 2.5X_0$. For the faintest bonding force in Table I, the critical distance of separation between molecules $X_e = 2.5 \times 3.6 = 9.0$ Å. Beyond 9 Å, the electronic interactions between molecules decrease rapidly to nearly zero. Closer than 9 Å, the molecular contact begins, that is, the penetration of electron clouds, overlapping of molecular orbitals, or mathematically substantial exchange integrals.[13,14]

Table I Maximum Attractive Force of Chemical Bonds

Bond	Bond Length X_0 (Å)	Bond Energy E_0 (kcal/mol)	Bond Force[a] F_m (dyn/bond × 10^{-4})
C≡N	1.16	213	34.3
C≡C	1.20	200	31.2
C=O	1.23	179	27.2
O—H	0.96	111	21.6
C=C	1.34	146	20.4
C—H	1.07	99	17.3
N—H	1.01	93	17.2
C—F	1.36	116	16.0
O—Cr	1.57	102	12.2
C—O	1.43	86	11.2
O—Si	1.50	88	11.0
C—C	1.54	83	10.1
C—N	1.47	73	9.28
C—Cl	1.77	81	8.56
N—O	1.24	53	8.00
C—S	1.81	65	6.72
C—Si	1.94	69	6.65
N—N	1.12	39	6.51
C—P	1.87	63	6.30
S—S	2.04	54	4.95
O—O	1.48	35	4.42
Hydrogen bond	2.70	6	0.415
Van der Waals	3.60	2	0.010

[a] The bond force was calculated by eq. (9). It indicates the overall bond strength. Note that the O—H single bond is stronger than the C=C double bond.

The above deduction can be summarized in the following statement: Strong adhesion between materials is governed by two criteria: Intimate molecular contact of closer than 9 Å (necessary condition), and maximum attractive force with minimum potential energy (sufficient condition). Under these conditions the two sets of molecules experience the strongest bonding force (maximum attraction) yet remain the most stable state (lowest energy). This statement conforms to the key elements of most prior theories, vide infra, and hence may constitute a consolidated theory of adhesion, a dichotomy theory. The key elements of the seven prior theories and their relevance to the criteria are briefly summarized below. More details are discussed in the section on Merging of Prior Theories.

Intimate Molecular Contact
- Adsorption theory: Surface energy and rule of spreading/wetting.
- Diffusion theory: Physical contact, compatibility, and permeation.
- Interlocking theory: Flow, wicking, interpenetration, and cohesion.
- Weak boundary theory: Lack of intimate molecule contact.

Maximum Attractive Force
- Chemical bonding theory: Origin of intermolecular attractive forces.
- Acid-base theory: Hydrogen bonding force is the key of adhesion.

- Electrostatic theory: Electronic attractive forces.
- Weak boundary theory: Defects and lack of covalent/hydrogen bondings.

The reverse treatment offers a few advantages over the traditional treatment: (1) The nature of molecular contact is quantitatively defined; (2) the bond strength is defined with one parameter (maximum attractive force, such as lb/in.^{2}) instead of two parameters (bond length and bond energy); (3) the mathematical barrier of quantum mechanics, such as Hamiltonian operators, differential equations, and matrices, is bypassed without sacrificing its concept; (4) the criteria combine the central thoughts of most prior theories; and (5) the consolidated concept can be used to interpret/rationalize the ensuing guidelines on adhesion.

3.2 Adhesion of Polymers (Practice)

Attractive forces exist between any molecules when their distance of separation is closer than 9 Å. The attractive forces are usually insufficient for strong bonding when the molecules are small, but the attractive forces become spectacular when the molecules are large. Consequently, all binders/adhesives are made of macromolecules. The adhesion strength of polymers to wood, metal, ceramics, and plastics can be assessed by the afore-derived criteria:

Intimate Molecular Contact

Heat, pressure, solvents, surfactants and surface treatments have been used to promote intimate molecular contact. Solvents are routinely used in organic coatings/adhesives to carry polymers and spread on substrates. This practice is usually without problem for high energy surfaces (metals/oxides, 500-10,000 ergs/cm²), but not so easy for low energy surfaces (plastics/paraffins, less than 200 ergs/cm²), because, in the latter case, an increase in total surface energy is involved.[15, 16] Of course, all systems favor the lowest energy state available.

For low energy surfaces, the rule of spreading has been prescribed by the adsorption theory that spreading/wetting (intimate molecular contact) will occur when the critical surface tension of the solid is higher than the surface tension of the liquid. The critical surface tension of solid surfaces can be measured by the Zisman plot.[17] The measured critical surface tension of all plastics (18-50 dyn/cm) are well below the surface tension of water (72.8 dyn/cm); therefore, all plastic surfaces are hydrophobic.[18]

Maximum Attractive Force

The adhesion of polymers involves various types of chemical bondings. Typical attractive forces often encountered in organic coatings/adhesives are illustrated in Figure 1, where:

- **The covalent bond O—Cr** between polyester and chromate pretreated metals:

    $$X_0 = 1.57 \text{ Å} \quad (\text{known bond length}) \qquad (4)$$

    $$E_0 = 102 \text{ kcal/mol} \quad (\text{known bond energy}) \qquad (5)$$

    $$A = 1.59 \times 10^{-105} \text{ erg cm}^{12} \qquad (6)$$

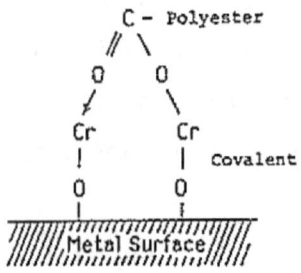
Adhesion of polyester to chromate treated aluminum.

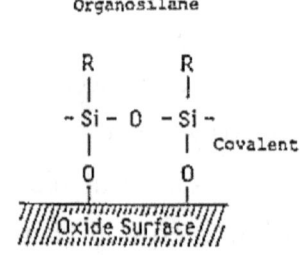
Adhesion of organosilane to glass, minerals, ceramics, cement or semiconductors.

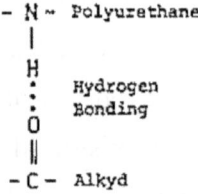
Adhesion of polyurethane to alkyd basecoat.

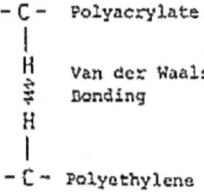
Adhesion of polyacrylate to polyethylene plastic.

Figure 1 Chemical bondings frequently in coatings

$$B = 2.12 \times 10^{-58} \text{ erg cm}^6 \tag{7}$$

$$X_m = 1.74 \text{ Å} \tag{8}$$

$$F_m = -2.69, E_0/X_0$$

$$= 12.2 \times 10^{-4} \text{ dyn/bond} \tag{9}$$

$$X_e = 1.40 \text{ Å} \tag{10}$$

Note that 1 kcal/mol/Å = 6.94×10^{-6} dyn/bond.

- **The covalent bond O—Si** between organosilane and fiberglass:

$X_0 = 1.50$ Å

$E_0 = 88$ kcal/mol

$F_m = 11.0 \times 10^{-4}$ dyn/bond

- **The hydrogen bond** between polyurethane clearcoat and alkyd basecoat:

    $X_0 = 2.7$ Å

    $E_0 = 6$ kcal/mol

    $F_m = 4.15 \times 10^{-5}$ dyn/bond

- **The van der Waals bond** between polyacrylate and polyethylene

    $X_0 = 3.6$ Å

    $E_0 = 2$ kcal/mol

    $F_m = 1.04 \times 10^{-6}$ dyn/bond

Covalent bonds are formed from chemical reactions between polymer and substrate, which are rather rare in conventional practice. Van der Waals attractions are always present. Consequently, hydrogen bondings are the only force technologists can put to work for the required adhesion.

The best achievable adhesion strength of a polymer is the cohesion strength of the polymer itself. Cohesion is a special case of adhesion where the two sets of molecules are identical. Equation (9) can also be used to estimate the cohesion strength of polymers. The cross-section area taken by a single polymer chain ranges 15-30 Å (Fig. 2), based on X-ray diffraction data.[19,20] Referring to the bond forces in Table I, the tensile strength of polyethylene adipate fiber ranges between 7.9×10^6 psi (based on C—C bonds) and 7.9×10^3 psi (based on end-to-end van der Waals bonds). The measured tensile strength is in-between (1.3×10^4 psi) due to numerous flaws and hydrogen bondings. Polymer fibers/films are highly oriented; hence their tensile/shear strengths are remarkably anisotropic.

4. EXPERIMENTAL

In order to draw up a set of guidelines on adhesion and supplement the missing pieces of information, 21 model polymers of varied functionality were synthesized to study (1) the effect of polar groups on adhesion and (2) the effect of polymer conformations on adhesion. The model polymers are acrylics, epoxies, or urethanes which have been widely used in organic coatings because of their low permeability to air/water hence impeding corrosions. Nine of them, six urethanes and three epoxies, were insoluble; therefore, only 12 polymers were included in the experiments.

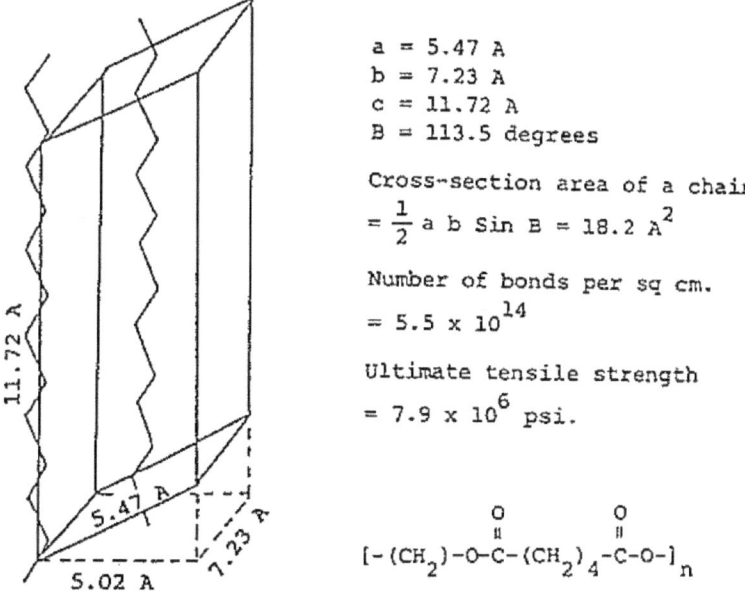

Figure 2   polyethelene adipate chains in a monoclinic unit cell determined by x-ray diffraction

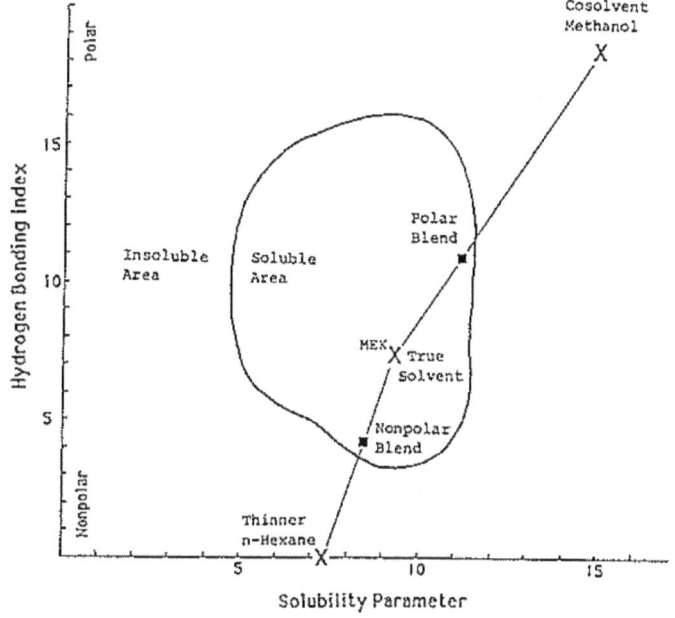

Figure 3   Solubiity map and solvent blends

4.1 Techniques and Procedures
Polymer Characterization

The glass transition (T_g) points were measured with the DuPont thermal analyzer 1090 in DSC mode. The molecular weights were determined by use of the Perkin-Elmer LC Series-10 in GPC mode. Polymer solubility was checked by placing 0.5_g solid polymer into 9.5_g solvent, standing overnight with occasional shaking (ASTM-D 3132). Eleven solvents of varied solubility parameters were chosen to trace the solubility maps.

The polymer conformation in solution may affect its adhesion performance. In polar solvents, the polar groups of the polymer are protruding, but, in nonpolar solvents, the polar groups are buried. The effects of polymer conformation on adhesion were studied by using three solvent systems for each polymer as shown in Figure 3. The polar/nonpolar blends were made in such a way that an excess of a few drops of the nonsolvent would change the polymer solution from clear to cloudy.

Adhesion Performance

The contact angles of sessile drops were measured with a Rame-Hart telegoniometer. Films of about 1-1.5 mil dry thickness were made from polymer solutions of 20% by weight. No additives nor cross-linkers were added. The films dry essentially by solvent evaporation: brief flash off then bake (180°F/30 min). The polymer conformation, i.e., extended/coiled chains and exposed/buried polar groups, was studied by measuring the hydrodynamic volume of the dissolved molecules. An Ubbelohde capillary viscometer in a 25°C thermostatted bath was set up to determine the intrinsic viscosity from which the hydrodynamic volume was calculated according to the Flory theory.[21] The adhesion strength was measured by the Pull-off Test (ISO-4624 and ASTM-D4541, Elcometer or Instron) and occasionally rechecked by the Cross-hatch Tape Test (ASTM-D3359, simple Gardner Cutter).

4.2 Results and Comments
Polymer Characterization and Spreading

The polymer compositions, molecular weights, and glass transition points are listed in Table II. Since the drying temperature was above the T_g's, the maximum intrinsic mobility should be developed. All molecular weights are in regular range, under 300,000.[22] The molecular weights of soluble poly-urethanes are on the low side. Polymers for conventional coatings must be soluble. The solubility of polymers in various solvents are presented in Table III. The data under CS_2 and xylene indicate that the solubility parameter alone cannot predict compatibility. The hydrogen bond index and dipole moment are also important parameters. Figure 3 shows that nonsolvents hexane and methanol were used to change the conformation for hydrodynamic volume measurements.

The surface tension of common solvents for coatings are below 30 dyn/cm, such as hexane = 18.4, methanol = 22.6, MEK = 24.6, and xylene = 28.9 dyn/cm. The critical surface tension of polyethylene is 31 dyn/cm (a reference point). That of fluorocarbons and silicones are lower. The critical surface tensions of all other commercial plastics such as SMC, nylon, LOMOD, etc., are higher due to the presence of N, O, Cl, or rings in the molecules.[17] Therefore, the rule of spreading/wetting was satisfied, and no wetting problem was encountered.

Effect of Polar Groups on Adhesion

According to the maximum attractive forces in Table I, the ratios of bond forces between covalent, hydrogen, and van der Waals bondings are 1000/40/1. The 1000/1 ratio would be the best, but covalent bondings are accessible only under unique conditions. Although the 40/1 ratio is the second best, hydrogen bondings are the most practical and should be the workhorse for adhesion. Numerous adhesion tests were run. The key data reflecting the effect of polar groups on adhesion strength are condensed in Table IV. The two-component (2K) systems are commercial products. All others are experimental systems without optimization.

Table II Characteristics of Polymers

Polymer Composition	T_g (°C)		Mol Wt × 10^{-3}		
	Onset	Inflection	M_n	M_w	P_d
MMA/BA	14	35	38.8	84.5	2.2
MMA/BA/DMAEMA	16	41	16.2	45.0	2.8
MMA/BA/glycidyl methacrylate	19	38	66.8	279.0	4.2
MMA/BA/hydroxyethyl acrylate	20	45	37.8	122.9	3.2
MMA/BA/acrylic acid	22	43	54.8	193.2	3.5
MMA/BA/styrene	34	55	36.0	63.0	1.8
MMA/BA/NMEMA	19	45	35.1	109.2	3.1
MMA/BA/vinyl pyrrolidone	45	69	45.6	117.0	2.5
MMA/BA/DMAPMA	29	62	17.2	51.0	3.0
Desmodur/bisphenol-A	41	84	4.4	9.1	2.1
Desmodur/1,6-hexanediol	49	86	14.3	31.2	2.2
Desmodur/CHDM	29	72	14.1	24.3	1.7

[a] MMA = methyl methacrylate, BA = butyl acrylate, DMAEMA = dimethylaminoethyl methacrylate, NMEMA = 2-N-morpholinoethyl methacrylate, DMAPMA = dimethylaminopropyl methacrylamide, CHDM = 1,4-cyclohexanedimethanol.

Table III Solubility of Polymers

Polymers	Hex	CyH	MAK	BAc	Xyl	Chf	MEK	CS$_2$	Acet	DMSO	MeOH
Solubility parameter	7.3	8.2	8.5	8.5	8.8	9.3	9.3	10.0	10.0	12.9	14.5
Hydrogen bond index	0	0	7.7	8.8	4.5	1.5	7.7	0	9.7	7.7	18.7
Dipole moment	0	0	2.7	1.9	0.4	1.2	2.7	0	2.9	4.0	1.7
MMA/BA plus											
Straight	X	X	O	O	O	O	O	X	O	O	X
DMAEMA	X	X	O	O	O	O	O	X	O	—	O
GMA	X	X	O	O	O	O	O	X	O	O	X
HEA	X	X	O	O	X	O	O	X	O	O	X
Acrylic acid	X	X	O	O	X	O	O	X	O	O	X
Styrene	X	X	O	O	O	O	O	O	O	X	X
NMEMA	X	X	O	O	O	O	O	X	O	X	X
Vinyl Pyrr	X	X	—	—	O	O	—	X	—	O	X
DMAPMA	X	X	—	—	O	O	—	X	—	O	O
Desmodur plus											
Bisphenol-A	X	X	X	X	X	O	—	X	—	O	X
Hexanediol	X	X	X	X	X	O	X	X	X	—	X
CHDM	X	X	X	X	X	O	X	X	X	X	X

X = insoluble, O = soluble, — = cloudy. All polymer solutions are 5% by weight. Hex = n-hexane, BAc = butyl acetate, MEK = methylethyl ketone, CyH = cyclohexane, Xyl = xylene, Acet = acetone, MAK = methyl amyl ketone, Chf = chloroform, DMSO = dimethyl sulfoxide.

Table IV Effect of Polar Groups on Adhesion

Polymer System	SMC Plastic (psi)	LOMOD Plastic (psi)	Bonderite Steel[a] (psi)
Different backbones			
Urethane, 2K	690	600	> 700
Epoxy, 2K	420	380	> 420
Urethane	300	300	> 540
Acrylic	350	260	> 350
Different side chains			
MMA/BA	120	110	> 350
MMA/BA + NR$_2$	350	260	> 350
MMA/BA + OH	250	150	> 350
MMA/BA + COOH	140	160	> 350
MMA/BA + C$_6$H$_5$	80	170	> 350
MMA/BA + epoxy	80	150	> 350
MMA/BA + \>N—	50	50	> 350

[a] Ruptures took place at film/glue or glue/dolly, but not film/steel interfaces, due to the strong adhesion between polymer and Bonderite.

The CONH and epoxy groups in the polymer backbone gave higher adhesion than the pendant NR_2 ($CONH_2$) and epoxy groups. Different pendant groups on the same backbone (MMA/BA) imposed only secondary effect on adhesion. The data seem to indicate that polar groups in backbone are more effective than those in side chains. Not all nitrogen-containing groups promoted adhesion, depending on their ability to form hydrogen bonds with the substrate. Note that the triple bonds — N < in Table IV has no hydrogen for bonding. Crosslinking in two-component (2K) systems exhibited fast dry, firm adhesion, and superb durability due to their excellent molecular contact and possible covalent bonding through residual reactive groups.

Commercial plastics contain ester/carboxyl, ether/ hydroxyl, epoxy/phenyl, or amine/amide/ imide, groups, which are potential bonding sites. In order to effect hydrogen bondings, the presence of both hydrogen donors (— OH, >NH, >— CH, — SH) and acceptors (>C=O, — NR_3, — OH, — NO_2, — CN, —CCl_3, > C = S) in the same backbone may maximize adhesion. Inter/intra bondings could be controlled by properly spacing these groups.

Effect of Polymer Conformations on Adhesion

Solvent blends are routinely used in coatings. Usually only one component in the blend is true (active) solvent. Other components under the names of co-solvent (latent solvent) or thinner (diluent) are nonsolvent. The cosolvent/thinner can be either polar like alcohols or nonpolar like hydrocarbons. In either case, the polymer molecules in solution will be more or less coiled.

When both the solvent and polymer have a similar solubility parameter and hydrogen bond index, the solvent is a true solvent for the polymer. All data in Table V indicate that true solvents enhance adhesion because of larger hydrodynamic volume and better mobility.[23] Alcohol cosolvent, if in the right amount, induces synergetic effect and boosts adhesion because of protruding polar groups. Hydrocarbon thinners tend to bury the polar groups and tighten up the polymer coils, and hence degrade adhesion. The maximum amount of nonsolvent was used in this study as shown in Figure 3 to reveal its effect on conformation/adhesion.

Table V Effect of Polymer Comformations on Adhesion

Polymer	M_n ($\times 10^{-3}$)	Solvent[a] Composition (wt %)	Intrinsic Viscosity (dL/g)	Hydrodynamic Volume ($\times 10^{-18}$ mL)	Adhesion[b] Strength (psi)
MMA/BA/HEA	37.8	MEK	0.435	5.87	380
		80 MEK/20 hexane	0.420	5.67	300
		20 MEK/80 MeOH	0.273	3.69	380
MMA/BA/DMAEMA	16.2	MEK	0.310	1.79	350
		50 MEK/50 hexane	0.225	1.30	50
		MeOH	0.188	1.09	180
MMA/BA/Styrene	36.0	MEK	0.260	3.34	300
		50 MEK/50 hexane	0.237	3.05	280
		60 MEK/40 MeOH	0.224	2.88	250
MMA/BA/DMAPMA	17.2	Toluene	0.220	1.35	120
		90 Tol/10 heptane	0.248	1.52	100
		MeOH	0.276	1.70	50
MMA/BA/NMEMA	35.1	MEK	0.330	4.14	100
		80 MEK/20 hexane	0.301	3.77	10
		30 MEK/70 MeOH	0.215	2.69	20
MMA/BA/vinyl pyrr.	45.6	Toluene	0.390	6.35	50
		90 Tol/10 heptane	0.394	6.42	10
		30 Tol/70 butanol	0.500	8.14	30
Desmodur/bisphenol-A	4.4	Chloroform	0.245	0.39	300
		90 Chf/10 hexane	0.110	0.17	10
		60 Chf/40 MeOH	0.174	0.27	40
Desmodur/1,6-hexanediol	14.3	Chloroform	0.480	2.45	220
		90 Chf/10 hexane	0.445	2.27	60
		60 Chf/40 MeOH	0.448	2.29	100
Desmodur/CHDM	14.1	Chloroform	0.404	2.03	220
		90 Chf/10 hexane	0.430	2.17	50
		60 Chf/40 MeOH	0.358	1.80	100

[a] Three solvent systems (true, polar, and nonpolar blends) were used for each polymer (Fig. 3). All films for adhesion tests were made from polymer solutions with 20% solids by weight.
[b] Adhesion data from pull-off tests of clear film on plastics. 100 psi or lower indicates poor adhesion.

Adhesion Tests

The Pull-off Test is an international standard (ISO) method. The Cross-hatch Tape Test is practiced widely in the coatings industry. Both were used in the present study. No correlation was found between the results of these tests. Superficially, the Pull-off Test measures tensile strength, and the Cross-hatch Tape Test estimates shear strength. But in reality, the induced stresses in both tests are likely non-uniform and superimposed. Most observed loci of failures were not interfacial except cases of poor adhesion (less than 100 psi). When the Pull-off Tests ran above 300 psi, cohesive failure of SMC often took place due to the weak bonding between the SMC (polyester) and its imbedded fiberglass.

Adhesion of polymer to pretreated steel is much stronger than its adhesion to bare aluminum (the dolly in the Pull-off Test). Therefore, the rupture in the Pull-off Test takes place most likely in the film/glue, glue/aluminum interfaces, or mixed mode, but hardly in the film/steel interface as shown in Table IV.

The adhesion of urethane to both steel and aluminum was excellent. In one incidence, when the Bonderite steel and the aluminum dolly were joined with a black urethane paint, cohesive failure of urethane itself took place. Since both the separated surfaces were black, the evidence was clear-cut.

Solids never rupture as a rigid body. Viscoelastic effects, surface structures, and fracture mechanics are mixed in an intricate manner, such that no ideal adhesion tests exist. Consequently, the value of adhesion strength obtained by different test methods may not be directly comparable.[24,25] Internal consistency of test data is what we can expect.

5. APPLICATIONS

The consolidated theory serves two purposes: to tie prior insights together such that the adhesion theories become unified and to interpret the guidelines on adhesion which were scattered and never put together due to the lack of a coherent/unified adhesion theory.

5.1 Merging of Prior Theories

The prior theories of adhesion overlap to certain degrees. None of them are really incompatible with others. The seven prior theories are briefly reviewed below with comments on how they mesh with the consolidated theory.[26-28]

Adsorption Theory[29-31]

The adsorption theory starts from surface tension equilibrium and arrives at the rule of spreading/ adsorption of liquids. It predicts adhesion strength from thermodynamic work of adhesion, and stresses that the intrinsic adhesion arising from van der Waals forces alone may lead to strong adhesion. Its extension beyond liquid/solid adsorption to solid/ solid adhesion is less convincing. Adhesion involves the boundary region of finite thickness. Obviously the boundary is not really a two-dimensional surface. The rule of spreading/ wetting from the adsorption theory is adopted in this study for assessing the necessary condition of intimate molecular contact.

Chemical Bonding Theory[22,32]

The molecular orbital theory considers the origin of attractive forces between molecules. Adhesion involves all types of chemical bondings: covalent, hydrogen, van der Waals, metallic, and ionic. Only the first three types are operative in organic coatings. Soldering involves metallic bondings; ion implantation/plating in the electronics industry may involve ionic bondings. The chemical bonding theory does not particularly recognize the prerequisite of intimate molecular contact for strong adhesion. It uses two parameters, bond energy and bond length, to characterize bond strength, while only one parameter, bond force (F_m), is necessary to rank bond strength. The reverse treatment of intermolecular attractions avoided the mathematical barrier of quantum mechanics and simplified the chemical bonding theory.

Electrostatic Theory[33]

The electrostatic theory claims that the attractive forces are the electrostatic effects at the interface. It postulates that all adhesion phenomena are charge transfer across the interface giving rise to electric double layers. The evidence was the well-known darkroom demonstration of electrical discharges when adhesive tapes are stripped rapidly from glass. The condenser discharge energy was correlated with the measured work of adhesion. It explains the adhesion of fine particles to surfaces such as Xerography.[34] Further development led to the DLVO theory popular in colloid science.[12] The DLVO theory contains too many parameters which are difficult to measure; hence it is not practical for industrial applications. Furthermore, the DLVO theory draws a lot of information from the chemical bonding theory which is stressed and simplified here.

Acid-Base Theory[35,36]

The acid-base theory applies the Bronsted acid-base concept to predict the relative magnitude of hydrogen bonding between polymers and oxide surfaces in the presence of moisture. The acid-base strength of an organic compound is judged by its pK value. The acid-base strength of an oxide surface is measured by its isoelectric point from the zeta potential experiment using aqueous suspension of powdered oxides. A low isoelectric point such as SiO_2 = 2 indicates acidic surface where amino materials should adhere well. A high isoelectric point such as MgO = 12 indicates a basic surface where carboxylic materials should adhere well. In general, optimum adhesion should be obtained when there is a substantial difference between pK and isoelectric point. The "acid-base force" is almost a synonym of the hydrogen bonding force, which is one of the inter-molecular forces emphasized in the consolidated theory.

Mechanical Interlock Theory[37-39]

The lock-key idea explains intuitively the adhesion of polymers to porous substrates (wood, paper, etc.). It is also a major factor for the strong adhesion of organic coatings to pretreated metals, and metal platings to pretreated plastics (metallization). Mechanical interlocking reinforces adhesion with cohesion; hence it is resistant to hydrolytic and thermal degradation. Penetration of polymer into pores forming interlocks is the macroscopic view. Microscopically, the adhesion between polymer and the wall of the pores still involves intermolecular forces. The discussion of intermolecular attractive forces is valid for both adhesion and cohesion.

Diffusion Theory[40-42]

The diffusion theory asserts that adhesion is due to intermolecular diffusion and entanglements across the interface. It is particularly useful for under-standing the adhesion of polymer to polymer, especially when both polymers are thermodynamically compatible and above their glass transition points. In concludes that dissimilar polymers are less adherent than similar polymers. Since diffusion is a dynamic process, it may take some time to develop the expected adhesion strength due to diffusion kinetics. Diffusion theory overlaps the interlocking theory. Mechanisms of diffusing, wicking, anchoring, and adhering certainly involve intermolecular attractive forces.

Weak Boundary Theory[43-45]

It represents the mechanical engineers' view of adhesion. Rupture always takes place at the weakest link. The weak spots may be inherent flaws such as bubbles, voids, crevices, or microcracks in the interface region. The weak spots may also be created during service due to stress/corrosion triggered by the permeated gas/water/ions. It is useful to interpret the locus of adhesion failure, and explain the big difference between ultimate and practical adhesion strength. However, it does not consider why things adhere. The weak spots are defects where neither molecular contact nor chemical bonding exists. The weak boundary theory is a theory of de-bonding rather than bonding or adhesion.

5.2 Guidelines of Adhesion

A set of practical guidelines on adhesion is condensed from the adhesion theory, this study, and prior work. These generalized conclusions may help the practitioners to get a perspective view without wading through voluminous literature. The complex behaviors of polymers certainly defy simple rules. Only the general trends were sought.

Effect of Functional Groups on Adhesion

The adhesion strength between coatings and substrates comes mainly from van der Waals forces, but substantially reinforced through hydrogen bondings and covalent bondings. Almost all polymers for coatings contain polar groups. As little as 0.1-1.0 mol % of functional groups can drastically increase the adhesion strength.[4] Although the effectiveness of functional groups is quite specific with regard to the surface of the substrate, experimental data of this study appear to indicate that polar groups are more effective in the backbone than in side chains. Different pendant groups on the same backbone impart lesser effects. Hydrogen donors and acceptors in the same backbone may maximize adhesion. Popular functional groups and the polarity of common materials arranged approximately in order of decreasing tendency for hydrogen bonding are compiled in Table VI. Note that poly (vinyl chloride) is nonpolar, poly (vinyl acetate), medium polar, but poly (vinyl alcohol), highly polar. The mobility and spacing of the polar groups are crucial for their ability to reach matching sites. Large molecules have higher cohesive strength but lower mobility. Except 2K systems, a polymer molecule of M_w, = 100,000–300,000, with a small number of long branches, is probably a starting point before optimization.' An excessive number of short branches adversely affect both spreading and adhesion.[46]

Table VI   Polar Groups and Polarity of Materials

- Polar groups in backbone

$$-\overset{O}{\underset{H}{\overset{\|}{C}}}-\overset{}{\underset{}{N}}-, \quad -\overset{}{\underset{CN}{CH}}-\overset{O}{\overset{\|}{C}}-O-, \quad -NH-, \quad -\overset{O}{\overset{\triangle}{CH-CH}}-$$

Amide   Cyanoacrylate   Amine   Epoxide

$$-\overset{O}{\overset{\|}{C}}-O-, \quad -O-, \quad -S-S-, \quad -Ring-$$

Carboxyl   Ether   Disulfide   5/6 members

- Polar groups in branches/ends

—NCO,   —CONH$_2$,   —NH$_2$,   —OH,
Isocyanate   Amide   Amine   Hydroxyl

$-\overset{O}{\overset{\triangle}{CH-CH_2}}$,   —CN,   —N(CH$_3$)$_2$,   —COOH,
Epoxide   Nitrile   t-Amine   Acid

—COOR,   —NO$_2$,   —Si(OCH$_3$)$_3$,   —Cl
Ester   Nitro   Silane   Chloro

- Polarity of matierals

Highly Polar: poly(vinyl alcohol), polyesther, polyurethane, polyurea, polyaminde, polyester, melamine, cellulosic, protein, startch, fabric, and sodiumm silicate

Medium Polar: acrylics, poly (vinyl acetate), polyvinylbutyral, epoxy, acrylonitrile, polysulfide, silicone, and metal oxides

Nonpolar: fluorocarbons, hydrocarbons, chlorocarbons, natural rubbers, and metals

Effect of Solvents on Adhesion

The conformation of the polymer molecules in solution depends upon the compatibility between polymer and solvent. Extended molecules have better mobility to reach bonding sites on the substrate. It also favors diffusion and entanglements. Experimental data of this study clearly indicate that true (active) solvents always promote adhesion. Alcohol cosolvents (latent solvents) help exposing polar groups by changing the polymer conformation, thereupon realizing a synergetic effect and boosting adhesion. Hydrocarbon thinners (diluents) tighten up the polymer coil and bury its polar groups, and thus reduce adhesion. Note that the cosolvents and thinners are usually nonsolvent.

Effect of Pigments on Adhesion

As a rule of thumb, pigments/fillers in polymer solution/emulsion increase its viscosity when wet, increase its cohesive strength when dry, and increase its adhesion when below CPVC (critical pigment volume concentration). A high PVC (less amount of polymer) gives high adhesion; a low PVC (higher amount of polymer) gives high cohesive strength.[2,5] The types of pigments make some difference but do not override the above trend. The acid-base theory provides a convenient tool for matching polymers to pigments for good adhesion.

Promotion of Molecular Contact

Heat, pressure, solvent, surfactant, flame, plasma, corona discharge, radiation energy, catalyst, and chemical treatments are among the industrial techniques to promote adhesion.[1,4,8] The first four improve molecular contact (necessary condition) between the two sets of molecules. The last six create new bonding sites (sufficient condition) on the substrate. Heating polymer above its T_g increases its adhesion. Pressure speeds up flow and eliminates air pockets. True solvents always enhance adhesion. Surfactants may reduce water resistance, and hence should be used sparingly.

DeBruyne's rule of adhesion[47] states that strong joints can never be made to polar adherents with nonpolar adhesives, or to nonpolar adherents with polar adhesives. This is usually true because of the hydrophobic nature of nonpolar adhesives toward moisture-covered polar adherents, and the high surface tension of polar adhesives toward the low critical surface tension of the nonpolar adherents. Intimate molecular contact is hampered in both cases.

Promotion of Bonding Forces

Among the three types of bonding forces, van der Waals forces are always present and contribute the major attractive force in coatings/adhesives. Hydrogen and covalent bondings are far superior in terms of durability, especially in the presence of water. Even a small number of them can substantially reinforce adhesion strength. For coatings applications, 300 psi is good adhesion, but for construction adhesives 500 psi is about the lower limit. Two-part crosslinking systems promise fast-dry, firm adhesion, and superb durability due to the presence of residual reactive groups for covalent bondings.

For nonpolar substrates such as hydrocarbons, fluorocarbons, and chlorocarbons, surface treatments[48] are necessary to impart polar groups which raise their surface tension and create bonding sites. With good wetting, rough surface shows better adhesion because it has interlocking cavities, large contact area, and redistributed (randomized, averaged, hence not maximum) stress.

Causes of Adhesion Loss

Some coatings/joints fail too soon. The likely causes of adhesion failures are: inadequate surface preparation (contaminated, or nonpolar surfaces), lack of intimate molecular contact (poor spreading/wetting), structural defects (air bubbles, voids, crevices, and other flaws), rigid molecular structure (not enough mobility for alignment to bonding sites), inefficient distribution of polar groups (on side chain or uneven spacing), internal stress/strain (too thick, high shrinkage, odd thermal expansion), corrosion in harsh environments (heat, cold, water, salt, radiations, and fumes), etc. These causes should be kept in mind when formulating coatings.

Adhesion in Harsh Environments

Coatings/adhesives for harsh environments must be specially designed to achieve adhesion. According to the current research trend,[7] polymers of fused/ joined/connected rings such as polyimides, polybenzimidazoles, and polyquinoxalines are suitable for high temperature (up to 1000°F) applications. Cyanosilicones and fluoroelastomers can stand low temperatures (down to -100°F). Polyphenyl-*as*–triazines, fluorine-modified polyimides, and chelating agents have good water resistance. Alodining and anodizing pretreatments can endure salt water.

6. CONCLUSIONS

The consolidated theory of adhesion integrates seven diverse theories into a coherent concept. Thermodynamically, spreading of a liquid on a solid will occur if the surface (free) energy of the system is thereby reduced. To ensure intimate molecular contact, the rule of spreading/wetting is adopted from the *adsorption* theory. The driving force of adhesion is the electronic interactions between molecules. When the maximum attractive force is near the minimum potential energy, *chemical bondings* are established. Covalent bonds need special design. Van der Waals bonds are ubiquitous. Only hydrogen bonds are practical for adhesion promotion.

Various chemical elements or groups have different electron drawing power, i.e. electronegativity, which is the source of attractive forces for adhesion. The *electrostatic* force, the acid-base force, and the intermolecular force are similar concepts in different terms. The best achievable adhesion of the polymer is equal to the cohesion strength of the polymer it-self. *Diffusion*, entanglement, and *interlocking* of polymers across the boundary reinforce the adhesion strength up to their cohesion strength. Adhesion/cohesion are intermolecular hetero/homo attractions, respectively.

Adhesion rupture always takes place at the weakest link. The formation of a *weak boundary* is due to the following mishaps: No molecular contact (crevices, flaws), inadequate molecular contact (poor spreading/wetting), or weak intermolecular attraction (lack of covalent/hydrogen bondings). These conditions may exist originally or develop during service.

The consolidated theory capitalizes on the insights of prior theories. The set of guidelines on adhesion attempts to put the practical aspects of polymer adhesion in a nutshell. Adhesion is a complex and evasive field beyond the reach of any single study. The model polymers and the experiments in this study have only a very limited scope. Further work is in progress to resolve some of the uncertainties. Future advances could make the adhesion theory more quantitative and closer to engineering.

7. REFERENCES

Alexander, L. E. *X-Ray Diffraction Methods in Polymer Science*. Wiley-Interscience, New York, 1969.

Azaroff, L. V. *Introduction to Solids*. McGraw-Hill, New York, 1960, pp. 214, 275.

Benson, S. W. *The Foundations of Chemical Kinetics*. McGraw-Hill, New York, 1960, p. 213.

Bikerman, J. J. *J. Colloid Sci.*, 2, p. 163 (1947).

Bikerman, J. J. *Recent Advances in Adhesion*. L. H. Lee, Ed. Gordon & Breach, New York, 1973, p. 351.

Bikerman, J. J. *The Science of Adhesive Joints*, 2nd ed., Academic, New York, 1986.

Bolger, J. C. *Adhesion Aspects of Polymeric Coatings*. K. L. Mittal, Ed., Plenum, New York, 1981.

Bolger, J. C.; and A. S. Michaels. *Interface Conversion*, P. Weiss and D. Cheevers, Eds. Elsevier, New York, 1969, Chap. 1.

Borroff, E. M.; and W. C. Wake. *Trans. Inst. Rubber Ind.*, 25,

190, 199, 210 (1949).

Brewis, D. M.; and D. Briggs. *Industrial Adhesion Problems*. Wiley, New York, 1985.

Clark, D. T.; and W. J. Feast. *Polymer Surfaces*. Wiley, New York, 1978.

Colins, E. A.; J. Bares; and F. W. Billmeyer, Jr. *Experiments in Polymer Science*. Wiley, New York, 1973.

Damico, D. J. *Adhesives Age*, 25 (Oct. 1987).

Davison, N. *Statistical Mechanics*. McGraw-Hill, New York, 1960, p. 321.

DeBruyne, N. A.; and R. Houwink. *Adhesion and Adhesives*. Elsevier, New York, 1951.

DeBruyne, N. A. *Aircraft Eng*. XVIII (12), 53 (1939).

Debye, P. J. W. *Adhesion and Cohesion*, P. Weiss, Ed. Elsevier, New York, 1962, p. 1.

Deryaguin, V. *Research*, 8, 70, 365 (1955).

Donald J., D. K. *Adhesives*, 4, 233 (1972).

Eley, D. D. *Adhesion*. Oxford University Press, London, 1961.

Fowkes, F. M. *Am. Chem. Soc. Adv. Chem. Ser.*, 43, 54 (1964).

Fowkes, E. M. *Treatise on Adhesion and Adhesives*. Dekker, New York, 1967, Vols. *i*, II, Chap. 4, 9.

Gardon, J. L. *J. Phys. Chem.*, 67, 1935 (1963).

Grayson, M.; and D. Eckroth, Eds. *Encyclopedia of Chemistry*, 3rd ed. Wiley, New York, 1982, Vol. 18, p. 407.

Grimley, H. *Aspects Adhesion*. 7, 11-27 (1973).

Heimanz, P. C. *Principles of Colloid and Surface Chemistry*. Dekker, New York, 1977, pp. 364, 419.

Henderson, A. W. *Aspects Adhesion.* 1, 33 (1965).

Katz, I.; and C. V. Cagle. *Adhesive Materials: Their Properties and Usage.* Foster, Long Beach, CA, 1971.

Kauzmann, W. *Quantum Chemistry.* Academic, New York, 1957, p. 375.

Kinloch, A. J. *J. Materi. Sci.* 15, 2141 (1980); 17, 617 (1982).

Lee, L. H. *Adhesive Chemistry: Development and Trends.* Plenum, New York, 1984.

Lifshitz, E. M. *Dokl. Akad. Nauk. SSSR*, 97, 643 (1954).

Mark, H. P. *Adhesion and Cohesion*, P. Weiss, Ed. Elsevier, New York, 1962, p. 240.

McBain, J. W.; and D. G. Hopkins. *J. Phys. Chem.* 29, 88 (1925).

Mittal, K. L. *Adhesion Aspects of Polymeric Coatings.* Plenum, New York, 1981, p. 355.

Salomon, G. *Adhesion and Adhesives*, 2nd ed. Elsevier, Armsterdam, 1965, Vol. 1, p. 6.

Schneberger, L. *Adhesives Age*, 10 (May 31, 1985).

Sickfeld, J. *J. Oil Coll. Chem. Assoc.*, 61, 292 (1978).

Skeist, I. *Adhesive Handbook*, 2nd ed. Reinhold, New York, 1977, p. 12.

Smith, J. R. *Mater. Sci. Engi.*, 83, 169-234 (1986).

Taylor, D.; and J. Rutzler. *Ind. Eng. Chem.*, 50, 928 (1958).

Vasenin, R. M. *Adhesion: Fundamentals and Practice, Ministry of Technology.* MacLaren, London, 1969.

Voyutskii, S. S. *Autohesion and Adhesion of High Polymers.* Wiley-Interscience, New York, 1963, p. 138.

Wake, W. C. *Polymer*, 19, 291 (1978).

Weaver, C. *Adhesion, Fundamentals and Practice*. McLaren, London, 1969.

Weiss, P. *Adhesion and Cohesion*. Elsevier, New York, 1962, p. 89.

Wu, S. H. *Polymer Interface and Adhesion*. Dekker, New York, 1982.

Zisman, W. A. *Adhesion and Cohesion*, P. Weiss, Ed. Elsevier, New York, 1962, p. 176.

2.

Three Laws of Quantitative XRD & XRF Analyses

> **Vision without action is daydream. Action without vision is nightmare.**
>
> — Aristotle

ABSTRACT

New insights led to the three laws of quantitative X-ray Diffraction (XRD) and X-ray Fluorescence (XRF) analyses: 1. the law of zero matrix effects, 2. the law of constant slope, and 3. the law of binary mixtures. The first two laws change the current <u>concept</u> of quantitative XRD & XRF analyses. The third law greatly simplifies the current <u>practice</u> of quantitative XRD & XRF analyses. Consequently, one decoding formula unifies and quantifies both XRD & XRF. One X-ray scan of one sample preparation determines all components in any mixtures. That means chemical compounds by XRD or chemical elements by XRF. This new procedure reduces some 80% lab work required by current practice. Its precision has been evaluated statistically to be ± 5% or better

1. INTRODUCTION

All current textbooks on quantitative X-ray diffraction (XRD) and X-ray fluorescence (XRF) analyses teach the Internal Standard and/or the Spiking methods. Both methods require calibration lines from standards, which is tedious and time-consuming. New insights lead to the three laws of quantitative XRD & XRF analyses: 1. the law of zero matrix effects; 2. the law of constant slope of calibration line; and 3. the law of binary mixtures. The first two laws clarify the current concept of XRD & XRF analyses. The third law transforms any complex mixture into a set of simple binary mixtures. Moreover, one of the two components (the added reference standard) in each binary mixture has known % weight. The % weight of the other component is simple and obvious. These three laws unify and grossly simplify the quantitative XRD and XRF analyses of mixtures.

2. UNIFIED THEORY OF QUANTITATIVE XRD & XRF

Examine the fundamental equations relating intensity (I_i) to concentration (X_i), (Klug & Alexander, 1974; Herglotz & Birks, 1978; Jenkins et al., 2002; Grieken & Markowicz, 2002), we found that with a fixed experimental setup, the fundamental intensity-concentration equations of both XRD & XRF becomes $I_i = k_i X_i$, where k_i is an overall constant including the absorption coefficient (μ), density (ϱ), thickness (d) of the specimen and other instrumental parameters. Note that the μ & ϱ factors vary with the composition and the d factor varies with specimen preparation. However, all components in the same specimen have identical μ,& d factors, which can be exactly cancelled mathematically.

- According to Klug and Alexander (1974) for XRD with pellet specimen:

$$I_i = K_i X_i / \mu = k_i X_i \qquad (1)$$

Where K_i is a proportional constant, k_i includes K_i / μ factor.

- According to Herglotz & Birks (1978) for XRF:

$$I_i = K_i X_i (1 - e^{-\mu \varrho d}) / \mu = k_i X_i \qquad (2)$$

Where k_i includes $K_i (1 - e^{-\mu \varrho d}) / \mu$ factor

With pellet specimen, where d = , $e^{-\mu \varrho d}$ = 0, we have (Chung, 2016):

$$I_i = K_i X_i / \mu = k_i X_i \qquad (3)$$

Where k_i includes the K_i / μ factor.

With thin-film specimen (Chung et al., 1974), for which d is very small. We have

$$e^{-\mu d} = 1 - \mu \varrho d$$

$$I_i = K_i X_i \varrho d = k_i X_i \qquad (4)$$

Where k_i includes the $K_i \varrho d$ factors.

Bundling the K_i and the µ, ϱ & d factors into an overall constant (k_i) creates six effects: 1. Leave the µ, ϱ & d factors out of the decoding formula. 2. Cancel the µ, ϱ & d factors in the k_i/k_s ratios. 3. Make the k_i factor a characteristic constant (i.e. constant slope of calibration line) of each component sought. 4. One simple equation decodes both XRD & XRF, 5. It works for full range of concentrations, and 6. No restriction on the amount of reference standard (X_s). Note that in the Internal Standard method, all samples must have identical X_s (Klug and Alexander, 2nd Ed. P. 537, 1974), which is very tedious.

Let us implant known amount of a reference standard (X_s) into the totally unknown mixture and homogenize the admixture. Couple the reference standard (s) with each of the original components (i). Now we have a pair of simultaneous equations for each original component. Note that each component in the same specimen has identical µ, ϱ, & d factors, hence these matrix factors are neatly cancelled in the ratio k_i / k_s in Equations (5) & (6) for both XRD & XRF.

$$I_i = k_i X_i \text{ and } I_s = k_s X_s$$

$$\frac{I_i}{I_s} = \frac{k_i}{k_s} \cdot \frac{X_i}{X_s},$$

$k_s = 1.00$ by definition (see section on Specific Intensity)

$$\frac{I_i}{I_s} = k_i \cdot \frac{X_i}{X_s}, \quad k_i = \text{slope of a straight line.} \tag{5}$$

Rearrange Equation (5), we have:

$$X_i = \frac{X_s}{k_i} \cdot \frac{I_i}{I_s} \quad (6)$$

The basic decoding formula.

Note that each individual k_i and k_s contains matrix factors (μ, ϱ & d). However, the ratio k_i / k_s is free from matrix factors, because all matrix factors (μ, ϱ & d) are exactly cancelled.

Equations (5) & (6) indicate k_i = slope of calibration line = characteristic constant of component i. Consequently, the complex matrix factors and the tedious calibration work are eliminated. Equation (6) is the new basic decoding formula for both XRD & XRF. It is free from matrix effects. Yet it works for full range of concentrations.

The Basic Decoding Equation (6) contains two unknowns (X_i & k_i) and three measured data (X_s, I_i & I_s). Given composition X_i in synthetic mixture, the constant k_i can be calculated. Given the constant k_i in an unknown mixture, the composition X_i can be calculated. In other words, a single XRD or XRF scan quantifies the chemical compounds or elements in any complex mixtures. Because k_i is a characteristic constant of each chemical compound or chemical element, it needs to be measured only once for all.

The reference standard (X_s) plays multiple roles: 1. It puts all X-ray intensity data on the same scale. 2. It acts as a probe or spy and flashes back its findings in X-ray signals. 3. It defines the characteristic constant slope of calibration line. 4. It functions as an internal standard.

The physics of XRD and XRF are fundamentally different. However, both techniques share the same primary X-rays. Both diffraction and emission take place simultaneously. The X-ray instrument selectively collects the coded signals either by Bragg angle (XRD) or by wavelength (XRF). The primary X-rays, the diffracted X-rays, and the emitted X-rays should obey the same Adiabatic Principle of X-rays (Chung, 1974b).

3. APPLICATION OF THE DECODING FORMULAS
3.1 XRD: All components are crystalline:

No reference standard is required. We have n unknowns (X_i), which must satisfy the following (n + 1) equations. These (n+1) equations can be conveniently treated in terms of matrix algebra:

$$I_i = k_i \bullet X_i, \quad i = 1 \text{ to } n \quad \text{and} \quad \sum_1^n X_i = 1$$

$$\begin{pmatrix} k_1 & 0 & 0 & \ldots & 0 \\ 0 & k_2 & 0 & \ldots & 0 \\ 0 & 0 & k_3 & \ldots & 0 \\ \ldots & & & & \\ 0 & 0 & 0 & \ldots & k_n \\ 1 & 1 & 1 & \ldots & 1 \end{pmatrix} \bullet \begin{pmatrix} X_1 \\ X_2 \\ X_3 \\ \vdots \\ X_n \\ 1 \end{pmatrix} = \begin{pmatrix} I_1 \\ I_2 \\ I_3 \\ \vdots \\ I_n \\ 1 \end{pmatrix}$$

When the matrix equation, **KX = I,** has a solution, the solution is unique if and only if the rank of the (**K**) matrix is equal to the rank of the (**K, I**) matrix, which is also equal to the number of unknowns. The unique solution has the following symmetrical form:

$$X_i = \begin{bmatrix} k_i & I_i \\ I_i & k_i \end{bmatrix}^{-1} \quad \text{It implies the Law of Binary Mixtures} \quad (7)$$

Equation (7) involves only intensity (I_i) data. The k_i's are characteristic constants. The weight fractions (X_i) of all components can be determined with one single XRD scan of the original mixture. It establishes the Law of Binary Mixtures (See next sections), which leads to the simplest way to decode complex mixtures as demonstrated by the three sets of experimental data.

3.2 XRD: Some components are amorphous:

A reference standard is required for this case. Apply equation (6) to each identified component. We have

$$X_o = X_1 + X_2 + \cdots + X_n = \frac{X_s}{I_s}\left(\frac{I_1}{k_1} + \frac{I_2}{k_2} + \cdots + \frac{I_n}{k_n}\right) = \frac{X_s}{I_s}\frac{I_i}{k_i}$$

$$\Sigma\% = \bullet\, I_s \quad \begin{array}{c} I_i < X_o \\ \text{The discriminant equation for XRD} \\ k_i > X_s \end{array} \quad (8)$$

Where X_o = weight fraction (% wt) of original mixture.

< indicates presence of amorphous contents.
= indicates all components are crystalline.
> indicates wrong experimental data.

Equation (8) determines if amorphous contents are present and how much in the original sample.

3.3 XRF: Regardless Crystalline or amorphous

No matter the mixture is crystalline or amorphous. X-ray Fluorescence can assess all elements heavier than oxygen. A reference standard is always required for XRF. Apply Equation (6) to each element sought. Equation (7) may even work for metallic alloys (pending further research work). Equation (8) applies to XRD only.

Equations (6) (7) & (8) are the essence of the proposed Unified and Simplified Theory of XRD & XRF. Equation (6) is the basic decoding formula, applies to both XRD & XRF for any mixtures. It is the key to solve most problems in XRD & XRF. Equation (7) transforms any complex mixtures into a set of binary mixtures for both XRD & XRF (see section on Law of Binary Mixtures). Equation (8) works for XRD only. It determines if amorphous contents are present and how much.

For both XRD & XRF, only X-ray intensities (I_i & I_s) are to be measured from a single sample preparation. Other factors are either constants (k_i & k_s) or known quantities (X_s & X_o). Consequently, one single X-ray scan of one sample preparation quantifies all components with one decoding formula. Be sure the prepared sample is a homogeneous mixture. Certainly, the average intensity of repeated scans improves the precision.

4. SPECIFIC INTENSITY OF XRD & XRF

For easy comparison of the same property of different materials, usually a pure compound is assigned as reference standard. For example, the specific gravity of gold is 19.3, which means that the mass of gold is 19.3 times as heavy as the mass of water at 4°C. The ubiquitous water is assigned as reference standard. Note that both the density and specific gravity of water are defined to be exactly one (1.00). The word "specific" means per unit mass, such as the specific impulse of rocket fuel in space ship. In order to put all X-ray intensity data on the same scale, let us define a new term Specific Intensity (k_i) of any chemical element or chemical compound (i) as follows:

$$\text{Specific Intensity} = \frac{\text{XRD or XRF intensity of 1.00\% of component } (I_i)}{\text{XRD or XRF intensity of 1.00\% of reference standard } (I_s)} = \left(\frac{I_i}{I_s}\right)_{1\% \text{ each}} \quad (9)$$

Rearrange the basic decoding equation (6), we have:

$$k_i = \frac{I_i / X_i}{I_s / X_s} = \left(\frac{I_i}{I_s}\right)_{1\% \text{ each}} = \left(\frac{I_i}{I_s}\right)_{50/50} = \text{Specific Intensity} \quad (10)$$

= slope of Equation (5) & (6) for both XRD & XRF

A set of Specific Intensity (k_i) can be determined from a synthetic mixture for either compounds (XRD) or elements (XRF).

The Specific Intensity (k_i) indicates how strong is the diffracted or emitted X-rays relative to that of the reference standard (Y_2O_3 or Al_2O_3 for XRD, Y in Y_2O_3 for XRF). We have $k_s = 1.00$ by definition. Yttrium Y is a rare element, most likely it is not present in the original sample. High purity chemical grade Y_2O_3 powder is white, stable, inexpensive, and readily available.

Equations (9) & (10) display five roles of the specific intensity (k_i): 1. It puts all X-ray intensity data on the same scale. 2. It is the slope of the basic decoding formula. 3. It is a characteristic constant of each chemical element or compound. 4. It allows simultaneous determination of a set of k_i with one sample preparation. 5. It unifies and simplifies quantitative XRD & XRF analyses.

Three new discoveries led to the 1. Eliminate the matrix effects; 2. Discard the calibration lines; and 3. Establish the Law of Binary Mixtures. Consequently, a single X-ray scan of one sample preparation quantifies all components sought in any mixture. It reduces over 80% lab work required by current practice. Three sets of experimental data are presented to illustrate lab practice. Its precision has been evaluated statistically to be ± 5% or better, which is sufficient to solve most problems.

5. THREE LAWS OF QUANTITATIVE XRD & XRF

Equations (6), (7) & (8) are the key conclusions of the Matrix Flushing Theory, which establishes the three laws of quantitative XRD & XRF analyses:

1. The law of zero matrix effects: The matrix effects have bothered quantitative XRD & XRF analyses of mixtures

ever since the pioneering days. However, all matrix effects can be neatly cancelled mathematically. There are no matrix effects from the very beginning.

2. The law of constant slope: The slope of calibration line of any chemical element or compound is a characteristic unique constant. A set of these constants can be easily determined. No calibration work is necessary for quantitative XRD & XRF analyses from the very beginning.

3. The law of binary mixtures: The plot of X-ray intensity ratio to the weight ratio of any binary mixture is a straight line passing through the origin with a slope equal to the ratio of corresponding specific intensities. This intensity-concentration relationship of any binary mixture is not perturbed by the presence or absence of other components. In other words, each pair of components in any complex mixture form an isolated (adiabatic) unit, immune (adiabatic) from the coexistence of any other components.

The first and second laws correct and clarify the current concept of quantitative XRD & XRF analyses. The third law unifies and simplifies the current practice of quantitative XRD & XRF analyses.

The decoding formula Equation (6), leads to the unique solution Equation (7). Equation (7) establishes the Law of Binary Mixture. Implant known amount of a reference standard (Y_2O_3 or Al_2O_3) into any complex mixture and couple it with each component sought. The complex mixture becomes a set of binary mixtures, the simplest mixture. Moreover, one of the two components in each binary mixture is the known reference standard. It makes each binary mixture "simpler than the simplest mixture", a strange but hard fact. It empowers the Unified Theory of XRD & XRF as the simplest way to quantify any mixtures. The term Law of Binary Mixture is more intuitive than the esoteric term Adiabatic Principle (Chung, 1974b).

Equation (6) and (7) eliminates the matrix effects, the calibration lines, and the certified standards. Just one single XRD or XRF scan gives the weight fractions of all components sought. Moreover, the weight fraction is expressed in terms of ratios, I_i / I_j and k_i / k_j (i, j = 1, 2, ..., n). Therefore, many disturbing factors are mutually cancelled, including matrix factors, specimen thickness, specimen density, residual stress, micro-absorption, preferred orientation, crystallinity, extinction, particle size and size distribution, specimen position, instrumental drifting, and others.

6. NUMERICAL EXAMPLE FOR QUANTITATIVE XRD ANALYSIS

The key point of the unified theory is to measure the Specific Intensities (k_i) of components in a synthetic mixture by Equation (6), then determine the weight fractions of components sought (X_i) in unknown mixtures by Equations (6) (7) or (8). Note that Equations (7) & (8) are derived from the basic Equation (6).

Three numerical examples are presented to illustrate the applications of the three working equations (6), (7) & (8). The compositions and X-ray data are summarized in the corresponding Tables. The strongest resolved X-ray peak was chosen for intensity measurements. In order to avoid repetitions, the detailed lab procedures are referred to the cited references. The numerical calculations are detailed for easy references.

6.1 All Components Are Crystalline

Because the Specific Intensity (k_i) is the characteristic constant of each component i, the XRD intensities (I_i) from a single scan of one sample preparation are the only experimental data required to calculate X_i, The set of data in Table 1 illustrates the applications of the working Equations (6), (7) & (8):

Table 1. All components are crystalline.

Weight, g	% Weight X_i Known	XRD, cps I_i Found	Specific Intensity k_i, Constant slope		
ZnO	0.5554	32.54	32.5	4036	4.27
CdO	0.2266	13.27	13.3	3029	7.85
LiF	0.3712	21.74	21.8	778	1.23
Al$_2$O$_3$	0.5539	32.45	---	943	1.00

- **Basic Decoding Equation (6):** Let the ZnO, CdO & LiF in Table 1 are the components (X_i) in the original mixture. The Al$_2$O$_3$ is added as the Reference Standard (X_s). Apply the Basic Decoding Equation (6): Giving I_i & k_i, we can calculate X_i. Given I_i & X_i we can calculate k_i. We have:

Given I_i & k_i, calculate X_i   **Given I_i & X_i, calculate k_i**

$$X_{ZnO} = \frac{32.45}{4.27} \cdot \frac{4036}{943} = 32.5 \quad \text{or} \quad k_{ZnO} = \frac{32.45}{32.54} \cdot \frac{4036}{943} = 4.27$$

$$X_{CdO} = \frac{32.45}{7.85} \cdot \frac{3029}{943} = 13.3 \quad \text{or} \quad k_{CdO} = \frac{32.45}{13.28} \cdot \frac{3029}{943} = 7.85$$

$$X_{LiF} = \frac{32.45}{1.23} \cdot \frac{778}{943} = 21.8 \quad \text{or} \quad k_{LiF} = \frac{32.45}{21.74} \cdot \frac{778}{943} = 1.23$$

$$X_{Al2O3} = \frac{32.45}{1.00} \cdot \frac{943}{943} = 32.5 \quad \text{or} \quad k_{Al2O3} = \frac{32.45}{32.45} \cdot \frac{943}{943} = 1.00$$

- **Adiabatic Equation (7):** Let all four chemicals in Table 1 are components of the original mixture. No Reference standard is required. One single XRD scan of the original mixture quantifies its composition. Apply the Adiabatic Equation (7). We have:

$$X_{ZnO} = [1 + \frac{4.27}{4036}(\frac{3029}{7.85} + \frac{778}{1.23} + \frac{943}{1.00})]^{-1} = 32.5$$

$$X_{CdO} = [1 + \frac{7.85}{3029}(\frac{4036}{4.27} + \frac{778}{1.23} + \frac{943}{1.00})]^{-1} = 13.3$$

$$X_{LiF} = [1 + \frac{1.23}{778}(\frac{3029}{7.85} + \frac{4036}{4.27} + \frac{943}{1.00})]^{-1} = 21.8$$

$$X_{Al2O3} = [1 + \frac{1.00}{943}(\frac{4036}{4.27} + \frac{3029}{7.85} + \frac{778}{1.23})]^{-1} = 32.5$$

- **Discriminant Equation (8):** Reference standard is required. Apply Equation (8) to check if amorphous content present and how much. Referring to the calculated data from Equation (6), we have:

$$\sum_{1/n} \frac{I_i}{k_i} \begin{matrix} < X_o \\ = \bullet \quad I_s \\ > X_s \end{matrix} \qquad (8)$$

Where $X_0 = 67.5\%$ and $X_s = X_{Al2O3} = 32.5\%$.

$$\text{Left side of Equation (8)} \sum_{1/n} \frac{I_i}{k_i} = \frac{4036}{4.27} + \frac{3029}{7.85} + \frac{778}{1.23} = 1963$$

Right side of Equation (8): $\dfrac{X_o}{X_s} \cdot I_s = \dfrac{67.5}{32.5} \cdot 943 = 1959$

The left side is almost equal to the right side, 1959 ~ 1963, which indicates no amorphous content.

6.2 Some Components Are Amorphous.

A reference standard Al_2O_3 is required in this case. Equation (6) gives the weight fractions of all identified crystalline components. Equation (8) determines whether amorphous materials are present and how much. Note that $k_{ZnO} = 4.35$, $k_{CaCO3} = 2.98$ in Table 2 are averages from prior work. A good estimate of amorphous content (Silica gel) is obtained as shown in Table 2.

Table 2. Some components are amorphous and/or unidentified

Component	Composition gram	Intensity I_i (cps)	Sp. Int. k_i	%wt.in admix. Known Found		% wt in orig. mix. Known Found	
ZnO	0.9037	4661	4.35	34.43	36.4	43.8	46.3
CaCO3	0.7351	2298	2.98	28.00	26.2	35.6.	33.4
Silica gel	0.4234	0	—	16.13	(16.0)	20.5	(20.4)
Added Ref. Std.							
Al2O3	0.5629	631	1.00	21.44	—	0	0

Total weight of original mixture = 2.0622 grams (no Al_2O_3).
Total weight of admixture = 2.6251 grams (with Al_2O_3).

According to basic decoding Equation (6):

$$X_{ZnO} = \frac{21.44}{4.35} \cdot \frac{4661}{631} = 36.4\%$$

$$X_{CaCO_3} = \frac{21.44}{2.98} \cdot \frac{2298}{631} = 26.2\%$$

Amorphous content = 100 − 36.4 − 26.2 − 21.4 = 16.0 % in admixture (with Al2O3), or apply Equation (8) we have: Xo = 78.56 %, Xs = 21.44 %

$$\sum \frac{I_i}{k_i} = \frac{4661}{4.35} + \frac{2298}{2.98} = 1842, \text{ and } \frac{X_o}{X_s} \cdot I_s = \frac{78.56 \times 631}{21.44} = 2312$$

The left side is much smaller than the right side of Equation (8), 1842 << 2312. It indicates the presence of amorphous contents, which can be determined by mass balance. Note that all the symbols in Equations (6) & (8) are referring to the admixture (original sample plus reference standard). Equation (7) has no reference standard. Equation (11) converts X_i to $X_i°$, the weight fraction of component (i) in original sample.

$$X_i^\circ = \frac{X_i}{1 - X_s} = X_i \cdot \frac{\text{Wt. of admixture}}{\text{Wt. of original mixture}} \quad (11)$$

$$X_{ZnO}^\circ = 36.4 \cdot \frac{2.6251}{2.0622} = 46.3\%$$

$$X_{CaCO3}^\circ = 26.2 \cdot \frac{2.6251}{2.0622} = 33.4\%$$

$$X_{Silica\ gel}^\circ = 16.0 \cdot \frac{2.6251}{2.0622} = 20.3\% = \text{Amorphous content in original mixture.}$$

Double check for admixture: 36.41 + 26.20 + 16.00 + 21.44 = 100% (with Al_2O_3).

Double check for original sample: 46.3 + 33.4 + 20.3 = 100% (without Al_2O_3).

7. Numerical Example for Quantitative XRF Analysis

A reference standard (Y_2O_3) is required in this case. It shows the application of Equations (6) for XRF, no matter crystalline or amorphous.

Add known amount of reference standard (Y_2O_3) into the synthetic sample. Grind and homogenize the admixture. Run XRF experiments. All XRF intensity data are from one single scan of one sample preparation. Average intensity from duplicate scans is used to improve precision. Just like the XRD analysis, given the measured intensity (I_i) and the constant specific intensity (k_i), the composition (X_i & X_i^0) can be calculated by Equations (6) & (11).

Table 3 lists the measured intensities (I_i) and the average specific intensities (k_i) of prior work. All other data are results of calculations. The experimentally "found % weights" from XRF agree well with the "known % weights" of the synthetic mixture.

The X-ray intensities (I_i) in Table 3 are the only experimental data required to quantify all components sought. The k_i's are the average Specific Intensities (characteristic constant slopes of calibration lines) from prior work. All other data in Table 3 are results of calculations. Equation (6) gives X_i. Equation (11) converts X_i (% weight in admixture) to X_i° (% weight in original synthetic sample).

Note that given the composition (X_i) and intensity (I_i) data of the synthetic mixture, a set of specific intensities (k_i) can be determined. All k_i 's are characteristic constants, which vary within experimental errors.

Table 3. Access all elements heavier than oxygen in mixture.

Composition, g	I_i, cps XRF Inten.	k_i, Spec. Inten. Constant slope from prior work	X_i, Weight % in Admixture Known	Found	X^0_i, Weight % in Original Mix. Known	Found
BaSO$_4$	0.0695					
Ba	124	1.96	2.60	2.55	2.83	2.78
S	0.97	0.0635	0.61	0.61	0.66	0.66
ZnO	0.3677					
Zn	1357	2.88	18.8	19.0	20.5	20.7
K$_2$HPO$_4$	1.0071					
K	464	0.646	28.8	28.9	31.3	31.5
P	11.3	0.0393	11.4	11.6	12.4	12.6
Reference Standard						
Y$_2$O$_3$	0.1277					
Y	159	1.00	6.40	---	0	0

Total weight of original sample (without Y$_2$O$_3$) = 1.4443 g. that of admixture (with Y$_2$O$_3$) = 1.5720 g.

$$X_{Ba} = \frac{6.40}{1.96} \cdot \frac{124}{159} = 2.55\% \qquad X_{Ba}^° = 2.55 \cdot \frac{1.5720}{1.4443} = 2.78\%$$

$$X_{Zn} = \frac{6.40}{2.88} \cdot \frac{1357}{159} = 19.0\% \qquad X_{Zn}^° = 19.0 \cdot \frac{1.5720}{1.4443} = 20.7\%,$$

$$X_{K} = \frac{6.40}{0.646} \cdot \frac{464}{159} = 28.9\% \qquad X_{K}^° = 28.9 \cdot \frac{1.5720}{1.4443} = 31.5\%,$$

$$X_S = \frac{6.40}{0.0635} \cdot \frac{0.97}{159} = 0.61\% \qquad X_S^\circ = 0.61 \cdot \frac{1.5720}{1.4443} = 0.66\%,$$

$$X_P = \frac{6.40}{0.0393} \cdot \frac{11.3}{159} = 11.6\%. \qquad X_P^\circ = 11.6 \cdot \frac{1.5720}{1.4443} = 12.6\%,$$

$$X_Y = \frac{6.4}{1.00} \cdot \frac{159}{159} = 6.40\% \qquad X_Y^\circ = 0.00\ (I_Y = 0.00,\ \text{No Yttrium})$$

If the XRF job were done by the book (Muller, 1972; Herglotz & Birks, 1978; Jenkins et al., 2002; Grieken & Markowicz, 2002), it would require 5 calibration lines for the 5 elements sought. Each calibration line needs 3 data points plus the sample itself. That means 5x4=20 times more sample preparations from certified standards, 20 times more X-ray scans, 20 times more intensity measurements, 20 times more calculations, and 20 times more plotting. This situation is much the same for the XRD analyses.

In summary, the Unified and Simplified Theory of Quantitative XRD & XRF has been practiced in our lab. It reduces by over 80% of the lab work required by current practice. Statistical evaluation (Chung, 1974a,b, 1975c, 2016; Chung & Smith, 2000) gives a precision of ± 5% or better, which is sufficient to solve most industrial problems. Note that the Rietveld refinement whole pattern fitting methods have the potential to approach the precision limit of ± 1% at the expense of much more time for data collection and processing (Chung & Smith, 2000; Hubbard et al. 1983; Snyder, 1992; Smith et al., 1987; Bish & Howard, 1988).

8. CONCLUSIONS

New insights establish three laws of quantitative XRD & XRF analysis, which unify, simplify, and change the current concept of quantitative XRD & XRF analyses:

1. The law of zero matrix effects: Every component in the same specimen has identical matrix factors (μ, & d), which can be exactly cancelled mathematically. There is no matrix problem from the very beginning.

2. The law of constant slope: The slope of calibration line for any chemical compound (XRD) or chemical element (XRF) is a characteristic constant (i.e. Specific Intensity, k_i). A set of Specific Intensities can be easily determined, thus eliminates all calibration work.

3. The law of binary mixtures: It transforms any complex mixture into a set of the simplest binary mixtures. That means each binary mixture has a component of known % weight. The % weight of the other component is obvious.

These three laws of XRD & XRF realize six advantages: 1. eliminate the matrix effects, 2. evade the calibration lines, 3. one decoding formula applies to both XRD & XRF, 4. It works for full range of concentrations, 5. one single X-ray scan of one sample preparation decodes all % weights, and 6. Reduce lab work by some 80% required by current practice. It amounts to a breakthrough in quantitative XRD & XRF analysis.

9. REFERENCES

Azaroff, L. V. (1968). *Elements of X-ray Crystallography*, p. 202. New York: McGraw Hill.

Bish, D. L. & Howard S. A. (1988). "Quantitative Phase Analysis Using the Rietveld Method." *J. Appl. Cryst.* 21, 86-91.

Bunn, C. W. (1961). *Chemical Crystallography*, 2nd ed. p. 223. London: Oxford University Press.

Chung, F. H. (1974a). "Quantitative Interpretation of X-Ray Diffraction Patterns of Mixtures. I. Matrix Flushing Theory for Multi-Component Analysis." *J. Appl. Cryst.* 7: 519-525.

Chung, F. H. (1974b). "Quantitative Interpretation of X-Ray Diffraction Patterns of Mixtures. II. Adiabatic Principle of X-Ray Diffraction Analysis." *J. Appl. Cryst.* 7: 526-531.

Chung, F. H. (1974c). "A New X-Ray Diffraction Method for Quantitative Multi-component Analyses." Advances in X-Ray Analysis, Vol. 17, pp. 106-115.

Chung, F. H. (1975). "Quantitative Interpretation of X-Ray Diffraction Patterns of Mixtures. III. Simultaneous Determination of a Set of Reference Intensities." *J. Appl. Cryst.* 8: 17-19.

Chung, F. H. (2017). "Unified Theory for Decoding the Signals from X-Ray Fluorescence and X-Ray Diffraction of Mixtures." Applied Spectroscopy, Vol. 71(5), pp. 1060-1068.

Chung, F. H. (2018). "Quantitative XRD & XRF Analyses of Mixtures Unified and Simplified". J. Appl. Cryst. 51: 789-795.

Chung, F. H., Lentz, A. J. & Scott, R. W. (1974). "A Versatile Thin-Film Method for Quantitative X-Ray Emission Analysis." X-Ray Spectroscopy. 3: 172-175.

Chung, F. H. & Smith D. K. (2000). *Industrial Applications of X-Ray Diffraction.* pp. 3, 13, 37 & 511. New York: Marcel Dekker.

Clark, G. L. & Reynolds D. H. (1936a). At University of Illinois, adapted the highly developed quantitative "Internal Standard" method of ultraviolet spectroscopy for quantitative XRD analysis.

Clark, G. L. & Reynolds D. H. (1936b). "Quantitative Analysis of Mine Dusts." Ind. Eng. Chem., Anal. Ed. 8. p. 36.

Cohen, L. H. & Smith D. K. (1989). "Thin-Specimen X-ray Fluorescence Analysis of Major Elements in Silicate Rocks." Anal. Chem. 61: 1837-1840.

Copeland, L. E. & Bragg, R. H. (1958). "Quantitative X-Ray Diffraction Analysis." Anal. Chem. 30(2), 196-201.

Cullity, B. D. & Stock, S. R. (2001). *Elements of X-Ray Diffraction*, 3rd Edition. New Jersey: Prentice Hall.

Grieken, R. V. & Markowicz A. A. (2002). *Handbook of X-ray Spectrometry*: Methods and Techniques, 2nd ed. New York: Marcel Dekker.

Herglotz, H. K. & Birks, L. S. (1978). *X-Ray Spectrometry*. New York: Marcel Dekker.

Hubbard, C. R. Robbins, C. R.& Snyder R. L. (1983). "XRD Quantitative Analysis Using the NBS Quant82 System." Adv. X-Ray Anal. 26:149-157.

Jenkins, R., Gould R. W. & Gedcke D. (2002). *Quantitative X-Ray Spectrometry*, 2nd ed. New York: Marcel Dekker.

Jenkins, R. & Snyder R. L. (1995). *Introduction to X-ray Powder Diffractometry*. New York: John Wiley and Sons. pp. 369 - 376.

Klug, H. P. & Alexander L. E. (1974). *X-ray Diffraction Procedures for Polycrystalline and Amorphous Materials*. New York: John Wiley & Sons.

MIT X-ray SEF List of Resources (2018). Textbooks on X-Ray Diffraction. Website: http://prism.mit.edu/xray/oldsite/resources.htm.

Muller, R. O. (1972). *Spectrochemical Analysis by X-Ray Fluorescence*. pp. 59-61. New York: Plenum Press.

Rietveld, H. M. (1969). J. Appl. Cryst.2, 65-71.

Scheibe G. (1933). *Chemissche Spektralanalyse*, Physikalische Methoden der Analytischen Chemie, Vol. 1, p.108 and pp. 125-130. Leipzig, Akademische Verlagsgesellschaft (in German).

Smith, D. K., Johnson G. G., Scheible, A., Wims, A. M., Johnson, J. L. & Ullmann, G (1987). "Quantitative X-ray Powder Diffraction Method Using the Full Diffraction Pattern." Powder Diffraction, 2 (2): pp.73-77.

Snyder, R. L. (1992). "The Use of Reference Ratios in X-ray Quantitative Analysis." Powder Diffraction. 7: pp. 186-193.

3.

Unified and Simplified Practice of Quantitative XRD & XRF

> You find beauty in ordinary things, do not lost this ability.
>
> — Unknown

1. ABSTRACT

Due to the complex matrix effects, the current quantitative XRD & XRF analyses of mixtures require calibration lines from standards, hence tedious and time-consuming. New insights reveal that both the matrix effects and the calibration lines can be eliminated mathematically. Any complex mixture can be transformed into a set of simplest binary mixtures. One simple formula decodes both XRD & XRF. A single XRD or XRF scan quantifies the chemical compounds or chemical elements in any mixture. The unified and simplified procedure reduces some 80% lab work of current practice. Five sets experimental data are presented to verify its applications. Statistical evaluation of this new procedure gives a precision of ± 5% or better, which is normally expected from XRD & XRF analyses.

2. INTRODUCTION

The best way to determine the quantitative composition of a non-routine and totally unknown mixture is through X-ray Diffraction

(XRD) for chemical compounds and X-ray Fluorescence (XRF) for chemical elements. The MIT List of Resources [1], 2018 edition, compiles a list of textbooks on XRD [2-6], which covers "the seminal volumes for powder diffraction." All these XRD textbooks teach the internal standard technique and the spiking technique[7-8], both developed in the 1930s. Current XRF textbooks [9-12] mainly teach the spiking technique. The spiking technique is a special case of the internal standard technique. The internal standard technique is adapted from ultraviolet spectroscopy. [13-14] The spiking and the internal standard techniques require calibration lines of standards to control the matrix effects, the use of which is tedious and time consuming.

For solving technical problems or for research and development, one often receives samples marked "Rush Order" for quantitative XRD or XRF analysis. Because a factory is on hold, a legal case is in court, a big sale is under negotiation, a contract is near signing, or a new product is pending for mass production, waiting for the analytical report to make critical decisions. Obviously, the work must be done as soon as possible, and as accurately as possible. In order to simplify this challenging work, a wish list of five ideals was conceived: 1. Eliminate the matrix effects and the use of calibration lines. 2. Derive a linear decoding formula for full range of concentration. 3. Transform the slope of the linear decoding formula from variable to characteristic constant. 4. Apply the linear decoding formula to both XRD & XRF. 5. Extract weight fractions of chemical compounds or chemical elements by a single XRD or XRF scan.

Recent developments[15-22] led to the unified theory of quantitative XRD & XRF trying to fulfill these demands. This was made possible due to the visions and studies of many prior workers. Certainly, further work is necessary to substantiate or modify the unified concept of XRD & XRF. Five sets of experimental data are presented to verify its application. Its precision has been evaluated statistically ± 5% or better.

3. UNIFIED THEORY OF QUANTITATIVE XRD & XRF

The physics of XRD and XRF are fundamentally different. However, both techniques share the same primary X-rays. Both diffraction and emission take place simultaneously. The X-ray instrument selectively collects the coded signals either by Bragg angle (XRD) or by wavelength (XRF). The primary X-rays, the diffracted X-rays, and the emitted X-rays should obey the same Adiabatic Principle of X-rays [16]. Examining the fundamental equations relating intensity (I_i) to concentration (X_i),[5,10-11] we found that under fixed experimental setup, the fundamental intensity-concentration equations of both XRD & XRF becomes $I_i = k_i X_i$, where k_i is an overall constant including the matrix factor (μ), density (ϱ), or thickness (d) of the specimen and other instrumental parameters. The μ ϱ factors vary with the composition (unknown) and the d factor varies with specimen preparation (unknown). However, every component in the same specimen has identical μ, ϱ & d factors, which can be cancelled mathematically.

XRD: According to Klug and Alexander [5] with a pellet specimen:

$$I_i = K_i X_i / \mu = k_i X_i \qquad (1)$$

Where k_i includes the μ factors

XRF: According to Birks [10]:

$$I_i = K_i X_i (1 - e^{-\mu \varrho d}) / \mu = k_i X_i \quad (2)$$

Where k_i includes the $(1-2^{-\mu \varrho d}) / \mu$ factor

With pellet specimen [21], where d=, $e^{-mpd} = 0$ we have:

$$I_i = K_i X_i / \mu = k_i X_i \quad (3)$$

Where k_i includes the μ factors

With thin-film specimen [21], where d = very small, $e^{-mpd} = 1 - mpd$, we have:

$$I_i = K_i X_i \varrho d = k_i X_i \quad (4)$$

Where k_i includes the ϱ d factors

Bundling the K_i and the μ, ϱ & d factors into an overall constant (k_i) creates four effects: (1) Leave the μ, ϱ & d factors out of the decoding formula. (2) Cancels the μ, ϱ & d factors by the k_i / k_s ratios. (3) Make the k_i factor a characteristic constant of each component sought. And (4) Unify XRD & XRF with the same decoding formula free from matrix factors.

By implanting known amount of a reference standard (X_s) into the totally unknown mixture, the reference standard can be paired with each of the original components (i). We now have a pair of simultaneous equations for each component. Note that every component in the same specimen has identical μ, ϱ & d factors, hence these parameters are neatly cancelled in Equations (5) & (6) for both XRD & XRF.

$I_i = k_i X_i$ and $I_s = k_s X_s$,

$$\frac{I_i}{I_s} = \frac{k_i}{k_s} \cdot \frac{X_i}{X_s}, \quad k_s = 1.00 \text{ by definition, hence} \quad \frac{I_i}{I_s} = k_i \cdot \frac{X_i}{X_s} \quad (5)$$

$$X_i = \frac{X_s}{k_i} \cdot \frac{I_i}{I_s} \quad \text{or} \quad k_i = \frac{X_s}{X_i} \cdot \frac{I_i}{I_s} \quad \text{The basic decoding formula.} \quad (6)$$

Where: X_i & X_s = Weight fractions (%wt) of component i and reference standard s.

I & I_s = X-ray intensities (cps) of component i and reference standard s.

k_i & k_s = characteristic constant of the component i and reference standard s. Both include identical µ, ϱ & d factors, which are cancelled in k_i / k_s ratios.

k_s = 1.00 by definition of Specific Intensity (see section on Specific Intensity).

µ, ϱ & d = Mass absorption coefficient, density and thickness of specimen.

Equations (5) & (6) indicate k_i = slope of calibration line = characteristic constant of component i. Consequently, the matrix factors and the tedious calibration work are eliminated (see subsequent sections). Equation (6) is the new basic decoding formula for both XRD & XRF. It is free from matrix effects and works for a full range of concentrations. It contains two unknowns (X_i, k_i) and three measurements (X_s, I_i, & I_s). Given X_i

in a synthetic known mixture, k_i can be calculated. Given k_i in unknown mixtures, X_i can be calculated. Because $k_s = 1.00$ by definition, and k_i is a characteristic constant of each chemical compound or chemical element, it needs to be measured only once for all.

The reference standard plays multiple roles: 1. It puts all X-ray intensity data on the same scale; 2. It acts as a probe, sending back its findings via X-ray signals; 3. It defines the constant slope of any calibration line; and 4. It functions as an internal standard.

3.1 Three Cases of Mixtures

For quantitative XRD & XRF analyses, all mixtures can be divided into three cases:

Case 1, XRD: All components are crystalline and identified.

No reference standard is required. We have n unknowns (X_i), which must satisfy the following (n + 1) equations:

$$I_i = k_i \cdot X_i, \quad i = 1 \text{ to } n \quad \text{and} \quad \sum_1^n X_i = 1$$

These (n+1) equations can be conveniently solved with matrix algebra:

$$\begin{vmatrix} k_1 & 0 & 0 & \cdots & 0 \\ 0 & k_2 & 0 & \cdots & 0 \\ 0 & 0 & k_3 & \cdots & 0 \\ \vdots & & & & \\ 0 & 0 & 0 & \cdots & k_n \\ 1 & 1 & 1 & \cdots & 1 \end{vmatrix} \cdot \begin{vmatrix} X_1 \\ X_2 \\ X_3 \\ \vdots \\ X_n \end{vmatrix} = \begin{vmatrix} I_1 \\ I_2 \\ I_3 \\ \vdots \\ I_n \\ 1 \end{vmatrix}$$

When the matrix equation, **KX = I**, has a solution, the solution is unique if and only if the rank of the (**K**) matrix is equal to the rank of the (**K, I**) matrix, which is also equal to the number of unknowns. In other words, the necessary and sufficient condition for the existence of a unique solution of the matrix equation is that all its characteristic determinants vanish. The unique solution has the following symmetrical form:

$$X_i = [\frac{k_i}{I_i} \sum_1^n \frac{I_i}{k_i}]^{-1} \quad (7)$$

It implies the Adiabatic Principle.

Equation (7) involves only intensity (I_i) data. The k_i's are characteristic constants. The weight fractions (X_i) of all components can be determined with one single XRD scan of the original mixture. It establishes the Adiabatic Principle of X-rays [16], which leads to the simplest way to decode complex mixtures as shown in the five sets of experimental data in Tables 1-5.

Case 2, XRD: Some components are amorphous and/or unidentified.
Reference standard is required. Applying equation (6) to each identified component, we have:

$$X_o = X_1 + X_2 + \cdots + X_n = \frac{X_s}{I_s}(\frac{I_1}{k_1} + \frac{I_2}{k_2} + \cdots + \frac{I_n}{k_n}) = \frac{X_s}{I_s} \sum_1^n \frac{I_i}{k_i}$$

Rearrange: equation

$$\sum_1^n \frac{I_i}{k_i} \leq \frac{X_o}{X_s} \cdot I_s \qquad \text{The discriminant formula} \qquad (8)$$

Where X_o & X_s = weight fractions (% wt) of the original mixture and the reference standard.

< indicates presence of amorphous and/or unidentified contents.

= indicates all components are crystalline and identified. > indicates wrong experimental data.

Equation (8) determines if amorphous components are present and how much are in the original sample. Its applications are illustrated for case 1: All components are crystalline and identified (Table 1), and for case 2: Some components are amorphous and/or unidentified (Table 4).

Case 3, XRF: Assess all elements heavier than oxygen in any mixture, regardless of whether they are crystalline or amorphous.

A reference standard is required in this case. Apply Equation (6) to each element sought. Equation (7) may even work for metallic alloys (pending further research work).

Equations (6) (7) & (8) are the essence of the unified theory. Equation (6) is the basic decoding formula, applies to all three cases of mixtures. It is the key to solve most problems in XRD & XRF. Equation (7) transforms any complex mixtures into a set of the simplest binary mixtures for both XRD & XRF (see section on Adiabatic Principle). Equation (8) applies to XRD only.

In all three cases of mixtures, only the X-ray intensities (I_i &

I_s) are to be measured. Other factors are either constants (k_i & k_s) or known quantities (X_s & X_o). Consequently, one single XRD or XRF scan quantifies the chemical compounds or chemical elements in any mixtures. All work with calibration lines is eliminated. Of course, duplicates improve the precision of the data.

3.2 Specific Intensity

For easy comparison of the same property of different materials, usually a pure compound is assigned as a reference standard. For example, the specific gravity of zinc oxide (ZnO) is 5.47, which means that the mass of ZnO is 5.47 times as heavy as the mass of water at 4°C. The ubiquitous water is assigned as reference standard in this case. Note that both the density and specific gravity of water are defined to be exactly one (1.00). The word "specific" means per unit mass, such as the specific impulse of rocket fuel. In order to put all X-ray intensity data on the same scale, let us define a new term Specific Intensity (k_i) of any chemical element or chemical compound (i) as follows:

$$\text{Specific Intensity} = \frac{\text{XRD or XRF intensity of 1.00\% component, } I_i}{\text{XRD or XRF intensity of 1.00\% reference standard, } I_s} = \left(\frac{I_i}{I_s}\right)_{1\% \text{ each}} \quad (9)$$

Rearranging the basic decoding Equation (6), we have:

$$k_i = \frac{I_i / X_i}{I_s / X_s} = \left(\frac{I_i}{I_s}\right)_{1\% \text{ each}} = \left(\frac{I_i}{I_s}\right)_{50/50} = \text{Specific Intensity} \quad (10)$$

= slope of decoding formula (5) or (6)
= Reference Intensity Ratio (RIR).

The Specific Intensity (k_i) indicates how strong is the diffracted or emitted X-rays relative to that of the reference standard (Y_2O_3 or Al_2O_3 for XRD, Y in Y_2O_3 for XRF). We have k_s = 1.00 by definition. Yttrium Y is a rare element, unlikely to be present in the original sample. High purity chemical grade Y_2O_3 powder is white, stable, inexpensive, and readily available. Any pure compound can be chosen as the reference standard so long as it is not in the original sample. In the case of peak overlapping, select the resolved strongest peak for intensity data as long as the same resolved peak is used for both k_i and X_i measurements.

The specific intensity (k_i) plays six roles: (1) It puts all X-ray intensity data on the same scale. (2) It is a characteristic constant of each chemical element or compound. (3) It is the slope of the basic decoding formula. (4) It allows the determination of a set of k_i with one sample preparation[16]. (5) It unifies and simplifies XRD & XRF. (6) It replaces RIR in XRD with much broader implications.

3.3 Adiabatic Principle

The word "adiabatic" literally means "no crossing" In technical usage, it means "no heat enters or leaves the system" in thermodynamics, "no change of quantum numbers" in quantum mechanics, and "no matrix effects" in quantitative XRD & XRF analyses.

The decoding formula Equation (6), leads to the unique solution Equation (7). Equation (7) establishes the Adiabatic Principle.[16] Adiabatic Principle prescribes: The plot of intensity ratio vs. the weight ratio of any two components is a straight line passing through the origin with a slope equal to the ratio of corresponding specific intensities. This intensity-concentration relationship between each and every pair of components in a multi-component system is not affected by the presence or absence of other components. In other words, each pair of components in a multi-component mixture forms an isolated (adiabatic) unit, immune (adiabatic) from the coexistence of any other components.

Implant a known amount of a reference standard (Al_2O_3 or Y_2O_3) into a complex mixture and couple it with each component sought. The complex mixture becomes a set of binary mixtures, the simplest mixture. Moreover, one of the two components in each binary mixture is the known reference standard. It makes each binary mixture "simpler than the simplest mixture" a strange but hard fact. The Adiabatic Principle empowers the unified theory of XRD & XRF as the simplest way to quantify complex mixtures.

Equation (6) eliminates the matrix effects, the calibration lines, and the certified reference standards. Just one single XRD or XRF scan gives the weight fractions of all compounds or elements sought. Moreover, the weight fraction is expressed in terms of ratios, I_i/I_j and k_i/k_j, where i, j = 1, 2, ..., n. Therefore, many disturbing factors are mutually cancelled, including matrix effect, specimen thickness, specimen density, residual stress, micro-absorption, preferred orientation, crystallinity, extinction, particle size and size distribution, specimen position, instrumental parameters, instrumental drifting, and others.

4. DETERMINATION OF SPECIFIC INTENSITY

The procedures for specimen preparation are detailed in References

#15 & #17 for XRD, and #20 & #21 for XRF. Generally, the synthetic mixtures are wetted with methanol or ether, ground and homogenized. The methanol or ether will evaporate during the grinding process. The dry mixture is made bulk pellet or thin-film of routine standard size. Then the XRD or XRF intensity measurements are taken.

Each chemical element or compound has a characteristic Specific Intensity (k_i). The basic Equation (6) is applied to determine Specific Intensity (k_i) with known composition (X_i). A set of k_i's can be determined simultaneously by one sample preparation. Because k_i is constant, once it is measured, it can be used for future XRD or XRF analyses.

Two sets of k_i were measured and calculated by Equation (6). The XRD results are summarized in Table 1, and the XRF results are summarized in Table 2. The detailed calculations are included for easy reference.

4.1 Specific Intensity of XRD A Alpha

Table 1. Determination of a set of k_i for XRD with a Alpha Al_2O_3 as the reference standard

Components	Weight g	% Weight X_i	XRD Intensity hkl	I_i, cps	Sp. Int. k_i	50/50 Mixture RIR
ZnO	0.4183	16.29	101	1564	4.27	4.35
CdO	0.3562	13.87	111	2418	7.76	7.62
LiF	0.6302	24.54	200	729	1.32	1.33
CaF_2	0.5395	21.00	220	651	1.38	1.41
Ref. Std. $\alpha\ Al_2O_3$	0.6242	X_S 24.30	113	546	1.00	1.00
Total	2.5684 g	100.0 %				

The first three columns in Table 1 show the composition (X_i) of the synthetic mixture. The next three columns present the measured XRD intensity (I_i) and the calculated Specific Intensity (k_i). The last column indicates the Reference Intensity Ratio (RIR) of 50/50 binary mixtures for comparison. Given X_i and X_s, calculate k_i by Equation (6):

$$k_{ZnO} = \frac{X_s}{X_{ZnO}} \cdot \frac{I_{ZnO}}{I_s} = \frac{24.30}{16.29} \cdot \frac{1564}{546} = 4.27$$

$$k_{CdO} = \frac{X_s}{X_{CdO}} \cdot \frac{I_{CdO}}{I_s} = \frac{24.30}{13.87} \cdot \frac{2418}{546} = 7.76$$

$$k_{LiF} = \frac{X_s}{X_{LiF}} \cdot \frac{I_{LiF}}{I_s} = \frac{24.30}{24.54} \cdot \frac{729}{546} = 1.32$$

$$k_{CaF2} = \frac{X_s}{X_{CaF2}} \cdot \frac{I_{CaF2}}{I_s} = \frac{24.30}{21.00} \cdot \frac{651}{546} = 1.38$$

Besides Specific Intensity (k_i), this set of data validates Equation (8): Its left side equals its right side, indicating that all components are crystalline, with no amorphous content.

$$\sum_{1}^{n}\frac{I_i}{k_i} = \frac{1564}{4.27} + \frac{2418}{7.76} + \frac{729}{1.32} + \frac{651}{1.38} = 1701$$

$$\frac{X_o}{X_s} \cdot I_s = \frac{75.7}{24.3} \times 546 = 1701$$

4.2 Specific Intensity for XRF

Table 2. Determination of a set of ki's for XRF with Y in Y2O3 as the reference standard

Components	Weight g	% Weight of Element X_i	XRF Intensity I_i in cps	Specific Intensity k_i
ZnO	0.6429	32.8% Zn	1849	k_{Zn} = 2.84
BaSO$_4$	0.3678	13.7% Ba	531	k_{Ba} = 1.95
		3.21% S	3.95	k_S = 0.0619
K$_2$HPO$_4$	0.4633	13.2% K	169	k_K = 0.644
		5.23% P	4.25	k_P = 0.0409
Element Y as Ref. Std.		X_S	I_s	
Y$_2$O$_3$	0.1005	5.03% Y	100	k_Y = 1.00

Total weight   1.5745 grams

The first three columns in Table 2 show the composition (X_i) of the synthetic mixture. The last two columns present the measured XRF intensity (I_i) and the calculated Specific Intensity (k_i).

The intensity of L line of Ba and K lines of Zn, K, S, P, and Y were counted. Given X_y & X_i, calculate k_i by Equation (6).

Mol. wt. of ZnO = 81.37. Zn in ZnO = 65.37/81.37 = 80.34%. Zn in admixture = 0.8034 x 0.6429/ 1.5745 = 32.80%. Mol. wt. of $BaSO_4$= 233.40. Ba in $BaSO_4$= 58.84%. Ba in admixture = 0.5884 x 0.3678/1.5745 = 13.7%. Mal. wt. of Y_2O_3 = 136.91. Y in Y_2O_3 = 78.74%. Y in admixture = 0.7874 x 0.1005/1.5745 = 5.03%. Total weight of admixture (original mixture + Y_2O_3 = 1.5745 g. We have:

$$k_{Zn} = \frac{X_Y}{X_{Zn}} \cdot \frac{I_{Zn}}{I_Y} = \frac{5.03}{32.8} \cdot \frac{1849}{100} = 2.84$$

$$k_{Ba} = \frac{X_Y}{X_{Ba}} \cdot \frac{I_{Ba}}{I_Y} = \frac{5.03}{13.7} \cdot \frac{531}{100} = 1.95 \quad \text{Repeat for elements S, K \& P.}$$

5. QUANTITATIVE XRD & XRF ANALYSES

The key point of the unified theory is to measure the Specific Intensities (k_i) of components in a synthetic mixture by Equation (6), then determine the weight fractions of components sought (X_i) in unknown mixtures by Equations (6) (7) or (8). Note that Equations (7) & (8) are derived from the basic decoding Equation (6). Three sets of Xi were measured and calculated for three cases of mixtures: Cases 1 and 2 for XRD are collected in Tables 3 and 4. Case 3 for XRF is collected in Table 5. The strongest resolved X-ray peak is chosen for intensity measurements. A notable feature of the unified theory is that one single XRD or XRF scan quantifies the chemical compounds or elements in any mixture. Of course, duplicates improve the precision. Detailed lab procedures can be found in the cited references. All specific intensity constants (k_i) are either adopted from averages of prior work or measured from synthetic mixtures as shown in the previous section.

5.1 Case 1 Mixtures for XRD

Table 3. Case 1 for XRD: All components are crystalline and identified.

Component	Composition (gram)	Intensity I_i (cps)	Sp. Int. k_i	% Weight Known	% Weight Found
ZnO	0.2236	610	4.35	9.87	9.26
NiO	0.5454	1412	3.81	24.06	24.5
CdO	0.6588	3303	7.62	29.07	28.6
KCl	0.8386	2207	3.87	37.00	37.6
Total	2.2664	---	---	100.0	99.96

No reference standard is required in this case. It shows the applications of Equation (7) for XRD. Very good agreement is achieved between the known and the estimated weight fractions. Note that the only input is the intensity data from a single X-ray scan. We have:

$$X_{ZnO} = \left[1 + \frac{4.35}{610}\left(\frac{1412}{3.81} + \frac{3303}{7.62} + \frac{2207}{3.87}\right)\right]^{-1} = 9.26\%$$

$$X_{NiO} = \left[1 + \frac{3.81}{1412}\left(\frac{610}{4.35} + \frac{3303}{7.62} + \frac{2207}{3.87}\right)\right]^{-1} = 24.5\%$$

$$X_{CdO} = \left[1 + \frac{7.62}{3303}\left(\frac{610}{4.35} + \frac{1412}{3.81} + \frac{2207}{3.87}\right)\right]^{-1} = 28.6\%$$

$$X_{KCl} = \left[1 + \frac{3.87}{2207}\left(\frac{610}{4.35} + \frac{1412}{3.81} + \frac{3303}{7.62}\right)\right]^{-1} = 37.6\%$$

Equation (7) gives the weight fractions of all crystalline components. Equation (8) determines whether amorphous materials are present and by how much, as shown in Case 2.

5.2 Case 2 mixtures for XRD

Table 4. Case 2 for XRD: Some components are amorphous and/or unidentified

Component	Composition gram	Intensity I_i (cps)	Sp. Int. k_i	% wt. in admix. Known	% wt. in admix. Found	% wt in orig. mix. Known	% wt in orig. mix. Found
ZnO	0.8090	4948	4.35	40.8	38.7	54.0	51.3
CdO	0.2825	3337	7.62	14.2	14.9	18.9	19.7
Resin powder	0.4057	0	—	20.5	21.9	27.1	29.0
Added Ref. Std. αAl_2O_3	0.4854	719	1.00	24.48	24.5	0	0

A reference standard is required in this case, which shows the application of Equations (6) and (8) for XRD. The method gives a good estimate of amorphous content in totally unknown mixture. Applying Equation (6), we have:

$$X_{ZnO} = \frac{24.48}{4.35} \cdot \frac{4948}{719} = 38.7\% \text{ in admixture by experiment}$$

$$X_{CdO} = \frac{24.48}{7.62} \cdot \frac{3337}{719} = 14.9\% \text{ in admixture by experiment}$$

Since $X_s = 24.48\%$, $X_o = 75.52\%$. Applying Equation (8), we have:

$$\sum_1^n \frac{I_i}{k_i} = \frac{4948}{4.35} + \frac{3337}{7.62} = 1575, \text{ and } \frac{X_o}{X_s} \cdot I_s = \frac{75.52}{24.48} \cdot 719 = 2218$$

$$\sum_1^n \frac{I_i}{k_i} \ll \frac{X_o}{X_s} \cdot I_s \; ; \quad \text{that is:} \quad 1575 \ll 2218$$

Because all peaks in the diffraction pattern are identified, Equation (8) indicates the presence of amorphous materials in the admixture. Its weight fraction can be determined by mass balance: 100 - 38.7 - 14.9 - 24.5 = 21.9%.

Note that all symbols in Equations (6) (7) & (8) are referring to the admixture (original sample plus the implanted reference standard Al_2O_3). Equation (11) converts X_i to X_i°, the weight fraction of component (i) in originals sample.

$$X_i^\circ = \frac{X_i}{1 - X_s} = X_i \cdot \frac{\text{Wt. of admixture}}{\text{Wt. of original mixture}} \tag{11}$$

$$X°_{ZnO} = 38.7 \cdot \frac{1.9826}{1.4972} = 51.3\%, \quad X°_{CdO} = 14.9 \cdot \frac{1.9826}{1.4972} = 19.7\%$$

$$X°_{Resin} = 21.9 \cdot \frac{1.9826}{1.4972} = 29.0\% = \text{Amorphous content in mixture.}$$

Double checks: 38.7 + 14.9 + 21.9 + 24.5 = 100% of admixture (with Al_2O_3).

51.3 + 19.7 + 29.0 = 100% of original sample (without Al_2O_3).

Table 5. Case 3 Mixtures for XRF: Assess all elements heavier than oxygen in any mixture regardless crystalline or amorphous.

	Ba	Zn	K	S	P	Y (ref. std.)
Intensity, cps	1680	343	126	12.8	3.00	113
k_i = Specific Intensity	1.96	2.88	0.646	0.0635	0.0393	1.00
Found % wt, admixture	39.6	5.50	9.01	9.31	3.53	---
True % wt, admixture	39.2	5.51	8.95	9.15	3.54	5.22
Found % wt, orig. mixture	42.4	5.89	9.60	9.97	3.77	0

Composition of synthetic Mixture: Mix 1.0511 g $BaSO_4$, 0.1082 g ZnO, 0.3147 g K_2HPO_4. Add 0.1047 g Y_2O_3 as reference standard. Total mixture is 1.5787 g.

A reference standard is required in this case, which shows the application of Equation (6) for XRF. All intensity data are from one single XRF scan, after the material was ground and homogenized. The XRF intensities of the six elements are listed in the first row in Table 5. The constant Specific Intensity (k_i's) are either known from averages of prior work, or measured from synthetic mixtures as shown in the previous section (Table 2). All other data in Table 5 are results of calculations as shown below:

$$X_{Ba} = \frac{5.22}{1.96} \cdot \frac{1680}{113} = 39.6\% \text{ in admixture (original mixture plus } Y_2O_3\text{)}.$$

$$X_{Zn} = \frac{5.22}{2.88} \cdot \frac{343}{113} = 5.50\% \text{ in admixture.} \quad \text{Convert } X_i \text{ to } X_i° \text{ by Equation (10), we have}$$

$$X°_{Ba} = X_{Ba} \cdot \frac{\text{Wt. of admixture}}{\text{Wt. of original mixture}} = 39.6 \times \frac{1.5787}{1.4740} = 42.4\% \text{ in original mixture}.$$

$$X°_{Zn} = 5.50 \times \frac{1.5787}{1.4740} = 5.89\% \text{ in original mixture}$$

The same calculations are run for other elements K, S & P. All intensity data are from a single XRF scan of one sample preparation. No certified standards are required, because no calibration lines are used. Each element has a characteristic constant slope (k_i), which equals its specific intensity. Good agreement is attained between the known and the found weight fractions.

If this XRF job were performed conventionally, [9-12] it would require five calibration lines for the five elements sought. Each calibration line needs three data points plus the sample itself. That means 5x4 = 20 times more sample preparations from certified standards, 20 times more X-ray scans, 20 times more X-ray intensity measurements, 20 times more calculations, and 20 times more plotting. The situation would be similar with the XRD analyses discussed in Cases 1 and 2. [2-6]

The unified theory of XRD & XRF has been practiced in our lab. It reduces some 80% lab work by current practice. The precision was evaluated statistically ± 5% or better, [15,20] which is typical and sufficient to solve most problems. The Rietveld and the whole pattern fitting methods for powder diffraction have the potential to approach the precision limit of ± 1%.[18, 23-26]

6. CONCLUSIONS

Three new insights unify and simplify the quantitative XRD & XRF analyses. First, every component in the same specimen has identical matrix factors (µ, ϱ & d), which can be neatly cancelled mathematically. Secondly, the calibration line of any chemical element or compound has a characteristic constant slope, which eliminates all calibration work. Thirdly, the adiabatic principle of XRD & XRF transforms any complex mixture into a set of the simplest binary mixtures. These insights led to four actions to realize the six ideals in the wish list.

1. **Define the Specific Intensity (k_i):** Each chemical compound or chemical element has a unique (constant) specific intensity, which puts all X-ray intensity data on the same scale. A set of specific intensities can be determined with one sample preparation.
2. **Derive the decoding formula:** Equation (6) applies to both XRD & XRF. It is free from matrix factors and works for the full range of concentrations. Its characteristic slope equals its Specific Intensity (k_i), eliminating the tedious work with calibration lines.
3. **Apply the Adiabatic Principle:** It breaks any complex mixture into a set of binary mixtures. Moreover, one component of each binary mixture is the known reference

standard, hence any complex mixture becomes a set of simplest binary mixtures.

4. **Decode the composition of mixtures:** One single XRD or XRF scan quantifies the chemical compounds or chemical elements in any mixtures with a precision of ± 5% or better, thus reducing some 80% lab work of current practice.

7. REFERENCES

Azaroff, L. V. (1968). *Elements of X-Ray Crystallography*, p. 202. New York: McGraw Hill.

Bish, D. L.; and S. A. Howard (1988). "Quantitative Phase Analysis Using the Rietveld Method." *J. Appl. Cryst.* 21: pp. 86-91.

Bunn, C. W. (1961). *Chemical Crystallography*, p. 223. London: Oxford.

Chung, F. H. (1974). "A New X-Ray Diffraction Method for Quantitative Multi-component Analyses." Advances in X-Ray Analysis, Vol. 17, pp. 106-115.

Chung, F. H. (1974). "Quantitative Interpretation of X-Ray Diffraction Patterns of Mixtures. I. Matrix Flushing Theory for Multi-Component Analysis." J. Appi. Crystallogr. 7: 519-525.

Chung, F. H. (1974). "Quantitative Interpretation of X-Ray Diffraction Patterns of Mixtures. II. Adiabatic Principle of X-Ray Diffraction Analysis." J. Appl. Crystallogr. 7: 526-531.

Chung, F. H. (1975). "Quantitative Interpretation of X-Ray Diffraction Patterns of Mixtures. III. Simultaneous Determination of a Set of Reference Intensities." J. Appl. Crystallogr. 8: 17-19.

Chung, F. H. (2017). "Unified Theory for Decoding the Signals from X-Ray Fluorescence and X-Ray Diffraction of Mixtures." Applied Spectroscopy, Vol. 71(5), pp. 1060-1068.

Chung, F. H.; A. J. Lentz; and R. W. Scott (1974). "A Versatile Thin-Film Method for Quantitative X-Ray Emission Analysis." X-Ray Spectroscopy. 3: 172-175.

Chung, F. H.; and D. K. Smith (2000). *Industrial Applications of X-Ray Diffraction*. pp. 3, 13, 37 & 511. New York: Marcel Dekker.

Clark, G. L.; and D. H. Reynolds (1936). At University of Illinois, adapted the highly developed quantitative "Internal Standard" method of ultraviolet spectroscopy for quantitative XRD analysis.

Clark, G. L.; and D. H. Reynolds (1936). "Quantitative Analysis of Mine Dusts." Ind. Eng. Chem., Anal Ed. 8. p. 36.

Cohen, L. H.; and D. K. Smith (1989). "Thin-Specimen X-ray Fluorescence Analysis of Major Elements in Silicate Rocks." Anal. Chem. 61: 1837-1840.

Copeland, L. E.; and R. H. Bragg (1958). "Quantitative X-Ray Diffraction Analysis." Anal. Chem. 30(2), 196-201.

Cullity, B. D.; and S. R. Stock (2001). *Elements of X-Ray Diffraction*, 3rd Edition. New Jersey: Prentice Hall, Inc.

Grieken, R. V.; and A. A. Markowicz (2002). *Handbook of X-ray Spectrometry*: Methods and Techniques, 2nd ed. New York: Marcel Dekker.

Herglotz, H. K.; and L. S. Birks (1978). *X-Ray Spectrometry*.

New York: Marcel Dekker.

Hubbard, C. R.; C. R. Robbins; and R. L. Snyder (1983). "XRD Quantitative Analysis Using the NBS Quant82 System." Adv. X-Ray Anal. 26: pp. 149-157.

Jenkins, R.; R. W. Gould; and D. Gedcke (2002). *Quantitative X-Ray Spectrometry*, 2nd ed. New York: Marcel Dekker.

Jenkins, R.; and R. L. Snyder (1995). *Introduction to X-ray Powder Diffractometry*. New York: John Wiley and Sons. pp. 369 - 376.

Klug, H. P.; and L. E. Alexander (1974). *X-ray Diffraction Procedures for Polycrystalline and Amorphous Materials*. New York: John Wiley & Sons.

MIT X-ray SEF List of Resources (2017). Textbooks on X-Ray Diffraction. Website: http://prism.mit.edu/xray/oldsite/resources.htm. It lists "the seminal volumes for X-Ray Powder Diffraction."

Muller, R. O. (1972). *Spectrochemical Analysis by X-Ray Fluorescence*. pp. 59-61. New York: Plenum Press.

Scheibe (1933). *Chemische Spektralanalyse*, Physikalische Methoden der Analytischen Chemie, Vol. *i*, p. 108, pp. 125-130. Leipzig, Akademische Verlagsgesellschaft (in German).

Snyder, R. L. (1992). "The Use of Reference Ratios in X-ray Quantitative Analysis." Powder Diffraction. 7: pp. 186-193.

Smith, D. K.; G. G. Johnson; et al. (1987). "Quantitative X-ray Powder Diffraction Method Using the Full Diffraction Pattern." Powder Diffraction, 2: pp. 73-77.

4.

Matrix Flushing Techniques for Quantitative X-ray Diffraction of Mixtures

> You find beauty in ordinary things, do not lost this ability.
>
> — Unknown

1. ABSTRACT

A matrix-flushing method for quantitative multi-component analysis by X-ray diffraction is reported. It is simpler and faster than, yet as reliable as, the conventional internal-standard method. In this new method, the calibration-curve procedure is shunted; a more fundamental 'matrix-flushing' concept is introduced. The matrix-flushing theory gives an exact relationship between intensity and concentration free from matrix effect. Contrary to most theoretical methods, the working equation is very simple, no complicated calculations are involved. The matrix-flushing theory and the analytical procedure are presented. Eight illustrative examples are drawn to demonstrate how this theory is applied to multi-component analysis and amorphous-content determination. A novel 'auto-flushing' phenomenon of binary systems was observed, which appears to make the analysis of any binary system a simple matter.

2. INTRODUCTION

Besides broad routine analysis X-ray diffraction is often the only technique available for distinguishing polymorphic structures and analyzing solid solutions. The results obtained are in terms of materials as they occur in the sample, not in terms of elements or ions present. The pigment industry is especially concerned about polymorphic forms and solid solutions. For example, the best red pigment quinacridone (Chung & Scott, 1971) has four polymorphic forms, only two of which have commercial value; the corrosion inhibiting pigment zinc molybdate (Kirkpatrick & Nilles, 1972) is composed of a solid solution of zinc oxide and molybdenum oxide. However, the only analytical technique, X-ray diffraction, dealing with these situations is hampered by a matrix effect. So far, the best method of quantitative X-ray diffraction analysis is still the internal-standard method developed before 1948 (Alexander & Klug, 1948). It involves the usual procedure of constructing a calibration curve from standards, which is rather tedious since each component sought needs a calibration curve, each calibration curve needs at least three standards, and each standard must contain exactly the same percentage of pure reference material chosen.

A new X-ray diffraction method for quantitative multicomponent analysis has been developed. It is much simpler and faster than, yet as reliable as, the conventional internal-standard method.

Since 1970, the Joint Committee on Powder Diffraction Standards has published a set of Reference Intensities (I/I_c) (Berry, 1970) of binary mixtures made with a pure material and synthetic corundum (α-Al$_2$O$_3$) by one-to-one weight ratio, where i and I_c are the intensities of the strongest lines of the pure material and corundum respectively. These Reference Intensities are intended for "rough quantitative X-ray diffraction analysis of mixtures." Occasionally there is a need for a rapid quantitative X-ray diffraction analysis with moderate accuracy in our industrial analytical laboratories; therefore, attempts were made to utilize these Reference Intensities. This thought led to a new X-ray diffraction method for quantitative multicomponent analysis. In this new method, the conventional calibration-curve procedure is shunted and a more fundamental "matrix-flushing" concept is introduced. The matrix-flushing concept gives an exact relationship between intensity and concentration since neither assumption nor approximation is made. Contrary to most theoretical methods, the working equation of this matrix-flushing method is very simple, no complicated calculations are involved. It is felt that one can utilize this method without understanding its theoretical basis.

The matrix-flushing theory and the analytical procedure are presented below. Experimental data are collected to demonstrate how this theory is applied to multicomponent analysis and amorphous-content determination. A novel "auto-flushing" phenomenon of binary systems emerged which makes the analysis of any binary system a simple matter.

3. MATRIX FLUSHING THEORY

In contrast with X-ray emission analysis, X-ray diffraction analysis has two factors favoring a more fundamental approach. Firstly, the primary and the diffracted X-rays are monochromatic and of the same wavelength. Secondly, the matrix effect consists of absorption only, no enhancement.

When the monochromatic primary X-rays impinge on a flat powder specimen, the intensities of diffracted rays are related to the percentage composition by equation (1) as derived by Klug & Alexander (1959),

$$I_i = K_i \frac{X_i/\varrho_i}{\sum \mu_i X_i} = K_i \frac{X_i/\varrho_i}{\mu_t}. \tag{1}$$

Hence we have:

$$\frac{I_i}{I_i^0} = X_i \frac{\mu_i}{\mu_t} \tag{2}$$

where:

I_i = Intensity of X-rays diffracted by (hkl) of component i.

I^o_i = Intensity of X-rays diffracted by (hkl) of pure compound i.

K_i = A constant which depends upon the geometry of the diffractometer and the nature of component i.

X_i = Weight fraction of component i.

d_i = Density of component i.

μ_i = Mass absorption coefficient of pure component i.

μ_t = Mass absorption coefficient of the total specimen exposed to primary X-rays, including component i, internal standard and reference material added, if any.

The μi and μ_t in equation (2) are called the 'absorption effect' which complicates X-ray diffraction analysis. Even μ_i, the mass absorption coefficient of pure compound i, has not been accurately measured. So the best way of approaching this problem is to flush all these μ factors out of the intensity-concentration equation. In order to achieve this, a flushing agent, which may be any pure compound *not* present in the sample, is added into a sample of n components. Let the weight fraction of the flushing agent and original sample be designated X_f and X_o respectively, that is

$$X_f + X_o = X_f + \sum_{i=1}^{n} X_i = 1 . \tag{3}$$

Then from equation (2):

$$\left. \begin{array}{l} \dfrac{I_i}{I_i^0} = X_i \dfrac{\mu_i}{\mu_t} \\[6pt] \dfrac{I_f}{I_f^0} = X_f \dfrac{\mu_f}{\mu_t} \end{array} \right\} . \tag{4}$$

We have:

$$\left(\dfrac{I_i}{I_f}\right)\left(\dfrac{I_f^0}{I_i^0}\right) = \left(\dfrac{X_i}{X_f}\right)\left(\dfrac{\mu_i}{\mu_f}\right) . \tag{5}$$

The individual values of $I^0{}_f$ and $I^0{}_i$ are dependent upon the atomic positions in their respective crystal structures. However, the ratio $I^0{}_f/I^0{}_i$ can be obtained through the Reference Intensities. In the meantime, all the matrix factors in equation (5) can be flushed out completely as shown below.

For a binary mixture of compound i and corundum c with one-to-one weight ratio, similar to equation (5), we have

$$\left(\frac{I_i}{I_c}\right)\left(\frac{I_c^0}{I_i^0}\right) = \left(\frac{\mu_i}{\mu_c}\right). \tag{6}$$

For the sake of simplicity, let k_i stand for the Reference Intensity ratio:

$$\frac{I_i}{I_c} = k_i, \tag{7}$$

then equation (6) becomes

$$\left(\frac{I_c^0}{I_i^0}\right) = \left(\frac{1}{k_i}\right)\left(\frac{\mu_i}{\mu_c}\right) \tag{8}$$

and similarly:

$$\left(\frac{I_c^0}{I_f^0}\right) = \left(\frac{1}{k_f}\right)\left(\frac{\mu_f}{\mu_c}\right). \tag{9}$$

Since $I^0{}_t$, $I^0{}_c$ and $I^0{}_f$ and are the intensities of the strongest lines of the X-ray diffraction patterns of respective pure materials, which are constants from the same diffractometer, we get:

$$\left(\frac{I_f^0}{I_i^0}\right) = \left(\frac{k_f}{k_i}\right)\left(\frac{\mu_i}{\mu_f}\right). \tag{10}$$

Substituting equation (9) into equation (5) we have:

$$\left(\frac{I_i}{I_f}\right)\left(\frac{k_f}{k_i}\right) = \left(\frac{X_i}{X_f}\right)$$

$$X_i = X_f \left(\frac{k_f}{k_i}\right)\left(\frac{I_i}{I_f}\right). \tag{11}$$

Equation (11) is a very important conclusion. First, it gives a simple relationship between intensity and concentration. Secondly, it is free from matrix effect; all absorption factors are neatly flushed out. Thirdly, it is an exact deduction; neither assumption nor approximation was made to reach the conclusion. This equation (11) can be used for quantitative multicomponent analysis. Any pure compound which is not a component of the sample can be used as the flushing agent. Yet one more simplification can be realized.

Since corundum (α-Al_2O_3) has been chosen for Reference Intensities by the Powder Diffraction File for its purity, stability and availability, it is convenient to choose the same corundum as a flushing agent for the same good reasons. Consequently $k_f = k_c = 1$, where the subscripts f and c stand for flushing agent and for corundum, and we obtain:

$$X_i = \left(\frac{X_c}{k_i}\right)\left(\frac{I_i}{I_c}\right). \tag{12}$$

This is the working equation for quantitative multicomponent analysis. It represents a straight line passing through the origin with a slope equal to X_c/k_i. It is free from matrix factors. No previous information as to the approximate concentration ranges of various components sought is required. Intensity ratios from the same scan are the experimental data needed. Since it uses intensity ratios of the same scan, the errors due to instrumental drift and sample preparation are minimized. Note that the working equation (12) not only comprehends the internal-standard method (Alexander & Klug, 1948; Klug & Alexander, 1959) but also prescribes the slope of calibration curve for *every* component in the sample, thus it is not necessary to actually work out these calibration curves. Furthermore, it can be used for amorphous-content determination.

When corundum is chosen as the flushing agent, applying equation (12) to equation (3), we get:

$$X_c + \sum_{i=1}^{n} X_i = 1, \tag{13}$$

$$X_c + \frac{X_c}{I_c} \sum_{i=1}^{n} \frac{I_i}{k_i} = 1, \tag{14}$$

$$\sum_{i=1}^{n} \frac{I_i}{k_i} = \left(\frac{X_o}{X_c}\right) I_c \quad \because 1 - X_c = X_o. \tag{15}$$

Equation (15) affords a means to check experimentally the correctness of the matrix-flushing theory, to appraise the reliability of intensity data, and to predict and assay the presence of amorphous materials in a sample.

As no restriction is imposed by the theory upon the weight fraction of flushing agent X_f in equation (11) and X_c in equations (12) and (15), if the specimen is made of exactly 50·00% original sample and 50·00% of corundum, then equations (12) and (15) are reduced to the following rather interesting relations:

$$X_i = \frac{1}{2k_i} \left(\frac{I_i}{I_c}\right) \tag{16}$$

$$\sum_{i=1}^{n} \frac{I_i}{k_i} = I_c. \tag{17}$$

For a binary system, a novel "auto-flushing" phenomenon exists. No flushing agent is needed. One component automatically serves as a flushing agent for the other component and *vice versa*, which means quantitative analysis can be done by merely grinding the original sample and scanning it.

Let the weight fraction of the two components of a binary mixture be X_1 and X_2 similar to the derivation of equation (10) we have:

$$\left. \begin{array}{l} X_1 + X_2 = 1 \\ \dfrac{I_1}{I_2} = \dfrac{k_1}{k_2} \cdot \dfrac{X_1}{X_2} \end{array} \right\} . \quad (18)$$

Solving these two simultaneous equations, we get:

$$X_1 = \frac{1}{1 + \dfrac{k_1}{k_2} \cdot \dfrac{I_2}{I_1}} . \quad (19)$$

Therefore, the quantitative composition of a binary system can be easily calculated from the intensity ratio of the strongest diffraction peak from each component in a single diffraction scan.

Mathematically, equation (19) represents a straight line passing through the origin with a slope $k=k_1/k_2$ when I_1/I_2 is plotted against X_1/X_2.

Note that this auto-flushing phenomenon is true for any binary system. It is interesting to observe that this linear relationship of binary systems of polymorphic crystals was recognized quite a long time ago. Prime examples are the determination of anatase in rutile TiO_2 (Spurr & Myers, 1957), quartz in cristobalite SiO_2 (Alexander & Klug, 1948), calcite in aragonite $CaCO_3$ (Azaroff & Buerger, 1958), *etc.* This is because in such special cases $\mu_1=\mu_2$ and inherently there are no matrix effects whatsoever in the polymorphic systems. Nevertheless, the slope of the straight line, k, of each of these systems was acquired through a calibration curve of standards.

The new way of acquiring the slope k for any binary system and the procedure of quantitative analysis for multicomponent systems are presented in the experimental section below.

4. EXPERIMENTAL

Eight synthetic samples were prepared to illustrate the application of the matrix-flushing theory. The chemicals used were certified reagent grade or better. The rutile and anatase titanium dioxide were pigment-grade samples containing 99.7% rutile and 99.8% anatase respectively. The flushing agent, synthetic corundum, was Linde semiconductor-grade -Al_2O_3 (1.0 micron, Union Carbide).

In order to ensure optimum particle size and sample homogeneity, all samples were ground for about 20 min with an automatic mortar grinder (Fisher Scientific Co.). To avoid preferred orientation, the free-falling method of sample preparation recommended by NBS Monograph 25 (1971) was employed.

The intensity data were collected by use of a Norelco diffractometer equipped with a solid-state scintillation counter, graphite-monochromatized Cu K_α radiation, integrated-circuit control panel, and pulse-height analyzer. A constant-time (40 s) counting technique was used for each peak and each background. The accumulated counts range from 20 000 to 250 000 counts for the period of 40 s after background correction. Duplicate readings were taken for each peak and each background. A strip-chart trace of the X-ray diffraction pattern of each sample was recorded for later reference.

4.1 Reference Intensities

The relative intensities of peaks in an X-ray diffraction pattern of a crystalline compound vary slightly depending on the diffractometer design and instrumental conditions. The Reference Intensities (I/I_c) obtained from different laboratories hardly agree beyond two significant figures. Consequently, the Reference Intensities in the Powder Diffraction File are quoted to only two digits. If quantitative results are required, it is necessary to determine the Reference Intensities of materials concerned using the same diffractometer under the same instrumental conditions.

Table 1. *Reference Intensities by Counting*

	PDF card	Intensity (c.p.s.) I_i	Intensity (c.p.s.) I_c	Reference intensity, I_i/I_c By counting	Reference intensity, I_i/I_c PDF
ZnO	5–664	8178	1881	4·35	4·5
KCl	4–587	4740	1223	3·87	3·9
LiF	4–857	3283	2487	1·32	1·3
$CaCO_3$	5–586	4437	1491	2·98	2·0
TiO_2 rutile	21–1276	2728	1040	2·62	3·4
TiO_2 anatase	21–1272	3573	1054	3·39	4·3

Six chemical compounds were used for the following illustrative examples. Their Reference Intensities (I/I_c) and relevant data were determined and are listed in Table 1. The instrumental conditions of the diffractomter are given in Table 2. All subsequent intensity data were collected under the same instrumental conditions.

Table 2. *Diffractometer Conditions*

Copper X-ray tube, broad focus	50 kV, 40 mA
Scintillation detector	839 V
Gain	128
Baseline	1·0 V
Window	3·0 V
Constant time	40 s

4.2 Multicomponent Analysis

Four samples were prepared to demonstrate the application of the matrix-flushing theory to multicomponent analysis. Sample

1 consisted of three components ZnO, KCl and LiF. A known quantity of flushing agent (~16% α-Al_2O_3) was added into this sample which was then ground to a homogeneous fine powder. All the peaks in the diffraction pattern of this mixture were clearly resolved, and the intensity of the strongest line of each component was measured. The working equation (12) was applied to each component; for example:

$$X_{ZnO} = \frac{17\cdot96}{4\cdot35} \times \frac{5968}{599} = 41\cdot14\% \text{ (Experiment)}$$
$$41\cdot49\% \text{ (True Value)}$$

The true composition, the intensity data, and the composition found are listed in Table 3. It is found that even the intensities read directly from the pattern on a stripchart, Fig. 1, give fairly good accuracy.

The correctness of the matrix-flushing theory is further scrutinized by equation (15):

$$\sum \frac{I_i}{k_i} = \frac{5968}{4\cdot35} + \frac{2845}{3\cdot87} + \frac{810}{1\cdot32} = 2721 \text{ (Experiment)}$$

$$\frac{X_o}{X_c} \cdot I_c = \frac{82\cdot04}{17\cdot96} \times 599 = 2736 \text{ (Theory)}.$$

Table 3. *Intensity and Composition Data*

Sample	Composition (g)		Intensity (c.p.s.)	% Composition Known	% Composition Found	$\sum \dfrac{I_i}{k_i}$	$\dfrac{X_0}{X_c} \cdot I_c$
1	ZnO	1·8901	5968	41·49	41·14		
	KCl	1·0128	2845	22·23	22·04		
	LiF	0·8348	810	18·32	18·40		
	Al$_2$O$_3$ Flushing	0·8181	599	17·96	–	2721	2736
2	ZnO	0·9532	2856	18·98	19·10		
	KCl	0·6601	1651	13·15	12·41		
	LiF	0·8972	765	17·87	16·86		
	Al$_2$O$_3$ Flushing	2·5114	1719	50·00	–	1662	1719
3	ZnO	0·6759	2408	24·38	25·36		
	TiO$_2$ (R)	0·4317	931	15·57	16·28		
	CaCO$_3$	1·1309	2558	40·79	39·36		
	Al$_2$O$_3$ Flushing	0·5341	420	19·26	–	1767	1761
4	ZnO	0·0335	120	1·38	1·35		
	TiO$_2$ (R)	0·0633	139	2·60	2·57		
	CaCO$_3$	1·9197	4756	78·96	77·36		
	Al$_2$O$_3$ Flushing	0·4147	352	17·06	–	1677	1711
5	ZnO	0·9037	4661	34·43	36·41		
	CaCO$_3$	0·7351	2298	28·00	26·20		
	SiO$_2$ (Gel)	0·4234	0	16·13	15·95		
	Al$_2$O$_3$ Flushing	0·5629	631	21·44	–	1842	2312
6	ZnO	1·4253	6259	71·22	72·07		
	TiO$_2$ (R)	0·5759	1461	28·78	27·93		
7	TiO$_2$ (R)	0·7418	1373	43·80	42·70		
	TiO$_2$ (A)	0·9518	2386	56·20	57·30		
8	KCl	2·4530	5371	74·90	74·64		
	LiF	0·8219	604	25·10	25·36		

This proves that the relationship between X-ray intensity and concentration derived from the matrix-flushing theory is real.

Sample 2 is composed of the same three components in different quantities. Exactly 50·00% α-Al$_2$O$_3$ was incorporated into the sample. Again, a good agreement between experimental values and true values is obtained and the interesting relationship that the sum of I_i/k_i, should be equal to the intensity of corundum I_c, is fulfilled. It also shows that the matrix-flushing effect is independent of the amount of flushing agent used.

Sample 3 is made of ZnO, TiO$_2$ and CaCO$_3$ with about 20% flushing agent α-Al$_2$O$_3$. In this example, not all of the strongest lines are resolved. A CaCO$_3$ line interferes with the strongest line of α-Al$_2$O$_3$. The intensity of the next strongest line of α-Al$_2$O$_3$ was measured and this is used to calculate the intensity of the strongest line of α-Al$_2$O$_3$ with reference to the pattern of pure α-Al$_2$O$_3$ obtained by use of the same diffractometer under the same conditions. Two other CaCO$_3$ lines overlap the first and second strongest lines of ZnO. The intensity of the third strongest line of ZnO was counted to calculate the intensity of the strongest line of ZnO as before. Very good agreement between the experimental values and the true values is shown in Table 3.

Sample 4 was prepared to check the sensitivity and detection limit of the method. It contains about 1% ZnO, % TiO$_2$, and 80% CaCO$_3$ with 17% α-Al$_2$O$_3$ as flushing agent. It is apparent that the same working equation (12) can be applied regardless of the concentration range. Note that the scintillation counter might be flooded if the concentration of a strongly diffracting component is extremely high (such as 90%), while the Reference Intensities are obtained at 50% concentration. In such situations, more α-Al$_2$O$_3$ should be added to bring the concentration of the major component down. The lower concentration level, however, is limited only by the minimum size of the peak above background which is usually taken as three standard deviations of the background intensity. A sensitivity of 5 c.p.s. or better is typical with the solid-state scintillation counter. The detection limit depends upon the nature of the sample. Generally 0·5% by weight should be detectable.

The precision of intensity measurement is always in our favor. For example, the TiO$_2$ peak in this case gives 7394 counts with a background of 1824 counts in a fixed time of 40 s. Since duplicate readings were taken the relative standard deviations (Jenkins & DeVries, 1967) is 1·7%, which is quite good in normal analysis.

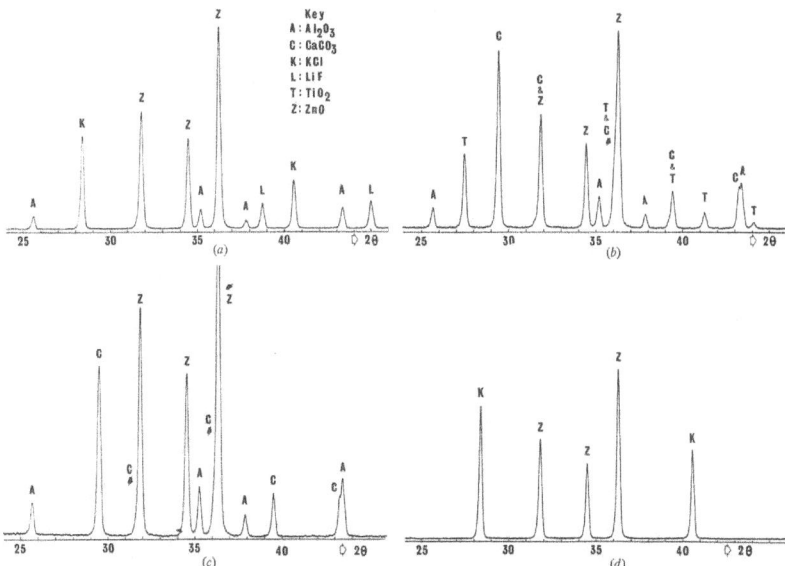

Fig 1. The X-ray diffraction patterns of (*a*) Sample 1 where all peaks are resolved, (*b*) Sample 3 where are some overlapping peaks, (*c*) Sample 5 which contains amorphous amterial, (*d*) the ZnO-KCl binary system for slope determination

$$\varepsilon\% = \frac{100\sqrt{2}}{\sqrt{T}} \frac{\sqrt{R_p + R_b}}{R_p - R_b} = \frac{100\sqrt{2}}{\sqrt{2 \times 40}} \frac{\sqrt{185 + 46}}{185 - 46} = 1 \cdot 7\%$$

4.3 Amorphous Content Determination

Traditionally, X-ray diffraction analysis provides information about crystalline components in the sample. The presence or absence of amorphous materials is generally ignored.

An interesting feature of this matrix-flushing method is that it can be used to detect and determine the total amorphous material in a sample.

Sample 5 contains three components ZnO, $CaCO_3$ and SiO_2 (silica gel, amorphous). A normal scan of this sample, Fig. 1, does not indicate the presence of amorphous material. However, the matrix-flushing method shows a large intensity imbalance as shown in Table 3.

In this case

$$\sum \frac{I_i}{k_i} \ll \frac{X_0}{X_c} I_c$$

which indicates the presence of amorphous materials. By the use of equation (12) and material balance, 15.95% amorphous material is found. Comparing this with the 16.13% silica gel actually put in, the recovery is felt to be very good.

4.4 Auto-flushing

For any binary system, an auto-flushing phenomenon emerges. Each component automatically serves as flushing agent for the other component. No flushing agent is needed. The percentage composition can be obtained from a regular scan of the binary mixture. Three examples are cited to depict the features of the auto-flushing phenomenon.

Sample 6 is a binary system of ZnO and TiO$_2$ which represents a general case. Substituting the intensity ratio of the strongest line of ZnO to that of TiO$_2$ into equation (19), we have:

$$X_{ZnO} = \frac{1}{1 + \frac{4.35}{2.62} \cdot \frac{1461}{6259}} = 72.07\%$$

$$X_{TiO2} = \frac{1}{1 + \frac{2.62}{4.35} \cdot \frac{6259}{1461}} = 27.93\%$$

These results compare very favorably with the true values as shown in Table 3.

Sample 7 is a special case of a binary system. It is composed of rutile and anatase TiO_2, a polymorphic binary system. There is no matrix effect whatsoever in such a system, hence a linear relationship between X-ray intensity and concentration exists for each component:

$$I_1 = k_1^* X_1$$
$$I_2 = k_2^* X_2$$
$$\frac{I_1}{I_2} = \frac{k_1}{k_2} \cdot \frac{X_1}{X_2}.$$

Note that k^*_1 and k^*_2 are *not* the Reference Intensities. They are unknown constants and have to be determined by calibration curves, which is what Alexander & Klug have demonstrated.

Fascinatingly, the k_1 and k_2 in equation (18) derived the matrix-flushing theory are indeed the Reference Intensities. The difference here is that k^*_1 and k^*_2 do contain absorption factors (μ) while the ratio (k_1/k_2) does not. It was previously shown that all μ factors are flushed out of the basic intensity-concentration equations (12), (15), (18) and (19).

Other polymorphic binary systems have been discussed in the literature, *e.g.* quartz and cristobalite SiO_2 (Alexander & Klug, 1948), calcite and aragonite $CaCO_3$ (Azaroff & Buerger, 1958), *etc.* The linear intensity-concentration relationship of all polymorphic systems was recognized and determined by calibration curves. The general nature of this linear relationship and the simplest way to obtain the slope was not discussed in the cited references.

Sample 8 is made if LiF and KCl to illustrate the simpler way to determine the slope $k = k_1/k_2$ in equation (19) without using the

Reference Intensities at all. The slope k is simply the corresponding intensity ratio of a 50/50 mixture of the same two components.

$$k = \left(\frac{I_{LiF}}{I_{KCl}}\right)_{50/50} = \frac{1846}{5583} = 0.331.$$

Note that:

$$\frac{k_1}{k_2} = \frac{1.32}{3.87} = 0.341$$

$$X_{LiF} = \frac{1}{1 + 0.331 \times \frac{5371}{604}} = 25.36\%.$$

Similarly for a ZnO-KCl binary system:

$$k = \left(\frac{I_{ZnO}}{I_{KCl}}\right)_{50/50} = \frac{4747}{3956} = 1.20$$

$$\frac{k_1}{k_2} = \frac{4.35}{3.87} = 1.13.$$

In conclusion, a general rule of auto-flushing for binary systems can be pronounced as follows:

For any binary system, the plot of weight ratio to X-ray intensity ratio is always a straight line passing through the origin. The slope of this line is equal to a singular value which is simply the corresponding intensity ratio of a 50/50 mixture of the two components.

In simple mathematics:

$$\frac{I_1}{I_2} = k \frac{X_1}{X_2}$$

$$k = \frac{I_1}{I_2} \text{ at } \frac{X_1}{X_2} = 1$$

$$\text{or } k = \left(\frac{I_1}{I_2}\right)_{50/50} = \text{slope}.$$

This rule, concise and precise, makes the quantitative X-ray diffraction analysis of any binary system a very simple matter.

It is not necessary to choose the strongest line of the diffraction pattern, any line obeys this rule as long as the same line is used consistently. Therefore, overlapping peaks do not present a problem.

Furthermore, neither component needs to be a good crystalline material as long as the perfection or imperfection in crystal structure of the component sought is of the same order as that of the reference material used to determine the slope k. The extreme case is the newly developed X-ray diffraction method for the determination of crystallinity of polymers (Chung & Scott, 1973) where one component is amorphous while the other component is a poor crystalline material. It was found that the slope of the straight line in this reference is another proof of equation (20).

The application of this rule of auto-flushing is not confined to strictly binary systems. A case in point is Sample 4 which contains four components when Al_2O_3 is counted as an unknown. However, this sample has only two major components $CaCO_3$ and Al_2O_3. The minor components contribute only 4% by weight. If it can be regarded as a binary system of $CaCO_3$ and Al_2O_3, then

$$X_{Al_2O_3} = \frac{1}{1 + \frac{1}{2 \cdot 98} \frac{4756}{352}} = 18 \cdot 1\%$$

(True value 17·1%;   Note 18·1 × 96% = 17·4%),

$$X_{CaCO_3} = \frac{1}{1 + 2 \cdot 98 \times \frac{352}{4756}} = 81 \cdot 9\%$$

(True value 79·0%;   Note 81·9 × 96% = 78·6%).

Fairly good results can be obtained this way and thus eliminate the addition of a flushing agent. If the percentage of the minor components is approximately known, closer results could be obtained.

5. DISCUSSION

In order to attain high precision and high accuracy of X-ray diffraction analysis by the matrix-flushing method, the following three factors deserve attention:

1. The material used to determine the Reference Intensity must have the same level of perfection or imperfection in crystal structure as the component sought. The level of perfection or imperfection is revealed by the width of their diffraction lines.
2. The grinding of the sample must be thorough enough to ensure optimum particle size and sample homogeneity.
3. The loading of the sample into the sample holder must be free-falling as recommended by NBS Monograph 25 (1971) to avoid preferred orientation and induced packing.

The features of this matrix-flushing method are multifold, the more important ones are compiled in the following:

1. No calibration curve is needed. All the matrix factors are flushed out of the intensity-concentration equation.
2. No previous information about the approximate concentration range is required.
3. All the intensity data are obtained from a single scan, hence the errors due to instrumental drift and sample preparation are eliminated.

4. It can be used for amorphous-content determination as well as multicomponent analysis.
5. The general rule of auto-flushing puts the analysis of any binary systems on its shortest route.
6. It is a most general method in its simplest form embracing the three cases of quantitative analysis put forth by Klug & Alexander (1959).

6. REFERENCES

Alexander, L.; and H. Klug. (1948). *Anal. Chem.* 20, 886-889.

Azaroff, L. V.; and M. J. Buerger (1958). *The Powder Method in X-ray Crystallography*, p. 205. New York: McGraw-Hill.

Berry, C. (1970). *Inorganic Index to the Powder Diffraction File*, p. 1189. Joint Committee on Powder Diffraction Standards, Pennsylvania.

Chung, F. H.; and R. W. Scott (1971). *J. Appl. Cryst.* 4, 506-511.

Chung, F. H.; and R. W. Scott (1973). *J. Appl. Cryst.* 6, 225-230.

Jenkins, R.; and J. L. De Vries (1967). *Practical X-Ray Spectrometry*, p. 96. New York: Springer-Verlag.

Kirkpatrick, T.; and J. J. Nilles (1972). U.S. Patent 3,677,783. Sherwin-Williams Company.

Klug, H. P.; and L. E. Alexander (1959). *X-Ray Diffraction Procedures*, p. 412. New York: John Wiley.

NBS Monograph 25 (1971). *Standard X-Ray Diffraction Powder Patterns*, p. 3. Washington, D.C.: U.S. Government Printing Office.

Spurr, R. A.; and H. Myers (1957). *Anal. Chem.* 29, 760-762.

5.

Adiabatic Principle of X-ray Diffraction of Mixtures

> **No one can change the direction of the wind, but anyone can adjust his sails to always reach his destination.**
>
> — Jimmy Dean

1. ABSTRACT

All the information relating to the quantitative composition of a mixture is coded and stored in its X-ray diffraction pattern. It has been the goal of X-ray diffraction analysts since the discovery of X-rays to retrieve and decode this information directly from the X-ray diffraction pattern rather than resort to calibration curves or internal standards. This goal appears to be attained by the application of the "matrix-flushing theory" and the now-proposed "adiabatic principle" in applied X-ray diffraction analysis. The matrix-flushing theory offers a simple intensity-concentration equation free from matrix effects which degenerates to "auto-flushing" for binary systems. The adiabatic principle establishes that the intensity-concentration relationship between each and every pair of components in a multi component system is not perturbed by the presence or absence of other components. A key equation is derived which conducts the decoding process. Both the matrix-flushing theory and the adiabatic principle are experimentally verified.

2. INTRODUCTION

X-ray diffraction analysis has been a very important tool in industrial laboratories for quality control and routine analysis. It stands out from most other techniques since the results are in terms of the materials as they occur in the sample, not in terms of elements, ions, functional groups, fractional species, or spin coupling. However, the X-ray diffraction technique is plagued by the matrix effect which makes quantitative X-ray analysis difficult and time-consuming.

Azaroff & Buerger (1958) wrote: "Unfortunately, the relations between the intensities of observed reflections and the percentage composition of a mixture are not simple and cannot be expressed by a usable mathematical expression. It is necessary, therefore, to use semi-empirical methods," which refers to the internal-standard method of Klug & Alexander (1954).

This internal-standard technique is so far the best method for quantitative X-ray diffraction analysis (Kaelble, 1967). It adopts the usual procedure of constructing a calibration curve from standards for each component sought which is rather tedious especially for multicomponent analysis.

In a recent paper (Chung, 1974*a*) and in Part I of this paper (Chung, 1974*b*) the derivation of a very simple intensity-concentration equation free from matrix effect is described. This theory comprehends and simplifies the three important cases of quantitative X-ray diffraction analysis put forth by Klug & Alexander. For binary systems, it degenerates to an auto-flushing phenomenon. Further development of this theory leads to an "adiabatic principle" in X-ray diffraction which is presented below. The adiabatic principle eliminates the flushing agent required in the matrix-flushing method. It makes the direct quantitative interpretation of the X-ray diffraction pattern of a mixture possible for the first time.

All information relating to the quantitative composition of a mixture is coded and stored in its X-ray diffraction pattern. The adiabatic principle is the key needed to retrieve and decode this information directly.

3. ADIABATIC PRINCIPLE OF AUTO FLUSHING

The well-known X-ray diffraction theory gives an expression for intensity of a reflection in a powder diffraction pattern which is well established and described elsewhere (Bunn, 1961; Azaroff, 1968; Nuffield, 1966):

$$I(hkl) = \left(\frac{I_0 \, e^4 \lambda^3 d}{32 \, \pi m^2 c^4 r} \right) \left(N^2 p F^2 \right) \times \left(\frac{1 + \cos^2 2\theta}{\sin^2 \theta \cos \theta} \right) TAV, \tag{1}$$

where:

I = Intensity of X-rays diffracted by (hkl) plane.

I_0 = Intensity of primary X-rays.

e, m = charge and mass of electron.

λ = X-ray wavelength.

d = slit width of detector.

c = Velocity of light.

r = Specimen-to-detector distance.

N = Number of unit cells per unit volume.

p = Multiplicity.

F = Structure factor.

θ = Bragg angle.

T = Temperature factor.

A = Absorption factor.

V = Volume of powder in the beam.

It consists of six factors as shown above, they are

(1) a constant factor for a particular diffractometer,
(2) a structure factor characteristic of the diffracting sample,
(3) an angle factor known as the Lorentz-polarization factor,
(4) a temperature factor to correct for thermal vibration of atoms,
(5) an absorption factor which attenuates the diffracted X-rays, and
(6) the volume of powder in the primary X-ray beam, in the case of a mixture V indicates the volume fraction.

For quantitative X-ray diffraction analysis of a mixture of n components by use of a diffractometer such as the Philips' unit, the weight fractions of each component are sought. The only unknown variable in the intensity equation (1) is the absorption factor A, all the other factors can be kept constant (K).

Let μ, and s be the linear absorption coefficient and thickness of the total sample, let V_i and X_i be the volume fraction and weight fraction of component i. Since the absorption of X-rays, just like the absorption of visible light, follows the well-known exponential law, for infinite specimen thickness, we have

$$A = \int_0^\infty \exp(-\mu_t s)\, ds = \frac{1}{\mu_t} \quad (2)$$

$$\mu_t = \sum_{i=1}^n V_i \mu_i = \sum_{i=1}^n \frac{X_i}{\varrho_i} \quad \mu_i = \sum_{i=1}^n X_i \left(\frac{\mu_i}{\varrho_i}\right) \quad (3)$$

where p_i and (μ_i/p_i) are density and mass absorption coefficient of component 2. Then the complicated intensity equation (1) is reduced to: At this moment, k_i is constant for a very small variation in X_i. Later on, it will be sown that k_i in the ratio k_i/k_j is constant for any X_i.

$$I_i = \left(\frac{K_i}{\varrho_i \mu_t}\right) X_i = k_i X_i \quad (4)$$

where k_i is a factor containing the mass absorbtion coefficient of the total sample

For the quantitative X-ray diffraction analysis of a mixture of n components, we have n unknowns (X_i, $i=1$ to n) which must satisfy the following ($n+1$) equations:

$$\left.\begin{array}{l} I_1 = k_1 X_1 \\ I_2 = k_2 X_2 \\ \cdot \\ \cdot \\ \cdot \\ I_n = k_n X_n \\ X_1 + X_2 + \ldots + X_n = 1 \end{array}\right\}. \quad (5)$$

This situation can be most conveniently treated in terms of matrix algebra:

$$\begin{bmatrix} k_1 & 0 & 0 & \ldots & 0 \\ 0 & k_2 & 0 & \ldots & 0 \\ 0 & 0 & k_3 & \ldots & 0 \\ \ldots & \ldots & \ldots & \ldots & \ldots \\ 0 & 0 & 0 & \ldots & k_n \\ 1 & 1 & 1 & \ldots & 1 \end{bmatrix} \begin{bmatrix} X_1 \\ X_2 \\ X_3 \\ \vdots \\ X_n \end{bmatrix} = \begin{bmatrix} I_1 \\ I_2 \\ I_3 \\ \vdots \\ I_n \\ 1 \end{bmatrix}. \quad (6)$$

If the above matrix equation, **KX = I**, has a solution, the solution is unique if and only if the rank of the (**K**) matrix is equal to the rank of the (**K, I**) matrix which is also equal to the number of unknowns.

In other words, the necessary and sufficient condition for the existence of a unique solution of equation (6) is that all its characteristic determinants vanish.

In order to satisfy these conditions, the unique solution of equation (6) has the following simple and symmetrical form:

$$X_i = \left(\frac{k_i}{I_i} \sum_{i=1}^{n} \frac{I_i}{k_i} \right)^{-1}. \quad (7)$$

Note that the unusual property of this unique solution is that the weight fraction of any component in a multicomponent system is expressed in terms of ratios like I_i/I_j and k_i/k_j $(i,j = 1, 2,\ldots,n)$.

From equation (4), each individual k_i contains an absorption factor. However, the ratio k_i/k_j contains no absorption factor, the absorption factor is flushed out.

From equation (5):

$$\frac{I_i}{I_j} = \frac{k_i}{k_j} \cdot \frac{X_i}{X_j} \tag{8}$$

$$\frac{k_i}{k_j} = \frac{I_i}{I_j} \quad \text{at} \quad \frac{X_i}{X_j} = 1 \tag{9}$$

$$\frac{k_i}{k_j} = \left(\frac{I_i}{I_j}\right)_{50/50} = \text{slope} = \text{constant}. \tag{10}$$

It can be shown easily (Chung, 1974a, b) that the ratio of k_i/k_j is the same ratio of respective Reference Intensities (Berry, 1972). Therefore, the correct solution Xi in equation (7) can be obtained by use of the intensity data directly from the X-ray diffraction pattern of the mixture and the corresponding Reference Intensities ($k_i = I_i/I_c$) of the components.

The foregoing deduction can be summarized in the following theorem:

The plot of intensity ratio (I_i/I_j) to the weight ratio (X_i/X_j) of any two components is a straight line passing through the origin with a slope equal to the ratio of corresponding Reference Intensities (k_i/k_j). This intensity-concentration relationship between each and every pair of components in a multi-component system is not perturbed by the presence or absence of other components. In other words, the auto-flushing phenomenon (Chung, 1974 a, b) of any binary system always holds true regardless of the coexistence of other components.

This theorem can be logically called the "adiabatic principle of auto-flushing," analogous to the adiabatic principle in molecular spectroscopy and in quantum mechanics.

The unique solution, equation (7), can also be arrived at in reverse by applying the adiabatic principle of auto-flushing to a multicomponent system.

In the matrix-flushing method, the only nuisance variable μ_t in equation (4) is flushed out, the constant term k_i is obtained through Reference Intensities. Thus a very simple intensity-concentration equation free from matrix effect is accomplished:

$$\frac{I_i}{I_c} = \frac{k_i}{k_c} \cdot \frac{X_i}{X_c},$$
(11)

where X_c and I_c are the weight fraction and the X-ray intensity of the flushing agent, corundum, k_i and k_c are Reference Intensities of component i and corundum respectively ($k_i=I_i/I_c$, $k_c=1$). The X-rays do not discriminate between a flushing agent and a component. So the same intensity-concentration relationship should hold true between each and every pair of components in a multicomponent system.

For an n-component system, there are $_nC_2 = n(n-1)/2$ combinations of pairs. There is an equation like equation (11) for each pair of components. However, only (n-1) out of these $n(n-1)/2$ equations are independent. But a normalization equation is available to make up n equations for the n unknowns Xi ($i=1,2,\ldots,n$). Therefore:

$$\left.\begin{array}{c} \dfrac{I_1}{I_2} = \dfrac{k_1}{k_2} \cdot \dfrac{X_1}{X_2} \\ \dfrac{I_1}{I_3} = \dfrac{k_1}{k_3} \cdot \dfrac{X_1}{X_3} \\ \cdot \\ \cdot \\ \cdot \\ \dfrac{I_1}{I_n} = \dfrac{k_1}{k_n} \cdot \dfrac{X_1}{X_n} \\ X_1 + X_2 + \ldots + X_n = 1 \end{array}\right\}. \tag{12}$$

Solving the n simultaneous equations we obtain n roots identical to that in equation (7).

Equation (7) is an exact deduction, since neither assumption nor approximation is made to derive it from the basic equation (1).

This concise yet precise conclusion has been found to simplify greatly applied X-ray diffraction analysis. It provides the key to retrieving and decoding the quantitative information coded and stored in X-ray diffraction patterns.

Another interesting consequence of the adiabatic principle of auto-flushing is that it offers a unified sensitivity expression. Let S be the unified sensitivity of an X-ray diffraction method for a certain sample. Then

$$S = \dfrac{I_i}{k_i X_i} = \sum_{i=1}^{n} \dfrac{I_i}{k_i} = \text{constant}. \tag{13}$$

The unit of the unified sensitivity S is c.p.s./1% Al_2O_3 no matter whether Al_2O_3 is a component of the sample *or not*. As indicated in equation (13), the unified sensitivity is a constant for any component in the same sample, but it is different for the same component in a different sample. This is because the unified sensitivity is a means to sense the presence of an absorption effect. The higher the absorption, the lower the S value. Their quantitative relationship is:

$$\frac{S_I}{S_{II}} = \frac{\mu_{II}}{\mu_I}, \quad (14)$$

where S_I and S_{II} are the unified sensitivities of two samples i and II and μ_I and μ_{II} are the total mass absorption coefficients of samples i and II respectively.

The sensitivity for a component i in a given sample is defined as the rate of change in net signal output, I_i (c.p.s.), to concentration increment, X_i (%), which is also the slope of a calibration curve. Hence the sensitivity for a component i in a sample by X-ray diffraction is:

$$S_i = \frac{I_i}{X_i} = k_i S. \quad (15)$$

The unit of S_i is c.p.s./% of component i.

The detection limit for a component i in a given sample is defined as the minimum concentration, X_i (%), required to give an observable signal, I_i (c.p.s.), which is usually taken as three standard deviations of the background intensity, $3\sqrt{N_b}$ (Birks, 1959). Hence the detection limit for component i in a sample by the X-ray diffraction technique is:

$$X_i(\text{d.l.}) = \frac{3}{k_i S}\sqrt{\frac{I_b}{t}} = \frac{3}{S_i}\sqrt{\frac{I_b}{t}}, \quad (16)$$

where
$S=$ Unified sensitivity for the sample.

S_i = Sensitivity for component i in the sample.

I_b = Background counting rate (c.p.s.).

t = Background counting time (s).

N_b = Background counts accumulated in t seconds.

Note the explicit relationship between sensitivity, S_i, and the detection limit, X_i (d.1.), in quantitative X-ray diffraction analysis.

The soundness and usefulness of this adiabatic principle of auto-flushing are demonstrated in the Experimental section.

4. EXPERIMENTAL

As discussed earlier in this report, the percentage composition of any mixture is coded and stored in its X-ray diffraction pattern. The adiabatic principle of auto-flushing offers a key to the retrieval and decoding of it. In order to demonstrate this point, four test mixtures of four components each were prepared with six chemical compounds of analytical reagent grade or better. The chemicals used are ZnO (certified A.C.S., Fisher Sci. Co.), KCl (Analytical Reagent, Mallinckrodt Chemical Works), LiF (Baker Analytical Reagent), $CaCO_3$ (Baker Analytical Reagent), -Al_2O_3 (Linde semiconductor grade, 1 micron, Union Carbide), and TiO_2 (99.7% rutile). Each test mixture was ground for about 20 min with an automatic mortar grinder (Fisher Scientific Co.) to achieve optimum particle size and sample homogeneity. The fine powder was introduced into the sample holder in accordance with the methods recommended by NBS Monograph 25 (1971) to avoid preferred orientation and induced packing. Then the X-ray diffraction pattern was obtained or specific peaks were counted to collect intensity data. A Norelco diffractometer equipped with a solid-state scintillation counter, graphite-crystal-monochromatized Cu K_α radiation, (integrated-circuit panel, and pulse-height analyzer was used for this work. The strongest line of each component is measured if it is resolved; otherwise, the strongest resolved line is measured, which is then converted to the intensity of its strongest line through direct comparison with the X-ray diffraction pattern of the pure compound.

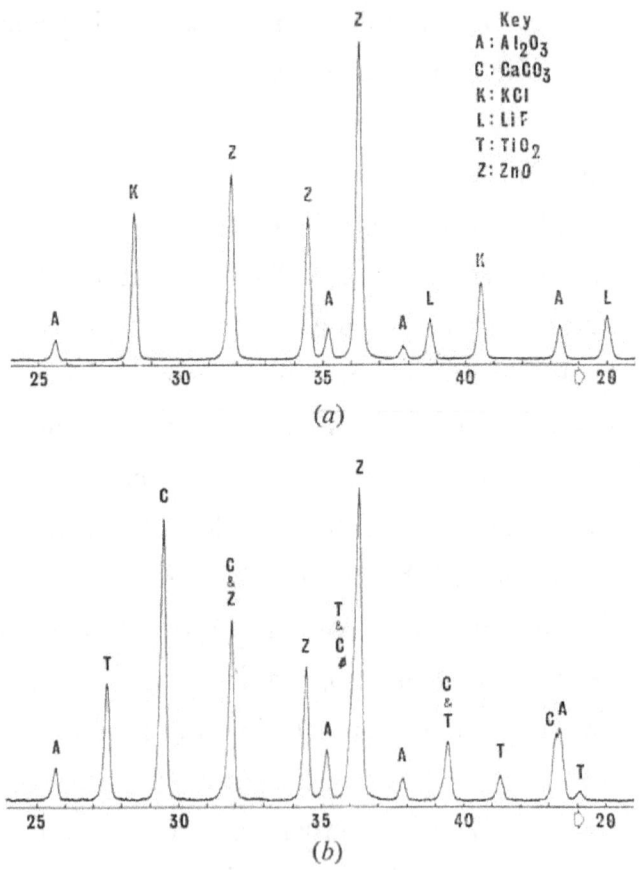

Figure 1. The X-ray differention patterns of (*a*) Sample 1 where all peaks are resolved, (*b*) Sample 3 where there are some overlapping peaks, (*c*) Sample 5 which contains amporphous material

(d) the ZnO—KCl binary system for slope determination.

Since the Reference Intensities ($k_i = I_i/I_c$) given in the Powder Diffraction File (Berry, 1972) have only two digits, the Reference Intensities of these six chemical compounds were determined by a procedure identical to that used for the actual analysis. The more detailed procedures are described in the previous paper (Chung, 1974b). Note that the concentrations of all components including Al_2O_3 are unknown. No flushing agent was used.

All the peaks are resolved in the X-ray diffraction patterns of samples 1 and 2, while some peaks are overlapping in that of samples 3 and 4 as shown in Fig. 1. Only about 1% ZnO and 2% TiO_2 were put into Sample 4 to check the detection limit.

The experimental data obtained are presented in Table 1. Sample 1 is used to present a typical calculation as shown below. From the key equation (7):

$$X_i = \frac{1}{\frac{k_i}{I_i} \sum_{i=1}^{4} \frac{I_i}{k_i}},$$

$$X_1 = \frac{1}{1 + \frac{k_1}{I_1}\left(\frac{I_2}{k_2} + \frac{I_3}{k_3} + \frac{I_4}{k_4}\right)}.$$

$$X_{ZnO} = \frac{1}{1 + \frac{4\cdot 35}{5968}\left(\frac{2845}{3\cdot 87} + \frac{810}{1\cdot 32} + \frac{599}{1\cdot 00}\right)} = 41\cdot 32\%,$$

Table 1. *Intensity and composition data*

These data are the same set used by Chung (1974a,b) except that the % Al_2O_3 is counted as unknown here.

Sample	Composition (g)		Intensity I (c.p.s.)	Reference Intensity k_i	% Composition, X_i		$\sum X_i$	$S = \dfrac{I_i}{k_i X_i}$
					Known	Found		
1	ZnO	1·8901	5968	4·35	41·49	41·32		33·2
	KCl	1·0128	2845	3·87	22·23	22·12		33·2
	LiF	0·8348	810	1·32	18·32	18·48		33·2
	Al_2O_3	0·8181	599	1·00	17·96	18·05	99·97	33·2
2	ZnO	0·9532	2856	4·35	18·98	19·42		33·8
	KCl	0·6601	1651	3·87	13·15	12·61		33·8
	LiF	0·8972	765	1·32	17·87	17·15		33·8
	Al_2O_3	2·5114	1719	1·00	50·00	50·84	100·02	33·8
3	ZnO	0·6759	2408	4·35	24·38	25·32		21·9
	TiO_2	0·4317	931	2·62	15·57	16·23		21·9
	$CaCO_3$	1·1309	2558	2·98	40·79	39·22		21·9
	Al_2O_3	0·5341	420	1·00	19·26	19·19	99·96	21·9
4	ZnO	0·0335	120	4·35	1·38	1·36		20·3
	TiO_2	0·0633	139	2·62	2·60	2·61		20·3
	$CaCO_3$	1·9197	4756	2·98	78·96	78·66		20·3
	Al_2O_3	0·4147	352	1·00	17·06	17·36	99·99	20·3

$$X_{KCl} = \frac{1}{1 + \dfrac{3·87}{2845}\left(\dfrac{5968}{4·35} + \dfrac{810}{1·32} + \dfrac{599}{1·00}\right)} = 22·12\%,$$

$$X_{LiF} = \frac{1}{1 + \dfrac{1·32}{810}\left(\dfrac{5968}{4·35} + \dfrac{2845}{3·87} + \dfrac{599}{1·00}\right)} = 18·48\%,$$

$$X_{Al_2O_3} = \frac{1}{1 + \dfrac{1·00}{599}\left(\dfrac{5968}{4·35} + \dfrac{2845}{3·87} + \dfrac{810}{1·32}\right)} = 18·05\%.$$

As shown in Table 1, in each case, very close agreement between experimental data and true values is realized; hence the correctness and authenticity of the adiabatic principle is experimentally verified.

Note that the unified sensitivity $S = I_i/k_i X_i$ = constant for each component in the same sample, as is implied in the adiabatic principle.

The detection limit of the X-ray diffraction analysis depends on the minimum detectable peak intensity above the background. This minimum intensity is usually taken as three standard deviations of the background intensity (Birks, 1959). The background intensities of this set of experiments range from 1000 to 3000 counts in. 40 s; hence the standard deviation of the background intensity is equation counts, or equation c.p.s. Therefore, the required minimum peak intensity is equation c.p.s.

The detection limit of ZnO in Sample 4 is calculated as follows:

Let ZnO be component 1, Al_2O_3 be component 2, and apply the adiabatic principle:

$$\frac{I_1}{I_2} = \frac{k_1}{k_2} \cdot \frac{X_1}{X_2}.$$

We have I_1= 4.5 c.p.s. (minimum peak intensity), I_2= 352 c.p.s. (experimental intensity), X_2=17.06% (con-centration of Al_2O_3), k_1=4.35, k_2=1.00, thus the detection limit of ZnO is X_1= 0.05%. It is not necessary to choose Al_2O_3 as the second component. Choosing either TiO_2 or $CaCO_3$ would give the same detection limit, which is of course a natural conclusion if the adiabatic principle holds true. Alternatively by substituting the above known quantities into equation (16), the same detection limit for ZnO can be obtained.

5. DISCUSSION

It has been an ideal if not a dream of X-ray diffraction analysts since the discovery of X-rays to determine the quantitative composition of a mixture from its X-ray diffraction pattern directly without resorting to a calibration curve or internal standard. This ideal seem to be fully realized by utilizing the matrix-flushing theory and the adiabatic principle of auto-flushing. The conclusion is simple and the method is straight forward. It broadens the field of application of this well-known technique.

It is not necessary to choose the strongest line of the diffraction pattern for intensity measurement. Any line of the diffraction pattern obeys the adiabatic principle as long as the same line is used consistently, although using the strongest line improves the sensitivity and detection limit of the measurement. For a component of extremely high concentration, it might be beneficial not to use the strongest line in order to avoid the possibility of a flooded counter.

The components sought need not be good crystalline materials. The criterion is that the material used to determine the Reference Intensity must have the same level of crystallinity as the component in the sample. The line width of reflections in an X-ray diffraction pattern is a good indication of the crystallinity of the component.

Corundum has been chosen as the reference material because of its purity, stability and availability. However, many materials have these good qualities, e.g. CaF_2, MgO, KCl, *etc.* all of which have cubic symmetry (α-Al_2O_3 has trigonal symmetry), very simple patterns, and low *d* spacings. They are equally eligible for use as the reference material. By choosing a standard reference material and by standardizing the diffractometer design, the Reference Intensities can be quoted to three significant figures and documented systematically. This would benefit everyone in the field of applied X-ray diffraction analysis.

The emergence of the adiabatic principle of auto-flushing makes the quantitative interpretation of X-ray diffraction patterns of mixtures possible for the first time. However, it will not supersede the matrix-flushing method because of the unique features of the latter:

(1) The matrix-flushing method can predict and determine the presence of amorphous material in a sample while the adiabatic method cannot.

(2) The matrix-flushing method can be applied to samples containing unidentified components by analyzing for only those components of interest. The adiabatic method can be applied to samples where all the components must be fully identified, by analyzing either for all the components or for none of them.

6. REFERENCES

Azaroff, L. V. & Buerger, M. J. (1958). *The Powder Method in R-Ray Crystallography*, p. 200. New York: McGraw-Hill.

Azaroff, L. V. (1968). *Elements of X-Ray Crystallography*, p. 202. New York: McGraw-Hill.

6.

Simultaneous Determination of Reference Intensities

> Smart people always choose comfort over luxury.
> — Secrets of Jewish wealth

1. ABSTRACT

A set of reference intensities, k_i are required for the quantitative interpretation of X-ray diffraction patterns of mixtures. Each k_i was heretofore determined individually from binary mixtures of a one-to-one weight ratio. A procedure for the determination of all k_i's of interest simultaneously is presented. The X-ray diffraction patterns of multicomponent mixtures usually contain overlapping peaks. This overlapping problem can be avoided by choosing an arbitrary reference material already present in the mixture and/or using the strongest *resolved* reflections directly. These concepts are substantiated by ten examples. The maximum standard deviation of the matrix-flushing method has been estimated to be 8% relative.

2. INTRODUCTION

Any practical quantitative X-ray diffraction analysis may be put under one of the two categories: First, all components are crystalline and identified; secondly, some components are non-crystalline and/or unidentified. The former can be analyzed by the application of the "adiabatic principle" (Chung, 1974*a*). The latter can be dealt with by the use of the "matrix-flushing" method (Chung, 1974*b*). Either case requires the characteristic Reference Intensity, k_i, of the component i concerned. To determine the Reference Intensities individually one by one is simple. Nevertheless, an even simpler way for the determination of a set of k_i's simultaneously is established. It further eases the work of quantitative X-ray diffraction analysis.

3. MULTIPLE k_i DETERMINATION

From the adiabatic principle of X-ray diffraction analysis of mixtures, a linear relationship exists between the intensity ratio of diffracted X-rays of any two components in a mixture and the corresponding weight ratio of these two components. This intensity-concentration relationship is independent of the presence or absence of other components. Therefore, n Reference Intensities could be determined simultaneously from a mixture of ($n+1$) components of *equal weight* including corundum (Chung, 1974*a*). Moreover, this intensity-concentration relationship is also independent of the amount of each component present. Hence, it is not even necessary to make the mixture of ($n+1$) components equal in weight (Chung, 1974*b*). This concept has been verified and is reported herein.

The chemicals used and the procedures of analysis are the same as those described in previous reports (Chung, 1974*a*, *b*, *c*). Sample 1 was made of exactly 0·5000 g of each component. Sample 2 was made of the same set of components but of different amounts. Sample 3 was made of four of the five components in Samples 1 and 2.

In these three samples, the strongest reflection of each component is resolved. The overlapping problem is taken up later. The intensities of these strongest reflections were counted using constant time (40 s). The composition and intensity data are shown in Table 1. The Reference Intensities were calculated with the following equation:

$$k_i = \left(\frac{X_c}{X_i}\right)\left(\frac{I_i}{I_c}\right) \quad (1)$$

where i and x are the X-ray intensity and weight percentage of component i and corundum c. The calculated k_i's are in good agreement with those obtained individually from 50/50 binary mixtures (Chung, 1974a, c). The intensity data are averages of duplicates.

4. OVERLAPPING PROBLEM

In practical analysis, the sample may contain amorphous components, some components may be unidentified, and the strongest reflections may not be resolved. Samples 5 and 6 are typical examples of this nature. Sample 4 was prepared for the simultaneous determination of k_i's of components of interest in Samples 5 and 6. Note that only components of interest should be included in the sample for multiple k_i determination. Because of overlapping problems, the strongest resolved reflections are used directly without converting them to the strongest reflection of each component through the powder pattern of the respective pure compound. In this particular case, the strongest resolved reflections are the strongest reflection (110) of TiO_2, the second-strongest reflection (104) of Al_2O_3, the second-strongest reflection (211) of $BaSO_4$, and the third-strongest reflection (002) of ZnO. Their composition and intensity data are given in Table 2. All data are averages of duplicate determinations. Equation (1) is used to calculate the k_i's from Sample 4. The following matrix-flushing equation (2) is used to calculate the percentage composition for Samples 5 and 6 with the k_i's obtained from Sample 4:

$$X_i = \left(\frac{X_c}{k_i}\right)\left(\frac{I_i}{I_c}\right). \tag{2}$$

ARBITRARY REFERENCE

The X-ray diffraction pattern of a multicomponent sample generally exhibits a large number of diffraction peaks. The introduction of corundum into a sample either for analysis or for multiple ki determination may cause more overlapping problems. Therefore, it is beneficial and rational to choose an arbitrary reference material already present in the sample rather than always choosing corundum as the fixed reference since any internally consistent set of k_i's should give equally correct results of analysis.

Table 1. Composition and Intensity Data

Sample No.	Component	Composition g	% Wt.	Intensity hkl	c.p.s.	k_i Calculated	50/50 Mixture
1.	ZnO	0·5000	20·00	101	1736	4·14	4·35
	CdO	0·5000	20·00	111	3208	7·66	7·62
	LiF	0·5000	20·00	200	533	1·27	1·32
	CaF$_2$	0·5000	20·00	220	565	1·35	1·41
	Al$_2$O$_3$	0·5000	20·00	113	419	1·00	1·00
2.	ZnO	0·4183	16·29	101	1564	4·28	4·35
	CdO	0·3562	13·87	111	2418	7·76	7·62
	LiF	0·6302	24·54	200	729	1·32	1·32
	CaF$_2$	0·5395	21·01	220	651	1·38	1·41
	Al$_2$O$_3$	0·6242	24·30	113	546	1·00	1·00
3.	ZnO	0·5554	32·54	101	4036	4·27	4·35
	CdO	0·2266	13·27	111	3029	7·85	7·62
	LiF	0·3712	21·74	200	778	1·23	1·32
	Al$_2$O$_3$	0·5539	32·45	113	943	1·00	1·00

Table 2. Composition and Intensity Data

Sample No.	Component	Composition g	Intensity hkl	c.p.s.	k_i	% Wt. Known	Found
4.	ZnO	0·4133	002	879	2·15	27·45	(Master
	TiO$_2$ (R)	0·3601	110	1061	2·97	23·92	Reference
	BaSO$_4$	0·3992	211	820	2·07	26·51	for
	Al$_2$O$_3$	0·3330	104	330	1·00	22·12	multiple k_i determination)
5.	ZnO	0·4655	002	1034	2·15	28·93	28·9
	TiO$_2$ (R)	0·1961	110	617	2·97	12·19	12·5
	BaSO$_4$	0·4097	211	860	2·07	25·47	25·0
	SiO$_2$ Gel	0·2173	—	0	—	13·51	13·7
	Al$_2$O$_3$ Flushing	0·3202	104	331	1·00	19·90	—
6.	ZnO	0·2743	002	346	2·15	14·77	14·5
	BaSO$_4$	0·6894	211	894	2·07	37·13	38·8
	CdO	0·1869	—	—	—	10·07	—
	KBr	0·2620	—	—	—	14·11	—
	Al$_2$O$_3$ Flushing	0·4442	104	266	1·00	23·92	—

Table 3. Composition and Intensity Data

Sample No.	Component	Composition g	Intensity hkl	c.p.s.	k_i	% Wt. Known	Found
7.	ZnO	0·4062	002	587	1·00	18·05	(Master
	As$_2$O$_3$	0·5669	222	3056	3·73	25·20	Reference
	CdO	0·4192	200	2324	3·84	18·63	for multi-
	NiO	0·6281	200	1907	2·10	27·92	ple k_i
	PbO$_2$	0·2296	110	1349	4·07	10·20	determination)
8.	ZnO	0·4190	002	468	1·00	17·07	16·7
	As$_2$O$_3$	0·4423	222	1978	3·73	18·01	18·9
	CdO	0·4523	200	1984	3·84	18·42	18·4
	NiO	0·6163	200	1455	2·10	25·10	24·7
	PbO$_2$	0·5252	110	2428	4·07	21·40	21·3
9.	ZnO	0·6001	002	756	1·00	21·06	21·4
	As$_2$O$_3$	0·3145	222	1445	3·73	11·04	11·0
	CdO	0·8208	200	3977	3·84	28·81	29·3
	NiO	1·1135	200	2837	2·10	39·09	38·3
10.	ZnO	1·0050	002	985	1·00	35·12	35·4
	As$_2$O$_3$	0·6818	222	2413	3·73	23·83	23·2
	CdO	1·1747	200	4431	3·84	41·05	41·4

This concept is illustrated by the following examples. The percentage compositions of Samples 8, 9 and 10 are to be determined. Sample 7 was prepared to obtain an internally consistent set of k_i's based on the arbitrary reference chosen. The component ZnO is chosen arbitrarily as the reference material here which means $k_{ZnO}=1$. Equation (1) is used to calculate the k_i's with reference to ZnO. That is $X_c=X_{ZnO}$ and $I_c=I_{ZnO}$. The following equation (3) from the adiabatic principle is used to calculate the percentage composition X_i's:

$$X_i = \left(\frac{k_i}{I_i} \sum_{i=1}^{n} \frac{I_i}{k_i} \right)^{-1}.$$

(3)

Of course, any other component in the sample such as As_2O_3, CdO, etc., is also eligible for use as the arbitrary reference. The composition and intensity data are listed in Table 3. Note that the intensity of the strongest *resolved* reflections were measured and used directly.

6. PRECISION

The matrix-flushing method has been applied to the analysis of various real samples in our laboratory. In order to establish the precision of this method, a field sample containing aluminum powder, iron oxide and a clay was used for a statistical evaluation. Although the iron oxide (goethite, PDF 17-536) contains impurities and the clay (kaolinite, PDF 6-221) is a metastable mineral; hence not favorable for quantitative X-ray diffraction analysis, the matrix-flushing method can still be used since the original raw materials are available for making the reference.

The results of ten single determinations by the same analyst with the same instrument on different days are listed in Table 4.

Table 4. *Precisions for k_i and X_i*

	Reference Intensity			Composition, %		
	k_{Al}	$k_{Fe_2O_3}$	k_{clay}	X_{Al}	$X_{Fe_2O_3}$	X_{clay}
1	1·97	1	0·549	2·78	36·6	60·7
2	2·09	1	0·551	2·22	37·8	60·0
3	1·90	1	0·477	2·67	33·9	63·5
4	2·05	1	0·521	2·61	37·2	60·2
5	2·06	1	0·545	2·60	36·0	61·4
6	2·11	1	0·511	2·30	33·5	64·2
7	2·01	1	0·528	2·56	37·5	60·1
8	2·08	1	0·550	2·39	34·2	63·5
9	2·13	1	0·546	2·15	37·5	60·4
10	2·14	1	0·576	2·51	37·3	60·3
$\bar{X}$	2·05	1·00	0·535	2·48	36·2	61·4
S	0·07	—	0·026	0·21	1·7	1·7
$S/\bar{X}$	3%	—	5%	8%	5%	3%

$\bar{X}$ = Average value. S = Standard deviation.

Since all the three components in the sample are identified, the adiabatic principle can be applied and the iron oxide is chosen as an arbitrary reference so that $k_{Fe_2O_3} = 1.00$.

These data indicate that the variation in results of analysis is dependent on the content of component sought. The higher the content, the lower the standard deviation. The *maximum* standard deviation of each result of a single determination obtained by the same analyst using the matrix-flushing method on different days would be 8% relative. For duplicate determinations the standard deviations should be reduced by a factor of $1/\sqrt{2}$ (i.e. 0·707). Better precision is expected if the component sought has a more defined crystal structure and the reference material has a higher purity.

Note that in this particular application the content of aluminum powder cannot be easily determined by other elemental analyses because the component clay (kaolin, aluminum silicate) also contains aluminum.

7. CONCLUSIONS

In order to obtain a set of reference intensities simultaneously for quantitative multicomponent analysis by the matrix-flushing method, a synthetic reference mixture which contains all the components sought with or without corundum is required. It is convenient to designate this synthetic reference mixture as a "Master Reference" which dictates what component should be assigned as the reference, specifies what reflection of each component should be chosen for intensity measurement, and eventually prescribes a consistent set of reference intensities.

By the use of Reference Intensities, the standard curves required for conventional quantitative X-ray diffraction analysis are no longer required. By the method of multiple k_i determination the need for preparation of several binary mixtures of an exact one-to-one weight ratio is avoided. By the direct use of the intensity data of the strongest resolved reflections, the overlapping problem is solved. By the choosing of an arbitrary reference material, the need for corundum as a standard and its associated difficulties are eliminated. Therefore, the quantitative interpretation of X-ray diffraction patterns of mixtures seems to be radically simplified without sacrificing accuracy.

8. REFERENCES

Chung, F. H. (1974a). *J. Appl. Cryst.* 7, 519-526.

Chung, F. H. (1974b). *J. Appl. Cryst.* 7, 526-531.

Chung, F. H. (1974c). Advanc. X-ray Anal. 17, 106-115.

7.

Matrix Flushing Techniques for Quantitative X-ray Fluorescence Analysis

> A smart man only believes half of what he hears. A wise man knows what half
> — Jeff Cooper

1. ABSTRACT

For research and development or for solving technical problems, we often need to know the chemical composition of an unknown mixture, which is coded and stored in the signals of its X-ray fluorescence (XRF) and X-ray diffraction (XRD). X-ray fluorescence gives chemical elements, whereas XRD gives chemical compounds. The major problem in XRF and XRD analyses is the complex matrix effect. The conventional technique to deal with the matrix effect is to construct empirical calibration lines with standards for each element or compound sought, which is tedious and time-consuming. A unified theory of quantitative XRF analysis is presented here. The idea is to cancel the matrix effect mathematically. It turns out that the decoding equation for quantitative XRF analysis is identical to that for quantitative XRD analysis although the physics of XRD and XRF are fundamentally different. The XRD work has been published and practiced worldwide. The unified theory derives a new intensity-concentration equation of XRF, which is free from the matrix effect and valid for a wide range of concentrations. The linear decoding equation establishes a constant slope for each element sought, hence eliminating the work on calibration lines. The simple linear decoding equation has been verified by 18 experiments.

2. INTRODUCTION

The major problem of quantitative X-ray spectrometry (XRS) is the matrix effect. Currently there are two major approaches to solve the matrix problem: (I) plot calibration lines with calibration standards, and (2) make mathematical corrections by fundamental-parameter or regression equations. To plot calibration lines, a set of standards is needed for each element sought. The slope of the calibration line changes with sample composition. Therefore, the calibration process is tedious and time-consuming, especially for multiple element analysis of non-routine and totally unknown samples. To make mathematical corrections needs prior knowledge of the sample and a sizable computer to run a complex program. It is mainly used in the steel and alloy industry. The ideal of quantitative XRS is (1) to find a new linear intensity-concentration equation free from the matrix effect, yet valid for wide concentration range; (2) to transform the slope of the linear intensity-concentration equation from variable to constant, which is characteristic for each element sought regardless of sample compositions. A unified theory of XRS is presented here to realize these two demands. It eliminates the tedious work on calibration lines.

It is well known that a thin-film specimen is free from the matrix effect in quantitative X-ray fluorescence (XRF) analysis. Purdue and Williams developed a smear (thin-film) technique using Zr in ZrO_2

as reference standard to control film thickness (sample weight) and 26 primary standards to construct calibration lines for the analyses of 25 elements in coal, ores, fuel ashes, slag, etc.[1] Prior published papers on thin films were reviewed. There are plenty of recent articles and books discussing thin farns,[2-5] but none are relevant to laboratory procedures for quantitative XRS due to the difficulties in preparing thin films with reproducible thickness, density, and uniformity. Therefore, the thin-film method is not so popular in practice. It is barely mentioned in passing in the book entitled Quantitative *X-ray Spectrometry* and the *Handbook of X-ray Spectrometry*.[6,7]

3. CURRENT THEORY OF QUANTITATIVE X-RAY SPECTROMETRY

The unified theory extends and simplifies the current theory of quantitative XRS. For easy reference, the current theory is briefly reviewed here. The relationships among X-ray intensity, concentration, and matrix effect in XRF analysis were derived by Muller;[8] Jenkins and De Vries,[9] and Birks[10]

$$I_i = k_i X_i (1 - e^{-\mu \rho d})/\mu \tag{1}$$

with equation, where

- I_i = XRF intensity of element i, counts/s
- k_i = proportional constant of spectrometer for element i
- X_i = weight fraction of element i, %
- ρ = density of specimen, g/cm^3
- d = thickness of specimen, cm
- μ = mass absorption coefficient of specimen, cm^2/g

μ_1 & μ_2 = mass absorption coefficients for primary and secondary X-rays

θ_1 & θ_2 = incident and take-off angles.

For easier discussion, let h = μ d, where the h is a dimensionless positive number, since the unit of μ is cm²/g, the unit of ρ is g/cm³, and the unit of d is cm. Note that the $\mu\rho$ factors are fixed in each sample; only the d factor can be changed. Eq. 1 can be simplified by making the specimen either a bulk pellet or a thin film.

For a bulk pellet, d = ∞, e^{-h}= 0, so Eq. 1 is reduced to

$$I_i = k_i X_i / \mu \quad (2)$$

This is the most common situation in XRF analysis. Various quantitative XRF techniques such as in-type, standard addition, internal standard, fundamental, and others are essentially efforts to compensate, minimize, or calculate the matrix effect (μ).[6]

For a thin film, α is very small. By series truncation, we have e^{-h}=1 – h. Eq. 1 is reduced to

$$I_i = k_i X_i \rho d \quad (3)$$

for a thin-film specimen. Note that the matrix factor μ. has disappeared, but the thin-film density ( )and thickness ( ) remain. The   quantities of very thin films are difficult to measure accurately. It is also difficult to prepare thin-film specimens with reproducible thickness, density, and uniformity. Consequently, the thin-film technique is not so popular in practice.[7]

The proportional constant *ki* in Eqs. 2 and 3 depends on the design of the spectrometer and the emitting power of the element sought. For the same spectrometer, the value of *ki* depends on the matrix composition. To date, the slopes of these two linear equations, Eqs. 2 and 3, are variables. Each slope must be determined by plotting a calibration line with a set of standards, which is tedious and time-consuming.

Chung et al. derived criteria to define both thin-film and bulk pellet specimens:[11] the Linear thickness for thin film, $d_i = 0.135/\mu p$, below which the matrix effect vanishes, and the critical thickness for a bulk pellet, $c.4.62/\mu p$, above which the matrix effect reaches an upper limit. Note that $d_c = 34 d_i$. The matrix effect becomes complex when the specimen thickness is in between, i.e., $d_i < d < d_c$.

4. UNIFIED THEORY OF QUANTITATIVE X-RAY SPECTROMETRY

For easy comparison of the same property of different materials, usually a pure compound is assigned as standard. For example, the specific gravity of zinc oxide (ZnO) is 5.47, which means that the mass of ZnO is 5.47 times as heavy as the mass of water at 4°C. The ubiquitous water is assigned as standard in this case. Note that both the density and specific gravity of water are defined to be exactly one (1.00). The word "specific" means per unit mass, such as the specific impulse of rocket fuel. In order to put all X-ray intensity data on the same scale, let us define a term Specific Intensity of any element *i* as follows:

Specific Intensity of element *i*

$$= \frac{\text{XRF intensity per unit mass (1.00\%) of element } i}{\text{XRF intensity per unit mass (1.00\%) of yttrium (Y)}} \quad (4)$$

The element yttrium is assigned as standard here. The specific intensity indicates how strong the emitting power of element i is relative to that of the element yttrium. Yttrium is a rare element. Therefore, most likely yttrium is not present in the original sample. Pure chemical grade Y_2O_3 powder is stable, inexpensive, and easily available. Of course, any rare element can be assigned as standard.

In order to cancel the troublesome μ factors mathematically in Eqs. 2 and 3, a small quantity of yttrium (Y in Y_2O_3) is added into the original sample as a flushing standard. To make a thin film, dilute the admixture with a film-forming polymer solution, mix well in a mini ball-mill, and make a thin film by a gauged film applicator (see Experimental section). To make a bulk pellet, dilute the admixture with Li_2CO_3 or starch, mix well, and make a bulk pellet with press and die. Two sets of equations can be set up for each pair of elements in a thin film or bulk pellet

$I_i = k_i X_i \rho d$, for any element i in a thin film.

$I_Y = k_Y X_Y \rho d$, for element Y from the flushing standard Y_2O_3; X_Y is a known quantity.

$I_i = k_i X_i / \mu$, for any element i in a bulk pellet.
$I_Y = k_Y X_Y / \mu$, for element Y from the flushing standard Y_2O_3; X_Y is a known quantity.

Cancelling the μ d factors mathematically in each pair of the simultaneous equations, we have the merged equation for both thin-film and pellet specimens

$$\frac{I_i}{I_Y} = \frac{k_i}{k_Y} \cdot \frac{X_i}{X_Y} = k_i \cdot \frac{X_i}{X_Y} \tag{5}$$

where I_y = XRF intensity of flushing standard element Y and k_y = 1.00 by definition (Eq. 4) of specific intensity.

All terms in Eq. 5 are ratios; hence many experimental variables, such as film thickness, film density, matrix effect, particle size/distribution, specimen position, and instrumental drifting, are mutually cancelled. A set of specific intensity ki values of interest can be determined by one simple experiment with one sample preparation (see Experimental design).

Note that the decoding equation (Eq. 5) for XRF is identical to that for X-ray diffraction (XRD), although the physics of XRF and XRD are fundamentally different.[12-14] The results of XRD have been published and practiced worldwide.[15] Therefore, we have a unified theory for decoding the signals from XRF and XRD of mixtures. Rearranging Eq. 5, we have

$$k_i = \frac{I_i/X_i}{I_Y/X_Y}$$
$$= \frac{\text{XRF intensity per unit mass (1\%) of element } i}{\text{XRF intensity per unit mass (1\%) of yttrium (Y)}}$$
$$= \textbf{Slope of}$$
$$= \textbf{Characteristic Constant of} \quad (6)$$

$$X_i = \frac{X_Y}{k_i} \cdot \frac{I_i}{I_Y} \quad \text{decoding equation for thin film or pellet} \quad (7)$$

Note that Eqs. 5, 6, and 7 are the same equation in different forms. Eq. 5 is the new linear relationship between intensity and concentration, free from the μρd factors and valid for a wide concentration range in either thin-film or bulk pellet specimens. Eq. 6 reveals that the slope (k_i) of Eq. 5 is characteristic constant of the element sought and is equal to its specific intensity. Eq. 7 converts X-ray intensity (c/s) to the percentage weight of the element sought. There are five variables in these equations: two unknowns, Xi and

k_j, and three measured date X_y, I_y and I_j. Given k_j, the X_i can be calculated. Given X_i, h_i can be calculated. Eq. 5 eliminates the calibration lines. It vastly simplifies the work of quantitative XRS.

This deduction leads to a unified theorem. The plot of intensity ratio (I_i/I_j) versus the weight ratio (X_i/X_j) of any pair of elements in either thin-film or bulk pellet specimens is a straight line passing through the origin with a slope equal to the corresponding ratio of specific intensities (k_i/k_j). When yttrium (Y) is assigned as flushing standard, then $k_j = k_y = 1.00$. The slope of the decoding equation (Eq. 5) for element i is equal to the specific intensity of element i. This new linear intensity-concentration relationship between each pair of elements is not perturbed by the presence or absence of other elements. In other words, no matrix factor is involved in the decoding (Eq. 5). The μρd factors are flushed out mathematically.

The unified theorem holds true for any mixture. The flushing standard Y_2O_3 is different from the conventional internal standard.[16] It provides a known component (X_Y), such that Eq. 5 gives a unique solution for any element sought (X_i). The flushing standard acts like a spy, or a probe. It flashes back its findings in X-ray signals. Eq. 5 is the decoding formula.

The percentage weight of the element sought (X_i) in the admixture (original sample plus Y_2O_3) is converted to the percentage weight of the element sought in the original sample (X_i^o) by

$$\frac{X_i}{X_i^o} = \frac{1}{1+X_f} \quad X_f = W_f/W_o \quad W_a = W_o + W_f \tag{8}$$

$$X_i^o = X_i(1+X_f) = X_i \cdot W_a/W_o \tag{9}$$

where

X_f = weight fraction of Y_2O_3 in terms of the original sample
W_f = weight of flushing standard Y_2O_3 in admixture
W_o = weight of original sample in admixture
W_a = weight of the admixture.

For reference, an example of numerical calculation is provided in the Appendix at the end of this article.

5. PRACTICE OF THE UNIFIED THEORY

In theory, there is no matrix factor in the new intensity-concentration relation (Eq. 5). In practice, the task is to make a thin-film specimen, which must be stable, homogeneous, reproducible, and satisfy the linear thickness criterion: de = 0.135/mp.[11]

Consider a thin film of polymethyl methacrylate (PMMA) polymer containing 1% $PbCrO_4$ (Chrome Yellow pigment). Its µρ values are estimated to be 9 cm^{-1} for Pb and 76 cm^{-1} for Cr. According to the criterion for a thin film, we have d=0.135/9=0.015cm=6mils for Pb and d=0.135/76 = 0.0018 cm = 1 mil for Cr. Therefore, a 1-mil film is thin to X-rays for both Pb and Cr, and thus free from the matrix effect. Note that 1 inch = 1000 mils =2.54 cm, or 1 mil =0.0025 cm. The film thickness marked on the Bird film applicator is in mils, which is popular in the film industry.

Special attention is paid to the approximation, e^{-h} = 1 - h, where h = µρd, which is valid only when h is very small. In laboratory practice, it means minimized matrix effect (µ), highly diluted sample (ρ), and thin enough film (d). In reality, the attempt here is to make a monolayer of sample particles evenly buried and set in a thin film of a low-density polymer. A dried (cured) polymer film is stable and very easy to handle. It takes only one minute for the wet film to dry.

5.1 Experimental Design

A set of 18 synthetic samples were prepared to demonstrate and verify the three salient aspects of the unified theory.

1. **Slope constancy and range of concentration:** Samples 1 to 6 verify the linear relationship between intensity and concentration of five elements, Ba, Zn, K, S, and P, at extreme range of concentrations in harsh matrices. Calculate the k_i, value from known concentrations (X_i) and the measured intensities (I_i) using Eq. 6. Examine whether each element has a constant slope under these various conditions. Meanwhile, use the average value of these slope k_i values to calculate the X_i values. Then check whether the calculated concentrations agree with the true concentrations.

2. **Multiple element analysis:** Samples 7 to 12 illustrate the analysis of three elements, Cd, Ti, and Al, at varied concentrations in diverse matrices. Due to the same flushing standard and the constant slope (k_i) for any element i, only minimal extra work is required for multiple element analysis relative to a single element sought.

3. **Worst matrix effect:** Samples 13 to 18 examine cases of worst matrix effect, namely analyzing a heavy element, zirconium (Zr). in light matrices of Li_2CO_3. Sample 13 was used to determine the constant slope of the element Zr using Eq. 6. This constant slope (k_{Zr} = 0.255) was used to calculate the percent-

age weight of Zr in samples 14 to 18 using Eq. 7. This constant slope value (k_{Zr} = 0.255) can be used for any future analysis of Zr, because it is the characteristic constant of element Zr.

5.2 Thin-Film Preparation Procedure

In order to make the thin-film specimen homogeneous, reproducible, and satisfy the thin-film criterion $dl = 0.135/\mu$, the following procedure has been practiced in our laboratory:

1. Grind the sample into a fine powder with mortar and pestle, or with an automatic mortar grinder (Fisher Scientific Company). A few drops of methanol or ethanol added into the sample may help the grinding process.

2. To a 0.25-pint metal can (popular in paint laboratories), add 1.5-2.5 g of the powder sample, about 0.1 g of pure Y_2O_3 as flushing standard, about 25g of polymer solution, and roughly 25 mL of steel balls of 1/8 inch diameter. (The recipe for a film-forming polymer solution is as follows: Add two parts PMMA and 3 parts polybutyl methacrylate (PBMA) into 7.5 parts toluene. Mix well. The polymer solution contains about 42% solids. It can be stored for several years. A varnish from any paint store is a good substitute.) Only the weight of the sample and the flushing standard Y_2O_3 are to be measured accurately.

3. Cover the metal can tightly and shake it vigorously on a Red Devil shaker (a popular shaker in paint stores) for about 20 min to make a homogeneous dispersion. This mini ball-mill step is essential to achieve particle size re-

duction, ultimate dispersion, and sample homogeneity.

4. Make a 2-mil drawdown of the dispersion on a thick Mylar film with a gauged film applicator (Gardner Laboratory, Inc., Bethesda, MD). The wet film will dry in about one minute to less than one mil thick. The Mylar substrate gives low X-ray background.
5. Cut a disk of 1¼-inch diameter from the drawdown with a press and die. Insert the disk into the X-ray spectrometer, then scan or count the analytical emission lines of the elements sought and the flushing standard yttrium (Y).
6. Use Eq. 7 to calculate the percentage weight (X_i) of the element sought, or use Eq. 6 to calculate specific intensity (k_i) from a synthetic standard.

5.3 Laboratory Equipment

Red Devil shaker: used for one gallon or smaller cans, from Sears, eBay, or others.

Gauged film applicator: Bird film applicator, Ecometer 3550, from GTECH Systems, or others.

Metal can with lid: unlined from Home of Cans, or others.

Grinding balls: stainless steel balls, 1/8 inch diameter, available from Across International.

X-ray spectrometer: any wavelength dispersive X-ray spectrometer can be used. A Norelco Philips X-ray spectrometer equipped with an XRG-3000 X-ray generator and a NIM IC-2000 circuit panel was used for data collection. Energy-dispersive spectrometers may have problems in peak resolution, hence not proper for this technique.

6. RESULTS OF EXPERIMENTS

Laboratory experiments were carried out according to the Experimental design described above. The experimental data are presented in Tables 1 to 4, and plotted in Figures 1 and 2. These findings are discussed below.

Table 1. Composition data for slope constancy and concentration range.

Components of admixture in grams

Sample	BaSO$_4$	ZnO	K$_2$HPO$_4$	Y$_2$O$_3$[a]	Total
1	1.4139	0	0	0.1027	1.5166
2	0	1.3652	0	0.1040	1.4692
3	0	0	1.3902	0.1006	1.4908
4	1.0511	0.1082	0.3147	0.1047	1.5787
5	0.3678	0.6429	0.4633	0.1005	1.5745
6	0.0695	0.3677	1.0071	0.1277	1.5720

[a]The three compounds BaSO$_4$, ZnO, and K$_2$HPO$_4$ are components of synthetic samples. The Y$_2$O$_3$ is the added flushing standard, which provides a known quantity of X$_Y$ such that the decoding Eq. 5 gives a unique solution for any element sought (X$_i$). Note that the k_i = slope = specific intensity = characteristic constant of the element sought.

Table 2. Results of constant slope and concentration range. k_i = specific intensity = slope. Y = element Y from flushing standard Y$_2$O$_3$.

Sample	Ba	Zn	K	S	P	Y
1 c/s	2241	0	0	16.6	0	109
wt %	54.86	0	0	12.81	0	5.33
k_i	2.00	–	–	0.0634	–	1.00
2 c/s	0	4386	0	0	0	114
wt %	0	74.65	0	0	0	5.57
k_i	–	2.87	–	–	–	1.00
3 c/s	0	0	688	0	15.8	136
wt %	0	0	41.87	0	16.58	5.31
k_i	–	–	0.642	–	0.0372	1.00
4 c/s	1680	343	126	12.8	3.00	113
wt %	39.2	5.51	8.95	9.15	3.54	5.22
k_i	1.98	2.88	0.650	0.0646	0.0391	1.00
5 c/s	531	1849	169	3.95	4.25	100
wt %	13.7	32.8	13.2	3.21	5.23	5.03
k_i	1.94	2.84	0.644	0.0619	0.0409	1.00
6 c/s	124	1357	464	0.97	11.3	159
wt %	2.60	18.8	28.8	0.61	11.4	6.40
k_i	1.92	2.91	0.648	0.0640	0.0399	1.00
Average k_i	1.96	2.88	0.646	0.0635	0.0393	1.00
wt % range	2~55%	5~75%	8~42%	0.6~13%	3~17%	–

Concentration range: from low end 0.6% S to high end 75% Zn and others in between.
Matrix range: includes light, medium, and heavy matrices.

6.1 Slope Constancy and Concentration Range

In XRF analysis, the linear relationship between intensity and concentration usually holds for a limited concentration range, and the slope of the intensity-concentration equation varies with the sample composition. In order to study the constancy of the slope (k_i) and the range of concentration (X) in Eq. 5, six synthetic samples were prepared by the proposed thin-film preparation procedure. Samples 1, 2, and 3 are pure compounds: $BaSO_4$, ZnO, and K_2HPO_4. Samples 4, 5, and 6 are mixtures made of these three pure compounds. About 0.1 g of flushing standard Y_2O_3 was added into each sample. All chemicals were certified reagents or better. The X-ray intensities of the Lα line of Ba and Kα lines of Zn, K, S, P, and Y were counted. The weight compositions are listed in Table 1. The X-ray intensity (c/s), concentration (wt%), and calculated slope (k_i) of each element are presented in Table 2. These data are plotted in Figures 1 and 2. The results demonstrate that the slope (k_i) of each element is essentially constant indeed for extremely wide range of concentrations (X_i) in varied matrices. When the concentration of the element sought is very low, especially for light elements such as sulfur in Sample 6, i.e., X_S= 0.61%, a longer counting time was required to attain better intensity data and thus achieve better precision.

Table 3. Results of XRF analysis for multiple elements.

Sample	Cd $k_{Cd}=0.232$	Ti $k_{Ti}=3.52$	Al $k_{Al}=0.0117$	La $k_{La}=1.00$
7 c/s	11.3	750	19.5	42.5
wt % found	1.36	5.92	46.3	–
wt % known	1.35	5.91	46.1	1.18
8 c/s	565	4100	1.2	56.3
wt % found	48.45	23.2	2.04	–
wt % known	49.84	22.8	2.02	1.12
9 c/s	3.5	420	27.5	52.5
wt % found	0.32	2.57	50.6	–
wt % known	0.34	2.58	49.7	1.13
10 c/s	205	51.3	20.5	58.8
wt % found	16.4	0.27	32.5	–
wt % known	17.5	0.22	31.9	1.09
11 c/s	37.5	2800	12.0	57.5
wt % found	3.20	15.8	20.3	–
wt % known	3.21	15.5	20.4	1.14
12 c/s	150	1975	7.5	70.6
wt % found	10.2	8.82	10.1	–
wt % known	10.0	8.50	10.1	1.11
Conc. range, wt %:	0.3–50%	0.2–23%	2–50%	–

Matrix type: light to heavy matrices.
Element La is from the added flushing standard La_2O_3.

6.2 Multiple Element Analysis

Because of the complex matrix effect, quantitative XRF analysis has been essentially empirical, and multiple element analysis becomes extremely tedious. Although various methods have been devised to cope with the complex matrix effect, they are either restricted by type of sample or limited by range of concentration. The existing quantitative multiple element methods are mainly mathematical correction procedures applied to steel or alloys of known elemental composition.[17] In industrial laboratories, the sample types are diversified, the concentration ranges are generally unknown. Therefore, it is highly desirable to have a fast, dependable, versatile XRF technique to run non-routine and totally unknown samples for single or multiple elements. The unified theory and practice fills this role.

The validity of Eq. 5 for multiple element analysis is illustrated by another set of six samples, i.e., Samples 7 to 12. Each sample consists of heavy element cadmium (Cd), medium element titanium (Ti), and light element aluminum (Al) from pure CdO, TiO_2, and Al_2O_3 respectively. A proper amount of Li_2CO_3 was added as an inert component to make a total weight of about 2.5 g to balance the sample/polymer ratio. The rare element lanthanum (La in La_2O_3) instead of yttrium (Y in Y_2O_3) was chosen as flushing standard to demonstrate the flexibility of the matrix flushing technique. The concentration ranges are 0.3-50% of Cd, 0.2-23% of Ti, and 2-50% of Al. The matrices vary from light to heavy. The concentration and intensity data are presented in Table 3. The set of specific intensities (characteristic constant of each element) were determined from prior experiments: $k_{Cd} = 0.232$, $k_T = 3.52$, $k_{Al} = 0.0117$. The agreements between the measured percentage weight (wt%) and the true percentage weight (wt%) are quite satisfactory for multiple elements at wide concentration ranges in varied matrices.

Table 4.

Table 4. Composition and intensity data for the case of worst matrix effect.

Sample	Components, g			X-ray c/s		% Zr	
	ZrO_2	Li_2CO_3	La_2O_3	Zr	La	Known	Found
13	0.1513	2.6950	0.0729	144	314	3.84	(k_{Zr}=0.255)
14	0.2755	2.4706	0.1363	274	619	7.08	7.00
15	0.6124	2.1573	0.1619	702	804	15.46	16.13
16	1.2618	1.4892	0.0852	1115	349	32.92	32.45
17	2.0187	0.6942	0.1136	1680	419	52.87	53.93
18	2.8361	0	0.1464	3075	704	70.40	71.77

Concentration range: 3–71% Zr.
Matrix: the constant slope (k_{Zr} = 0.255) holds true for the case of worst matrix effect, which is a heavy element of wide concentration range in light matrices.

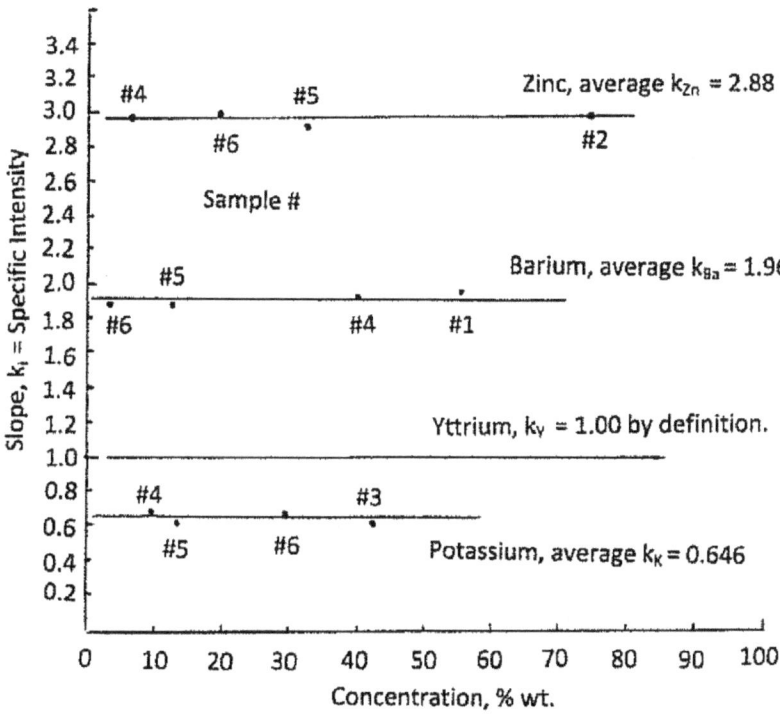

Figure 1. *Constant Slope (k_i) of heavy and medium elements, Ba, Y, Zn, and K, at wide concentration range in varied matrices. The horizontal lines indicate that the linear intensity-concentration relation (Eq. 5) has a constant slope for each element sought, regardless of concentration range and matrix composition.*

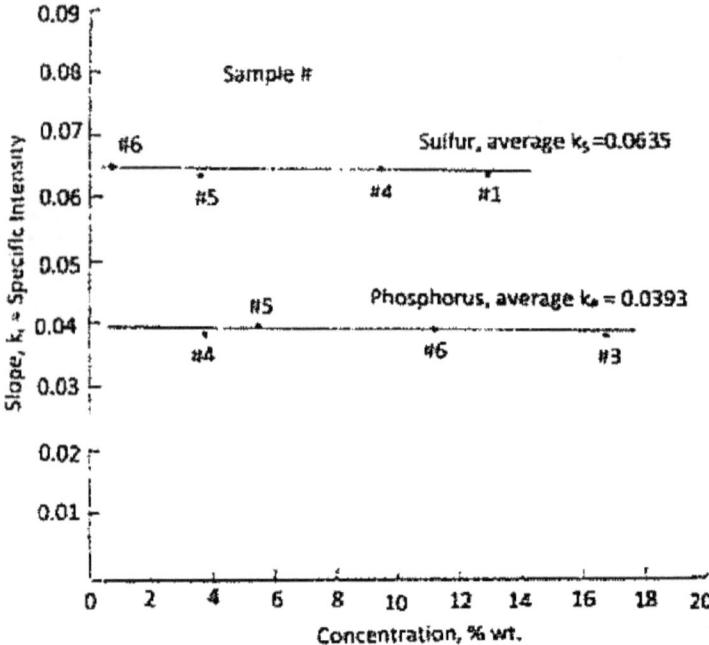

Figure 1. *Constant Slope (k_i) of light elementsm P and S, at wide concentration range and in harsh matricies. The horizontal lines indicate that the linear intensity-concentration relation (Eq. 5) has a constant slope for each element sought, regardless of concentration range and matrix composition.*

6.3 The Worst Matrix Effect

The conventional intensity-concentration lines for different elements at wide concentration ranges in various matrices are either concave or convex. Linear curves are exceptionally few. An extreme case is a heavy element in light matrices. Its intensity-concentration curve has the largest curvature due to self-absorption.° in order to study this extreme case with the unified theory and technique, six zirconium samples were prepared with pure ZrO_2 in a light matrix of Li_2CO_3. Lanthanum from La_2O_3 was chosen as the flushing standard. No effort was made to keep the percentage of La nearly constant. The intensity and concentration data are listed in Table 4. The specific intensity of zirconium, k_{Zr} = 0.255, was calculated using Eq. 6 with data from Sample 13. This k_{Zr} value was inserted into Eq. 7 to calculate the percentage weight of zirconium in Samples 14 to 18. No matrix effect was observed even in these worst cases. The matrix flushing equation (Eq. 5 or Eq. 7) is indeed valid for a wide concentration range and free from matrix effects.

6.4 Detection Limits

The detection limit is dependent on the sensitivity, the counting time, and the background intensity. The detection limits of the unified theory and technique for light elements found in this study are estimated to be 0.03 wt% of potassium (K), 0.11 wt% of sulfur (S), and 0.13 wt% of phosphorous (P). Cohen and Smith reported that the thin-film XRF technique is suitable for all major elements down to sodium (Na) for mineral/rock analysis.[18] The detection limit they found was 0.17wt% of sodium (Na). They adopted the thin-film specimen preparation procedure and plotted calibration lines with calibration standards. With the advent of the unified theory of quantitative XRF analysis, the time-consuming work on calibration lines could be eliminated.

6.5 Precision

The unified theory and technique has been used for multiple element analysis of non-routine and totally unknown samples in our laboratory. Some results were checked by use of atomic absorption and/or wet chemical analysis with good agreement. A real mineral sample was analyzed for barium (Ba) and phosphorous (P) seven times on different days by the same analyst. Lanthanum from pure La_2O_3 was chosen as the flushing standard in this case, such that the flow detector could be used for all the intensity data collection. The average values of the seven sets of data with corresponding standard deviation and precision are presented below.

Average, wt% $X_{Ba} = 26.9 \pm 0.50$ $X_P = 4.19 \pm 0.18$
Precision $0.50/26.9 = 1.9\%$ $0.18/4.19 = 4.3\%$
Average, slope $k_{Ba} = 1.010 \pm 0.046$ $k_P = 0.0181 \pm 0.0009$
Precision $0.046/1.010 = 4.6\%$ $0.0009/0.0181 = 5.0\%$

These results indicate that the precision of the matrix flushing technique is statistically estimated to be 5% relative or better.

7. CONCLUSIONS

The idea to cancel the matrix factor mathematically in quantitative XRS leads to the concept of specific intensity (k_i), which puts all X-ray intensity data on the same scale. The idea of matrix flushing with a flushing standard (the rare element Y in Y_2O_3) leads to a new linear intensity-concentration relation, i.e., Eq. 5. The highlights of the unified theory are summarized below:

1. The new linear intensity-concentration relation (Eq. 5) is valid for a wide range of concentration, independent of matrix composition, and requires no prior information about the sample.
2. The slope of the new intensity-concentration equation becomes a characteristic constant for each element sought,

thus eliminating the tedious work on calibration lines.
3. The thin-film preparation procedure controls the film thickness, film uniformity, particle size, and specimen homogeneity. It is easy, fast, dependable, and systematic.

8. REFERENCES

Bertin, E. P. Principle and Practice of X-ray Spectrometric Analysis. New York: Plenum Press, 1970, p.425.

Birkholtz, M. Thin Film Analysis by X-ray Scattering. Weinheim, Germany Wiley-VCH, 2006.

Birks, L. S. X-Ray Spectrochemical Analysis. New York: Interscience, 1959, 59, 63.

Chung, F. H. "Quantitative Interpretation of X-Ray Diffraction Patterns of Mixtures. 1. Matrix Flushing Theory for Multi-Component Analysis." *J. Appl. Cryst*allogr. 1974. 7: 519-525.

Chung, F. H. "Quantitative Interpretation of X-Ray Diffraction Patterns of Mixtures. II. Adiabatic Principle of X-Ray Diffraction Analysis of Mixtures." *J. Appl. Cryst*allogr. 1974. 7: 526-531.

Chung, F. H. "Quantitative Interpretation of X-Ray Diffraction Patterns of Mixtures. III. Simultaneous Determination of a Set of Reference Intensities." *J. Appl. Cryst*allogr. 1975. 8: 17-19.

Chung, F. H.; A. J. Lentz; and R. W. Scott. "A Versatile Thin-Film Method for Quantitative X-Ray Emission Analysis." X-Ray Spectrum, 1974. 3: 172-175.

Chung, F. H.; and D. K. Smith. Industrial Applications of X-ray Diffraction. New York Marcel Dekker, 2000, pp. 3, 13,

23, 37, 511.Cohen, L. H.; and D. K. Smith. "Thin-Specimen X-ray Fluorescence Analysis of Major Element in Silicate Rocks." Anal. Chem. 1989. 61: 1837-1840.

Grieken, R. V.; and A. A. Markowicz. Handbook of X-Ray Spectrometry: Methods and Techniques, 2nd ed. New York Marcel Dekker, 2002.

Jenkins, R.; and J. L. De Vries. Practical X-ray Spectrometry. London: MacMillan, 1967, pp. 20-21, 111, 133.

Jenkins, R.; K. W. Gould; and D. Gedcke. Quantitative X-Ray Spectrometry, 2nd ed. New York: Marcel Dekker, p.1995.

Lewis, G. J.; and E. D. Goldberg. "X-Ray Fluorescence Determination of Barium, Titanium, and Zinc in Sediments." Anal. Chem. 1956. 28: 1282-1285.

Mashin, N. I.; et al. "X-ray Fluorescence Method for Determining the Mass Absorption Coefficient in Two-Layer Thin-Film Ti/Ge and Ni/Ge Systems." Appl. Spectrosc. 2012. 79(2): 307-311.

Muller, R. O. Spectrochemical Analysis by X-ray Fluorescence. New York: Plenum Press, 1972, pp. 59-61.

Purdue, G. E.; and R. W. Williams. "Analysis of Powders by XRF Using the Smear Technique." X-Ray Spectrum. 1985. 14(3): 102-108.

Sahoo, N. K.; D. V. Udupa; and D. Bhattacharyya. Thin Films: Science and Technology. New York: Wiley, 2012. AIP Conference Proceedings, Vol. 1451.

Sherwin, T. (ed.). Langmuir Monolayers in Thin-Film Technology. New York: Nova Scientific Publishers,

9. APPENDIX OF NUMERICAL CALCULATIONS

All symbols and data in the unified theory refer to the admixture (original sample with added flushing standard Y_2O_3). There are two unknowns (X_i and k_i) and three measured data (X_Y, I_Y, and I_i) in the intensity-concentration equation (Eq. 5). Given X_i, the k_i can be calculated. Given k_i, the X_i can be calculated. For the convenience of laboratory work, the data of Sample 4 in Tables 1 and 2 are used as an example to show the three typical calculations: Eq. 6 for k_i, Eq. 7 for X_i, and Eq. 9 for $X_i°$, which is the percentage weight (wt%) of the element sought in the original sample.

1. **Constant slope (k_{Zn}):** Using Eq. 6 with given X_{Zn} = 5.51%
   Composition: 1.0511 g $BaSO_4$ + 0.1082 g ZnO + 0.3147 g K_2HPO_4
   Weight of flushing standard = W_f = 0.1047g = weight of Y_2O_3 added into the above sample
   Weight of original sample = W_o = 1.0511 + 0.1082 + 0.3147 = 1.4740g
   Weight of the admixture = 1.4740 + 0.1047 = 1.5787g
   Molecular weight of ZnO = 65.37 + 16.00 = 81.37%
   Zn in ZnO = 65.37/81.37 = 80.34%
   Molecular weight of Y_2O_3 = 2 x 88.905 + 3 x 16.00 = 225.81
   % Y in Y_2O_3 = 2 x 88.905/225.81 = 78.74%
   X_{Zn} = % Zn in admixture = 0.1082 x 0.8034/1.5787 = 5.51%
   X_Y = % Y in admixture = 0.1047 x 0.7874/1.5787 = 5.22%

I_{Zn} = measured intensity of Zn = 343 c/s
I_Y = measured intensity of Y = 113 c/s
Equation (6)

2. **Weight fraction (wt%) of element sought (X_{Zn}):**
Using Eq. 7 with given k_{Zn} = 2.876
Equation (7)

3. **Conversion of X_{Zn} to $X_{Zn}°$:**
Using Eq.9, given X_{Zn} = 5.51% and X_f = 7.10%
X_{Zn} is the weight percent of zinc found in the admixture (original sample plus flushing standard Y_2O_3). $X_{Zn}°$ is the weight percent of zinc found in the original sample. X_f is the weight% of Y_2O_3 added into the original sample.

$$\frac{X_{Zn}}{X_{Zn}^o} = \frac{1}{1+X_f} \tag{8}$$

$$X_f = \frac{\text{Weight of added } Y_2O_3}{\text{Weight of original sample}}$$
$$= \frac{0.1047}{1.4740} = 0.0710 = 7.10\% \, Y_2O_3$$
$$X_{Zn}^o = X_{Zn}(1+X_f)$$
$$= X_{Zn} \cdot \frac{W_a(\text{weight of the admixture})}{W_o(\text{weight of original sample})} \tag{9}$$

$X_{Zn}°$ = 0.1082 x 80.34% /1.4740 =5.90% Calculated from given composition.

8.

Thin-Film Technique of Quantitative X-ray Fluorescence Analysis

> It is how you deal with failure that determines how you achieve success.
>
> — Dale Carnegie

1. ABSTRACT

Liebhafsky and Zemany proposed two terms in X-ray emission analysis: one is the linear thickness of a specimen below which the absorption effect vanishes; the other is the critical thickness of a specimen above which the specimen becomes infinitely thick. The criteria for these two terms are defined here in quantitative terms. A procedure for making thin films of closely controlled thickness is described. The sample is dispersed into a film-forming polymer solution which contains an internal standard element. A thin film is made of this suspensoid by a film applicator. The internal standard element is used to correct for variation in actual film thickness. A *single* calibration curve can be used for samples of different matrices. Experimental data are presented to demonstrate the effectiveness of this thin-film method in eliminating matrix effects.

2. INTRODUCTION

The quantitative relationship between intensity and concentration in X-ray emission analysis is complicated by the absorption-enhancement effects of the matrix. It is well known that matrix effects substantially disappear in thin-film type specimens. Gunn[1] demonstrated the linear relationship between intensity and concentration of various elements in thin films. Application of Gunn's method requires first bringing the sample into solution which in turn involves wet chemistry. Once the sample is in solution atomic absorption spectroscopy might be easier to use for analysis. Finnegan[2] mixed a powdered dry sample with cellulose powder and then used butyl acetate to dissolve the cellulose and thus prepared thin films for X-ray analysis. Here constant film thickness of samples and standards were assumed. Rhodin[3] deposited oxide films on Mylar® and analyzed these films in terms of area density (g cm^{-2}) and thus avoided the problem of film thickness. McGinness, Scott and Mortensen[4] incorporated an internal standard into thin films to correct the variation in film thickness for the analysis of paints. Recently, Giauque et al.[5] reported a thin-film method but the film was not thin enough to X-rays, and absorption correction had to be determined by a separate experiment.

The question is just how thin must a film be for X-ray emission analysis before matrix effects can be ignored? Also, how can one prepare such a thin-film of controlled thickness for routine analysis? This paper will attempt to answer these questions.

3. CRITERIA FOR FILM THICKNESS

The general relationship between intensity, concentration, and matrix effect in X-ray emission analysis was derived by Muller,[6] Jenkins and De Vries.[7] and Birks.[8] If monochromatic primary radiation or an effective primary wavelength is assumed, it can be written in the following form,

$$I_i = k_i C_i \frac{1 - e^{-\mu \rho d}}{\mu} \quad (1)$$

$$\mu = \frac{\mu_1}{\sin \phi_1} + \frac{\mu_2}{\sin \phi_2} \quad (2)$$

where I_i = X-ray emission intensity of element i, counts s^{-1}; k_i = Proportional constant of the spectrometer for element i; C_i = percentage weight of element i in the specimen; = Density of specimen, g cm^{-3}; d = Thickness of specimen, cm. μ_1 and μ_2 = Mass absorption coefficients for the primary and secondary X-rays, cm^{2}g^{-1}; and ϕ_1 and ϕ_2 = Incident and take-off angles.

It is convenient for the following discussions to let h = $\mu \rho d$ and note that h is a dimensionless number, since the unit of mass absorption coefficient μ is cm^{2}g^{-1}, the unit of density is g cm^{-3}, and the unit of specimen thickness is cm.

1. **Thin films:** When the sample is very thin, then h is very small. By series expansion, equation Birks[8] reduced Eqn. (I) becomes

$$I_i = k_i C_i \rho d \quad (3)$$

This equation is independent of the matrix factor μ. It can be used for the thickness determination of thin films[9,10] wand for quantitative analysis of thin films.[3]

It is important to realize the limitation of Eqn. (3). It inherently contains an error due to the truncation of the infinite series in its derivation. The exact percent error can be expressed as follows:

$$\% \text{ Error} = [1-(1-h)e^h] \times 100 \tag{4}$$

The percent errors involved in various values of h are listed in Table 1.

Table. 1. Percent error introduced by the approximation $e^{-n} = 1 - h$

h	0.01	0.1	0.135	0.2	0.3	0.4	0.5	1.0
% Error	0.0	0.5	1.0	2.3	5.5	10.5	17.6	100

Table 2. Percent error introduced by the approximation $1 - e^{-h} = 1$

h	1	2	3	4	4.62	5	6	7
% Error	58.2	15.7	5.2	1.9	1.0	0.7	0.2	0.0

2. **Infinite thickness:** When the specimen is infinitely thick, then h = ∞, and 1 − e^-h 4.'4=1. Thus Birks[8] obtained

$$I_i = k_i \frac{C_i}{\mu} \tag{5}$$

This is the most common situation in X-ray emission analysis. Various quantitative X-ray methods such as standard addition (spiking), standard dilution, internal standard, etc., are essentially efforts to maintain a similar matrix and hence retain a nearly constant p factor. The exact percent error introduced by the approximation of infinite thickness can be calculated by the following equation

$$\% \text{ Error} = \frac{100}{e^h - 1} \quad (6)$$

The percent errors involved in various values of h are listed in Table 2.

Therefore, when h < 0.01, the specimen is genuinely thin to X-rays and the matrix effect can be ignored; when h > 7, the specimen is infinitely thick to X-rays and the thickness effect disappears. In the borderline situations, 0.4 <h< 2.4, either Eqn. (3) or Eqn. (5) would inherently contain at least 10 percent error, and their absorption corrections must be determined by separate experiments.

Liebhafsky and Zemany[9] defined two terms qualitatively in X-ray emission analysis: one is the linear thickness d_e, the point below which the absorption effect vanishes and the relationship of intensity to specimen thickness becomes linear; the other is the critical thickness d_c, the point above which the thickness effect disappears and the specimen becomes infinitely thick. As a result of the present treatment, the criteria for linear thickness and critical thickness can be quantified by equations (7) and (8) if 1 percent error from the approximations is tolerable.

$$d_e = 0.135/\mu\rho \quad (7)$$

$$d_c = 4.62/\mu\rho \simeq 34 d_e \quad (8)$$

In the above discussion, only the absorption effect of the matrix is included in the matrix factor µ; the enhancement effect of the matrix is not accounted for. However, for most analytical applications, the matrix enhancement effect is one to two orders of magnitude lower than the matrix absorption effect. The matrix enhancement effect in thin films is normally small and negligible.[5]

4. METHOD OF MAKING THIN FILMS

Consider a polymethylmethaerylate (PMMA) film containing 1 percent PbCrO$_4$ (Chrome Yellow Pigment). The $\mu\rho$ values are estimated to be 9 cm^{-1} for lead and 76 cm^{-1} for chromium by using the following information: An angle of 90° between the primary and the secondary (detected) radiation yields the highest peak-to-background ratio. The Philips X-ray spectrograph uses $\phi_1 \simeq b0°$ and $\phi_2 \simeq 30°$. An element can be excited to fluorescence most efficiently by radiation which has a wavelength of 0.2 ~ 0.3 Å shorter than the X-ray absorption edge of that element. The necessary data required for the estimation are listed in Table 3. For the sake of simplicity monochromatic primary radiation or an effective primary wavelength is assumed.

Table 3. Data for $\mu\rho$ calculation

Pb LII Edge = 0.815 Å	Cr K Edge = 2.07 Å
$\lambda_1 \simeq 0.7$ Å. for Pb	$\lambda_1 \simeq 1.7$ Å for Cr
λ_2 (PbLβ) = 0.98 Å	λ_2 (Cr Kα) = 2.29 Å
ρ (PMMA) = 1.19 g cm^{-3}	ρ (PbCrO$_4$) = 6.12 g cm^{-3}

Applying the criterion for thin films, Eqn. (7), to the above PMMA film, we find that a film thickness of 150 μm is thin for Pb analysis, and 18 μm is thin for Cr analysis if 1 percent error is tolerable (h = μρd = 0.135). It is interesting to note that a polymer is a better dispersing medium than water for X-ray emission analysis because the mass absorption coefficient of carbon is smaller than that of oxygen.

In order to take advantage of these facts for X-ray emission analysis, two practices must be followed: one is to disperse the sample homogeneously into a thin film, and the other is to make thin films of constant thickness (about 20 µm). To this end, the following procedure was devised.

A stock solution of film-forming polymer was made of 2 parts polymethyl methacrylate, 3 parts polybutyl methacrylate, 7.5 parts toluene, small amounts of necessary additives, and approximately 0.1 percent Zn, Cu or a rare element as an internal standard. Zinc isodecanoate and copper naphthenate are suitable substances for this purpose because they are soluble in the polymeric stock solution and hence homogeneity can be assured. The internal standard is used to measure the actual film thickness which can be normalized to constant film thickness.

About 24 grams of the stock solution was accurately weighed into a 1/4-pint metal can. The proper amount of sample in the form of homogeneous fine powder was blended into this resin solution so that the dried resin contained 0.5-1.0 percent of the element sought. An approximately equal volume of steel balls (1/8-inch diameter) were added into this mixture. Thus it essentially forms a mini-ball mill. After 5 ml. of toluene was added, the can was tightly covered and shaken vigorously on a Red Devil® shaker to obtain a homogeneous dispersion. A 2 mil drawdown on Mylar® film was made of this dispersion with a film applicator (Gardner Laboratory, Inc., Bethesda, Md.). The Mylar® substrate was chosen because it gives low background. The wet film will dry in about one minute to a film of less than 1 mil thick (thin film). A disc of 1 1/4-inch diameter was cut from the drawdown on Mylar® with a press and die. The disc was inserted into the X-ray spectrometer and then counted or scanned for the analytical emission lines of the element sought and the internal standard.

Table 4. Compositions and X-ray data for standards
(Cd–cy stands for cadmium cyclohexane butyrate)

Standard No.	Composition in g		% Cd	Resin Solution Used in g	% Cd in Thin Film	Normalized Intensity Ratio
1	TiO_2	2.5079	1.17	23.0469	0.251	0.32
	Cd–cy	0.1295				
2	TiO_2	2.3202	2.73	23.0012	0.582	0.67
	Cd–cy	0.2995				
3	TiO_2	2.0981	4.57	23.0150	0.967	1.12
	Cd–cy	0.4964				
4	TiO_2	2.0145	6.17	23.4267	1.334	1.59
	Cd–cy	0.7011				

Applying Eqn. (3) to the emission intensities of the element sought, I_i, and the internal standard, 4, we have

$$I_i = k_i C_i \rho d \qquad (9)$$

$$I_s = k_s C_s \rho d \qquad (10)$$

Since the concentration of the internal standard C_s is constant and the density ρ and thickness d of the film are the same for both elements, then

$$\frac{I_i}{I_s} = k C_i, \quad k = \text{constant}. \qquad (11)$$

The concentration of the element sought is proportional to the intensity ratio (I_i/I_s). The calibration curve of concentration versus intensity ratio should be a straight line passing through the origin. Note that the actual concentration of the internal standard is immaterial as long as it is low and constant, and the actual thickness of the film is also immaterial as long as it is thin to X-rays.

5. EXPERIMENTAL

In order to demonstrate the effectiveness of this thin film method in eliminating matrix effects, four standards were prepared to construct a calibration curve. Four synthetic samples of light, medium, heavy and mixed matrices were also prepared for analysis. Note that a *single* calibration curve can be used for the same element in various samples of different nature.

Table 5. Compositions and X-ray data for samples

Sample No.	Composition in g		Resin Solution Used in g	Normalized Intensity Ratio	% Cd Known	% Cd Found
1 (Light Matrix)	Al$_2$O$_3$ Li$_2$CO$_3$ Cd–cy	1.2035 1.2193 0.1517	22.9165	0.33	1.41	1.35
2 (Medium Matrix)	CaCO$_3$ TiO$_2$ Cd–cy	1.1323 1.2133 0.2956	22.8446	0.62	2.67	2.46
3 (Heavy Matrix)	PbSO$_4$ SrCO$_3$ Cd–cy	1.0045 1.1071 0.4509	23.2517	1.02	4.21	4.19
4 (Mixed Matrix)	Al$_2$O$_3$ TiO$_2$ PbSO$_4$ Cd–cy	0.6986 0.7098 0.5963 0.6056	23.1243	1.34	5.54	5.38

Table 6. Percentage of Pb in paints

	Paint I	Paint II	Paint III
X-Ray	1.06	0.41	0.36
AA	0.99	0.43	0.32

Cadmium from cadmium cyclohexane butyrate (SRM-1053, 23.9% Cd, NBS) was the element sought. The components used for making up the matrices are of certified reagent chemical grade or equivalent. The compositions of the standards and samples are shown in Tables 4 and 5. The non-volatile material (NVM) content of the stock solution of film-forming polymers is 42.0%. The internal standard is about 0.1% Zn from zinc isodecanoate.

Table 7. Percentage of Pb in alkyd house paint

Analysis	1	2	3	4	5	6	7	8	9	10	11	Average	Std. Dev.
X-Ray	1.06	0.97	1.04	1.18	1.06	1.08	1.15	0.96	1.03	1.05	1.04	1.06	0.07
AA	0.99	1.07	1.01	0.98	0.99	1.03	1.02	0.98	0.97	0.98	–	1.00	0.03

A Norelco Philips X-ray spectrometer equipped with a Mark

III electronic circuit panel was used to obtain the intensity data for CdK_α and ZnK_α. The instrumental conditions were set by normal PHA procedure. A fixed time of 200 seconds was chosen to accumulate about 10,000 to 16,000 counts after background correction. The intensity ratio I_{Cd}/I_{Zr}, normalized to constant film thickness were plotted against the % Cd in thin film standards to obtain the calibration curve. This same procedure was followed for the analysis of unknown samples. The calibration curve is shown in Fig. 1. The results of analysis are presented in Table 5. No matrix effect was observed. It appears to be practically eliminated.

The same procedure has been applied to the analysis of lead in paints. The calibration curve for lead is also shown in Fig. 1. Three paints of different formulation were analyzed by both the X-ray and the atomic absorption* methods. The results are listed in Table 6 for comparison.

Paint I is a typical alkyd house paint. It was used as the sample for a statistical evaluation of this thin film method. It was analyzed 11 times by the same analyst using the X-ray method on different days. It was also analyzed 10 times by another analyst using an atomic absorption spectrometer (Perkin-Elmer Model 403) on different days. The results are listed in Table 7. The data indicate that a standard deviation of 7% relative can be expected from this thin film X-ray method.

Other elements such as Ti, Ca, Sr, Ba, Cl and S in pigment powders have been successfully analyzed by this thin film X-ray method in our laboratory. For the two light elements Cl and S, we obtained 67.5 c s^{-1} for 1.03% Cl, 122 c s^{-1} for 2.05% Cl, and 11.5 c s^{-1} for 0.28% S, 15.0 c s^{-1} for 0.38% S, etc., in the presence of Ba, Cd, Ti and K. These data gave reasonably good calibration curves for Cl and S.

6. CONCLUSIONS

Solutions or fused pellets[11] are truly homogeneous samples. Most samples, however, require elaborate procedures to bring them into solution quantitatively and certain samples might decompose before melting. The next best approach is to disperse the sample uniformly into a light matrix. The present thin film X-ray method resulted from this concept in order to avoid tedious sample preparation.

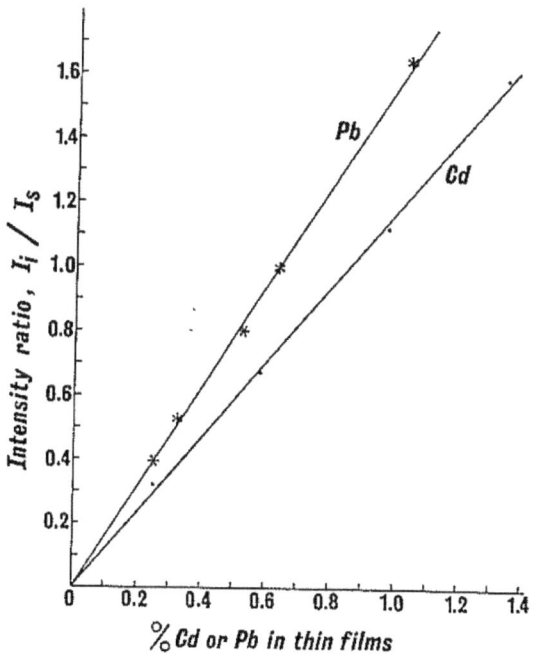

Fig. 1. Calibration Curves for percentage Cd and percentage Pb in Thin Films.

Although the thin film approach was suggested about two decades ago, its general application was deterred formerly by two factors. First, the film was too thin to make; secondly, the film thickness was beyond control.[7] With the criteria and method presented here, it is easy to make thin films of controlled thickness which are thin enough to X-rays; thus the matrix effect can be ignored, and single calibration curve can be used for samples of diverse constitution.

To achieve dependable and reproducible results of analysis by this thin film method the following two factors deserve attention. (1) In any dispersive grinding of a suspension of powders in a polymeric medium, the size of the particles will reach a limit. Further grinding has little effect on size reduction. This limit is generally around a few microns. The mini-ball mill step in the method of making thin films is devised to reach this limit. Therefore, a homogeneous dispersion of particles of similar size distribution can be realized. (2) The linear relationship between intensity and concentration holds only to a certain limit. For instance, the upper limit of lead content in the dried film of approximately 1 mil thickness is about 1 percent. This upper limit is higher for lighter elements. Samples and standards should be prepared in compliance with this limit. In case the concentration is beyond this upper limit, drawdowns of thinner films are necessary to bring the h value into tolerable range.

*Chicago Health Department, private communication, 1972.

7. REFERENCES

Birks, L. S. *X-Ray Spectrochemical Analysis*, Interscience, New York, 1959, 59, 63.

Cline, J. E. *Physical Measurement and Analysis of Thin Films*,

Plenum Press, New York, 1969, 83.

Croke, J. F.; W. R. Kiley; and E. E. Kaelble (Ed.). *Handbook of X-Rays*, Chapter 33. McGraw-Hill, New York, 1967.

Finnegan, J. J. *Adv. X-Ray Anal.* 5, 500 (1962).

Giauque, R. D.; F. S. Goulding; J. M. Jaklevic; and R. H. Pebl, *Anal. Chem.* 45, 671 (1973).E.L. Gunn, *Anal. Chem.* 33, 921 (1961).

Jenkins, R.; and J. L. De Vries. *Practical X-Ray Spectrometry*, MacMillan, London, 1967, 21, 133.

Liebhafsky, H. A.; and P. D. Zemany, *Anal. Chem.* 28, 455 (1956).

McGinness, J. D.; R. W. Scott; and J. S. Mortensen, *Anal. Chem.* 41, 1858 (1969).

Muller, R. O. *Spectrochemical Analysis by X-Ray Fluorescence*, Plenum Press, New York, 1972, 59-61.

Rhodin, T. N., *Anal. Chem*, 27, 1857 (1955).

9.

Quantitative X-ray Diffraction of Powder Samples

> The harder the conflict, the more glorious the triumph.
> — Thomas Paine

1. ABSTRACT

A unified matrix-flushing theory and its practical applications for quantitative multicomponent analysis by X-ray diffraction are presented. In this method, a fundamental "matrix-flushing" concept is introduced; the calibration curve procedure is shunted; the matrix (absorption) effect is totally eliminated; all components, crystalline or amorphous, can be determined.

2. INTRODUCTION

X-ray diffraction analysis is a standard technique used in industrial laboratories for quality control and routine analysis. It is often the only technique available for distinguishing polymorphic structures or for analyzing solid solutions (1). However, the X-ray diffraction technique is hampered by a matrix (absorption) effect which makes quantitative X-ray diffraction analysis rather tedious.

An X-ray diffraction method for quantitative multicomponent analysis has been developed in which the calibration curve procedure is shunted and a more fundamental "matrix-flushing" concept is introduced. This concept provides an exact relationship between intensity and concentration free from matrix effects. The unified matrix-flushing theory and its applications are presented below.

3. UNIFIED MATRIX FLUSHING THEORY

The intensity (energy per second) of a reflection (hkl) in a powder diffraction pattern is derived from the well-known X-ray diffraction theory and is described elsewhere (2, 3). The intensity equation is given below:

$$I(hk\ell) = \left(\frac{I_0 e^4 \lambda^3 d}{32 \pi m^2 c^4 r}\right) \left(N^2 pF^2\right) \left(\frac{1 + \cos^2 2\theta}{\sin^2 \theta \cos \theta}\right) TAV \quad (1)$$

where: I = Intensity of X-rays diffracted by (hkℓ) plane.
  I_0 = Intensity of primary X-rays.
  e, m = Charge and mass of electron.
  λ = X-ray wavelength.
  d = Slit width of detector.
  c = Velocity of light.
  r = Specimen-to-detector distance.
  N = Number of unit cells per unit volume.
  p = Multiplicity.
  F = Structure factor.
  θ = Bragg angle.
  T = Temperature factor.
  A = Absorption factor.
  V = Volume of powder in the beam.

This complete intensity equation can be conveniently separated into six factors: (1) a constant factor for a particular diffractometer, (2) a structure factor characteristic of the diffracting sample, (3) an angle factor known as the Lorentz-polarization factor, (4) a temperature factor to correct for thermal vibration of atoms, (5) an absorption factor which attenuates the diffracted X-rays, and (6) the volume of powder in the primary X-ray beam.

For a specific (hkl) reflection from a component i in an X-ray powder diffraction pattern of a mixture, the first four factors are constant, K_i. The absorption of X-rays, just like the absorption of visible light, follows the well-known exponential law. For infinite specimen thickness, we have the fifth factor:

$$A = \int_0^\infty e^{-\mu_t S} \, ds = \frac{1}{\mu_t} \quad (2)$$

where μt and S are the linear absorption coefficient and thickness of the total sample. The last factor V, in case of a mixture, indicates the volume fraction V_i of component i.

Let X_i and ρ_i be the weight fraction and density of component i; the complicated intensity equation (1) can be reduced to:

$$I_i = \frac{K_i}{\rho_i} \cdot \frac{X_i}{\mu_t} = k_i X_i \quad (3)$$

where equation is a characteristic constant of component i, and k_i is a factor containing the linear absorption coefficient of the total sample. At this moment, k_i is constant for a very small variation in Xi. Later on, it will be shown that k_i in the ratio k_i/k_j is constant for any Xi; the matrix factor μ_t is cancelled.

For the quantitative X-ray diffraction analysis of a mixture of n components, we have n unknowns (X_i, i = 1 to n) which must satisfy the following (n +1) nonhomogeneous equations:

$$\left. \begin{array}{l} I_i = k_i X_i, \quad i = 1 \text{ to } n \\ \displaystyle\sum_{i=1}^{n} X_i = 1 \end{array} \right\} \quad (4)$$

This situation can be most conveniently treated in terms of matrix algebra as follows:

$$\begin{pmatrix} k_1 & 0 & 0 & \cdots & 0 \\ 0 & k_2 & 0 & \cdots & 0 \\ 0 & 0 & k_3 & \cdots & 0 \\ \cdots & \cdots & \cdots & \cdots & \cdots \\ 0 & 0 & 0 & \cdots & k_n \\ 1 & 1 & 1 & \cdots & 1 \end{pmatrix} \begin{pmatrix} X_1 \\ X_2 \\ X_3 \\ \vdots \\ X_n \end{pmatrix} = \begin{pmatrix} I_1 \\ I_2 \\ I_3 \\ \vdots \\ I_n \\ 1 \end{pmatrix} \quad (5)$$

If the above matrix equation, **KX= I**, has a solution, the solution is unique if and only if the rank of the **(K)** matrix is equal to the rank of the **(K,I)** matrix.

In order to satisfy this condition, the unique solution of equation (5) has the following simple and symmetrical form:

$$X_i = \left(\frac{k_i}{I_i} \sum_{j=1}^{n} \frac{I_j}{k_j} \right)^{-1} \quad (6)$$

Note the unusual property of this unique solution is that the weight fraction of any component in a multicomponent system is expressed in terms of ratios like I_i/I_j and k_i/k_j (i, j = 1, 2,..., n). The use of an intensity ratio I_i/I_j, makes it immune to many sources of errors; the use of a k_i/k_j ratio, makes it free from matrix effect. But what is k_i? How can it be measured?

From equation (4) the following relationship between intensity and concentration holds true for any pair of components in a multi-component system:

$$\frac{I_i}{I_j} = \frac{k_i}{k_j} \cdot \frac{X_i}{X_j}$$

$$\frac{k_i}{k_j} = \frac{I_i}{I_j} \quad \text{at} \quad \frac{X_i}{X_j} = 1$$

$$\frac{k_i}{k_j} = \left(\frac{I_i}{I_j}\right)_{50/50} = \text{slope} = \text{constant} \tag{7}$$

If a compound j is a universal reference material such as corundum (α-Al_2O_3), then $k_j = k_c = 1$, where the subscript c stands for corundum, hence (4, 5):

$$k_i = \left(\frac{I_i}{I_c}\right)_{50/50} \tag{8}$$

This is the definition of "Reference Intensities Ratio" as given in the Powder Diffraction File (6) by the Joint Committee on Powder Diffraction Standards.

One interesting point is that the relationship between intensity and concentration of any two components as prescribed by equation (7) is independent of the existence of other components in the sample, which means equation (7) holds true whether the universal reference material corundum is indeed a component of the sample or not.

The foregoing deduction can be summarized in the following theorem: The plot of intensity ratio (I_i/I_j) to the weight ratio (X_i/X_j) of any two components is a straight line passing through the origin with a slope equal to the ratio of corresponding Reference Intensities (k_i/k_j). This intensity-concentration relationship between each and every pair of components in a multicomponent system is not perturbed by the presence or absence of other components.

The above derivation assumes that all the components in the mixture are crystalline and identified. However, in many situations some components in the mixture are amorphous and/or unidentified, quantitative data are often requested only for identified components or components of interest. In these cases, equation (6) cannot be applied, but equation (7) always holds true. In order to make use of this equation a flushing agent must be added into the sample to flush out the matrix effect. The flushing agent may be any pure compound not present in the sample.

Let the weight fraction of the flushing agent and the original sample be designated X_f and X_o respectively, that is:

$$X_f + X_o = X_f + \sum_{i=1}^{n} X_i = 1$$

$$\frac{I_i}{I_f} = \frac{k_i}{k_f} \cdot \frac{X_i}{X_f}$$

$$X_i = X_f \frac{k_f}{k_i} \cdot \frac{I_i}{I_f} \tag{9}$$

Equation (13) can be used for quantitative multicomponent analysis and in addition one more simplification can be realized. Since corundum (α-Al$_2$O$_3$) has been chosen for Reference Intensities by the JCPDS for its purity, stability, and availability, it is convenient to choose the same corundum as a flushing agent for the same good reasons. Consequently, $k_f = k_c = 1$, (k_c = relative intensities of corundum) from equation (13) we obtain:

$$X_i = \frac{X_c}{k_i} \cdot \frac{I_i}{I_c} \tag{10}$$

This is the working equation for quantitative multicomponent analysis. It represents a straight line passing through the origin with a slope equal to X_c/k_i. It is free from matrix factors. No previous information relating to the approximate concentration ranges of various components sought is required. Intensity ratios from a single scan are the only experimental data needed. Since intensity ratios from the same scan are used the errors due to instrumental drift and sample preparation are minimized. Note that the working equation (14) prescribes the slope of calibration curve for every component in the sample, thus it is not necessary to actually work out calibration curves.

Another interesting application of this matrix-flushing concept is that it can be used to detect and quantify the amorphous content in a mixture. When corundum is chosen as the flushing agent, substituting equation (14) into equation (11), we get:

$$\sum_{i=1}^{n} \frac{I_i}{k_i} = \frac{X_o}{X_c} \cdot I_c \tag{11}$$

where X_o/X_c is the weight ratio of the original sample to flushing agent corundum.

Equation (15) affords a means to experimentally check the correctness of this concept, to appraise the reliability of intensity data, and to predict and assay the presence of amorphous materials in a sample. That is to say the experimental data may fall into one of the following three cases:

$$\sum_{i=1}^{n} \frac{I_i}{k_i} \gtreqless \frac{X_o}{X_c} \cdot I_c \tag{12}$$

where: > indicates wrong data,
= indicates all components are crystalline,
< indicates the presence of amorphous material.

For a binary system, an "auto-flushing" phenomenon exists. No flushing agent is needed. One component automatically serves as a flushing agent for the other component and vice versa. Because:

$$\left. \begin{array}{c} X_1 + X_2 = 1 \\ \dfrac{I_1}{I_2} = \dfrac{k_1}{k_2} \cdot \dfrac{X_1}{X_2} \end{array} \right\}$$

$$X_1 = \dfrac{1}{1 + k \cdot \dfrac{I_2}{I_1}}$$

$$k = \dfrac{k_1}{k_2} = \left(\dfrac{I_1}{I_2}\right)_{50/50} \tag{13}$$

The slope k is simply the corresponding intensity ratio of a 50/50 mixture of the same two components, hence the quantitative analysis of a binary system can be performed without using the Reference Intensities at all. Of course, equation (18) is the result of degeneration of equation (6) when $n = 2$.

The soundness and usefulness of this matrix-flushing method for quantitative multicomponent analysis are demonstrated in the experimental section.

4. EXPERIMENTAL

In order to illustrate the simplicity and usefulness of the matrix-flushing theory, eight synthetic samples were prepared with pure chemicals of analytical reagent grade or better. The analytical procedure and the instrumental conditions used are described elsewhere (4, 5).

For the sake of clarity, the experimental data and their interpretation are divided into three categories:

4.1 All Components Are Crystalline and Identified

In this situation, the X-ray diffraction pattern of the original sample can be interpreted directly and quantitatively without using a flushing agent or an internal standard. Equation (6) is used to calculate the percentage composition from intensity data. The data of two examples No. 1 and No. 2 are given in Table 1. All intensity data refer to the strongest reflections of corresponding components.

Table 1. Intensity and Composition Data

Sample No.	Composition (grams)		Intensity I_i, c/s	Ref. Int. k_i	% Composition	
					Known	Found
1	ZnO	0.2236	610	4.35	9.87	9.2
	NiO	0.5454	1412	3.81	24.06	24.5
	CdO	0.6588	3303	7.62	29.07	28.7
	KCl	0.8386	2207	3.87	37.00	37.7
2	ZnO	1.8901	5968	4.35	41.49	41.3
	KCl	1.0128	2845	3.87	22.23	22.1
	LiF	0.8348	810	1.32	18.32	18.5
	Al_2O_3	0.8181	599	1.00	17.96	18.1

4.2 Some Components Are Amorphous And / Or Unidentified

In this case, a known quantity of flushing agent (α-Al_2O_3) was added to the sample. The sample was then ground to a homogeneous fine powder before the X-ray diffraction experiment was run. The working equation (14) was applied to each component sought to calculate its weight percentage.

In many practical situations the composition of an unknown sample is only partially determined. Even when all the peaks in the X-ray diffraction pattern are accounted for, one is still not sure whether there is any amorphous material present. Therefore, it is safe to use the flushing agent for the following reasons: First, the true values of x_c and k_c of the flushing agent are absolutely known; Secondly, the weight fraction X_i is dependent upon I_i and

k_i only, independent of any X_j, I_j and k_j where $j \ne i$; Thirdly, the weight fraction of a component sought is independent of the presence or absence of any other component, crystalline or amorphous. The data of two examples No. 3 and No. 4 are listed in Table 2. Note that Sample No. 3 is identical to Sample No. 2 except that the Al_2O_3 is an unknown component in Sample No. 2, but is a flushing agent in Sample No. 3.

Table 2. Intensity and Composition Data

Sample No.	Composition (grams)		Intensity I_i, c/s	Ref. Int. k_i	% Composition	
					Known	Found
3	ZnO	1.8901	5968	4.35	41.49	41.1
	KCl	1.0128	2845	3.87	22.23	22.0
	LiF	0.8348	810	1.32	18.32	18.4
	Al_2O_3 Flushing	0.8181	599	1.00	17.96	----

Table 2 (Cont.)

Sample No.	Composition (grams)		Intensity I_i, c/s	Ref. Int. k_i	% Composition	
					Known	Found
4	NiO	1.0743	4162	3.81	51.28	48.4
	CdO	0.1495	1160	7.62	7.14	6.7
	KCl	0.5410	2404	3.87	25.82	27.5
	Al_2O_3 Flushing	0.3304	356	1.00	15.77	----

It is found that even the intensity data read directly from the diffraction pattern on a strip-chart provide fairly good accuracy.

The authenticity of the matrix-flushing theory is further scrutinized by equation (15). Using the data of example No. 3, we have:

$$\Sigma \frac{I_i}{k_i} = \frac{5968}{4.35} + \frac{2845}{3.87} + \frac{810}{1.32} = 2721 \quad \text{(Experiment)} \quad (14)$$

$$\frac{X_o}{X_c} \cdot I_c = \frac{82.04}{17.96} \times 599 = 2736 \quad \text{(Theory)}. \quad (15)$$

Another interesting feature of this matrix-flushing theory is that it can be applied to detect and determine the total amorphous material in a sample.

Sample No. 5 contains an amorphous silica (silica gel) and Sample No. 6 contains an amorphous resin (Goodyear VPE). Their intensity data are presented in Table 3.

Table 3. Intensity and Composition Data

Sample No.	Composition (grams)		Intensity I_j, c/s	Ref.Int. k_j	% Composition		$\Sigma \frac{I_j}{k_j}$	$\frac{X_0}{X_c} \cdot I_c$
					Known	Found		
5	ZnO	0.9037	4661	4.35	34.43	36.4		
	$CaCO_3$	0.7351	2298	2.98	28.00	26.2		
	SiO_2 Gel	0.4234	0	----	16.13	15.9		
	Al_2O_3 Flushing	0.5629	631	1.00	21.44	----	1842	2312
6	ZnO	0.8090	4948	4.35	40.81	38.7		
	CdO	0.2825	3337	7.62	14.25	14.9		
	Resin (VPE)	0.4057	0	----	20.46	21.9		
	Al_2O_3 Flushing	0.4854	719	1.00	24.48	----	1575	2218

Normal scans of these two samples do not indicate the presence of amorphous material. However, the intensity data from the matrix-flushing method show a large intensity imbalance which indicates the presence of amorphous materials. By the use of material balance, the total amorphous content found in each case is in good agreement with the respective amount of silica gel and VPE resin actually put into the sample.

4.3 Binary Systems:

For a binary system no flushing agent is needed. The data of two examples, No. 7 and No. 8, are shown in Table 4. Equation (18) is used to obtain the % composition.

Table 4. Intensity and Composition Data

Sample No.	Composition (grams)		Intensity I_i, c/s	Ref. Int. k_i	% Composition	
					Known	Found
7	KCl	2.4530	5371	3.87	74.90	74.6
	LiF	0.8219	604	1.32	25.10	25.4
8	ZnO	1.4253	6259	4.35	71.22	72.1
	TiO_2 (R)	0.5759	1461	2.62	28.78	27.9

There are two different ways to obtain the slope k: $k = k_1/k_2$ the latter requires less work, gives better accuracy, and needs no Reference Intensities. For Sample No. 7:

$$k = \frac{k_{KCl}}{k_{LiF}} = \frac{3.87}{1.32} = 2.93 \quad \text{or} \quad k = \left(\frac{I_{KCl}}{I_{LiF}}\right)_{50/50} = \frac{5583}{1846} = 3.02 \quad (16)$$

An unusual example of auto-flushing is the determination of crystallinity of polymers (7) where one component is amorphous while the other component is a poor crystalline material. The slope of the straight line in this reference verifies equation (19) remarkably well.

5. CONCLUSION

Three working equations, (6), (14) and (16), for quantitative multicomponent analysis are derived from the well-established complete intensity equation (1). No assumption or approximation is made in the derivation. Since no calibration curves are required for this method, the work involved for quantitative X-ray diffraction analysis is greatly simplified.

6. REFERENCES

Azaroff, L. V. "Elements of X-Ray Crystallography," p. 202, McGraw-Hill, New York (1968).

Berry, L. G., Ed. "Inorganic Index to the Powder Diffraction File," p. 1421, JCPDS (1972).

Bunn, C. W. "Chemical Crystallography," p. 223, Oxford, London (1961).

Chung, F. H. "Quantitative Interpretation of X-Ray Diffraction Patterns of Mixtures. I. Matrix-Flushing Method for Quantitative Multicomponent Analysis." *J. Appl. Cryst.* in press (1973).

Chung, F. H. "Quantitative Interpretation of X-Ray Diffraction Patterns of Mixtures. II. Adiabatic Principle of X-Ray Diffraction Analysis of Mixtures." *J. Appl. Cryst.* in press (1973).

Chung, F. H.; and R. W. Scott. "A New Approach to the Determination of Crystallinity of Polymers by X-Ray Diffraction," *J. Appl. Cryst.* 6, 225-230 (1973).

Chung, F. H.; and R. W. Scott. "Vacuum Sublimation and Crystallography of Quinacridones." *J. Appl. Cryst.* 4, 506-511 (1971).

10.

X-ray Diffraction in U.S. Industry

> **Life is short, live it. Love is rare, grab it. Anger is bad, dump it. Fear is awful, face it. Memories are sweet, cherish it.**
>
> — Proverb

1. ABSTRACT

The discovery of X-rays by Roentgen in 1895 led a burst of innovations such as Quantum Theory, Relativity Theory, Atomic Theory, etc. The entire structure of science and philosophy was reconsidered and rebuilt. The XRD and XRF techniques were taking shape in Europe in the 1910s. These techniques were transferred to the New Continent in the same decade. Four institutes in the United States (MIT, Caltech, Cornell, and GE) started X-ray diffraction experiments. General Electric is the first industry in the world to apply X-ray diffraction (study the texture of tungsten filament), much later joined by other industries in USA and Europe. The current status of XRD in U.S. industries and future trends are reviewed including recent developments, single crystals vs. powder mixtures, chemical analyses, polymer crystallinity, residual stress, peak profile fitting, and relevant others.

2. INTRODUCTION

A burst of discoveries marked the start of this century. Rontgen discovered x-rays in 1895. Planck introduced the Quantum theory in 1900, and Einstein published the Theory of Relativity in 1905. The entire structure of science and philosophy was reconsidered and rebuilt. In 1912, Laue demonstrated the diffraction of x-rays by crystals at Munich. It proved the wave nature of x-rays and the periodicity of atomic packing within a crystal. In the same year, Bragg at Cambridge interpreted Laue's data with the now well-known Bragg equation, and published details of crystal structure determination by x-ray diffraction the following year (1913). Niels Bohr's first paper on atomic theory also appeared in 1913. In 1916, Debye and Scherrer discovered the powder method in Switzerland, which had enormous impact on industrial applications. In 1917, Hull invented the powder method independently in the New Continent.

In the United States, x-ray diffraction (XRD) experiments, in pioneer days, were carried out in four institutions: Burdick at the Massachusetts Institute of Technology (MIT), and later at Caltech, in 1916; Hull at General Electric Company (GE), in 1916; and Wyckoff at Cornell University, in 1917. Caltech, MIT, and GE were stimulated by direct contact with the Braggs. Cornell was inspired by the visit of Nishikawa from Japan.[1] GE was the first industry in the United States and in the world to apply x-ray diffraction. Much later, in Europe, GE used x-ray diffraction to study the texture of tungsten filament (1923 by Goucher), and Imperial Chemical Industries (ICI) studied the structure of nitrocellulose, lead azide (1930 by Miles), and polyethylene (1933 by Bunn). Soon afterward, x-ray diffraction techniques were practiced at Columbia University (Davis), Harvard University (Duane), University of Illinois (Clark), University of Chicago (Compton, Zachariasen, and Barrett), University of Minnesota (Gruner), Ohio State University (Blake), Penn State University (Pepinsky and Davey), Buffalo University (Harker), Oak Ridge National Laboratory (Levy and Busing), Naval Research Laboratory (Hauptmann and Karle), Dow Chemical (Hanawalt), Bell Laboratories (McKeehan), Eastman Kodak Laboratories (Huggins), Philips Laboratories (Parrish), and many prominent others.

3. PHYSICS VS. CHEMISTRY

Caltech, under the leadership of Linus Pauling, attained a dominant position as a center of crystal structure research and as a source of trained experts. Since the chemistry department was the center of x-ray diffraction activities at Caltech, crystal structure work in the United States has been more closely related to chemistry. In contrast, the x-ray diffraction technique in England has been more closely related to physics simply because both Braggs were physicists. Sir William Henry Bragg was Physics Chair at the University of Leeds (1912) and the University College, London (1915), then Director of the Royal Institution (1923). Sir William Lawrence Bragg was the Physics Chair at the University of Manchester (1919) and later Director of Cavendish Laboratories in Cambridge (1939). Naturally, new centers of x-ray/crystal study were leaning toward physics if the leaders were trained in Europe, albeit leaning toward chemistry (or mineralogy and material science) if the leaders were trained at Caltech.

Currently, the XRD workers in the U.S. are roughly divided into two large groups: the single-crystal group, mostly from academia, which usually attend ACA meetings, and the powder group, mostly from industry and government, which participate in Denver conferences (of course, many attend both).

4. A CAREER IN INDUSTRY

Government, academia, and industry are the three major employers in the United States. Government research is usually mission-oriented. Its purpose is to find more effective ways to defend (defense), to cure (health), or to protect (environment). Academic research is publication-oriented. Professors select topics of research with a reasonable prospect of success, so that the findings will be acceptable for publication and/or for students' theses. Industrial research is profit-oriented. Its attention is focused on the "bottom line" because the bottom line in the annual report is the sole indication of the performance of a company. Except for a few industrial giants, most industrial research laboratories chart their research with products in mind, hence make short-term plans, and are reluctant to wait for a potential breakthrough five years or more down the road.

Consequently, in the profession of x-ray diffraction, a career in industry is quite different from one in academia or government. In industry, the research effort must have business relevance matching the objectives of the company; it is usually a problem seeking solutions, never vice-versa, thus requiring one to be ready to accept assignments not related to x-ray diffraction at all, due to the shift in priority of company strategy.[2]

Table 1
Current hardware for XRD

Computers	Faster/smaller/less expensive computers for instrument control, data collection, profile fitting, search/match, and structure refinement
Detectors	Position-sensitive proportional/scintillation detectors and energy dispersive detectors
X-ray sources	High energy x-rays from synchrotrons, flash tubes, and lasers. Established synchrotron radiations include:

USA : Stanford, Cornell (CHESS), and Brookhaven.
UK : NINA
France : LURE
USSR : VEPP-3
Germany : DESY, DORIS
Italy : ADONE
Japan : National Lab

5. RECENT ACTIVITIES IN X-RAY DIFFRACTION

There are three factors contributing to the recent surge of activity in the x-ray diffraction field: faster/smaller/ less expensive computers, position-sensitive detectors, and stronger x-ray sources, as shown in Table 1.[3-10] Computers have been in existence for 40 years (Pepinsky produced XRAC in 1947), but only recently have computers been made faster, smaller, and more affordable, and hence are accessible to many laboratories for unattended programmed experiments and for tedious calculations that were not practical in the past. As a result, real-time observations of crystal growth or chemical reactions are practical. The old-fashioned film method of recording diffracted x-rays is indeed position-sensitive, but it is time-consuming to derive a consistent set of intensity data from x-ray photographs. The new position-sensitive detector is literally a digitized film; thus the x-ray position/intensity data can be easily stored/processed immediately. The detailed data of peak profile and line shift can be massaged at will. The new synchrotron x-ray sources have quite a few advantages: high intensity with a wide range of wavelengths; highly collimated and polarized without line spectra (K_α doublets); and each reflection is a single peak, thus particularly useful for stress analyses and thin-film studies.

Table 2

Single crystal vs powder

Single crystal	Powder
Drug	Chemicals
Electronics	Metals/alloys
Superconductors	Mineralogy/geology
Catalysts	Petroleum
Molecular biology	Ceramics/cement/glass
Life sciences	Plastics/fibers/films
	Semiconductors/thin films
	Aircrafts/spacecrafts
	Pigments/coatings
	Environmental science and technology

6. SINGLE CRYSTAL VS. POWDER

X-ray diffraction is a rather mature technique. Recent advances are mainly in speed, sensitivity, and convenience. Both single crystal diffraction and powder diffraction have industrial applications. In spite of the esoteric nature of single crystal structural analysis, it has earned niches in drug, electronic, catalyst, superconductor, and molecular biology industries. Much of our understanding of the properties of polycrystalline materials such as metals/alloys, plastics/fibers, and clays/minerals has been attained through the study of single crystal structures. Powder diffraction is simple, yet powerful, and particularly popular in industries such as those shown in *Table 2*. There is always some activity in every industry. Naturally, some institutions are more active in certain areas (see Table 3).

In the field of x-ray diffraction, powder implies all polycrystalline aggregates. It includes everything except single crystals. Hence metal sheets, wires, plastics, films, fibers, rubbers, ceramics, clays, and rocks are all treated as powders. Some unconventional powders and their typical particle sizes are listed in *Table 4*.

X-ray diffraction has been a basic technique in numerous laboratories. The major applications of XRD in U.S. industries include: chemical analysis, polymer crystallinity, stress analysis, thin-film studies, and ceramic structures, including superconductors. The current theory and practice of x-ray diffraction in the more active areas in recent years are briefly summarized in this paper.

Table 4

Unconventional particles

Powder	Particle	Typical size
Pigments	Discrete crystallites	0.01-100 μm
Metals	Grains/precipitates	1 mm-1 μm/~500 Å
Polymers	Crystalline domains	100-400 Å
Fibers	Micelles/microvoids	15-200 Å

Table 3

Applications of XRD in U.S.

Activities	**Selected representative institutions**
Chemical analysis	Industries: GE, Du Pont, Dow, USS, Gulf, IBM, AT&T, Sherwin-Williams
	Government: NBS, EPA, USGS, TVA, Los Alamos, Sandia, Oak Ridge, Argonne, Lawrence Livermore, Naval Research
	Academia: Denver, Penn State, Iowa State, SUNY, UCLA, Florida, North Dakota State, South Dakota State, South Dakota School of Mines
Stress	Northwestern U., Denver U., Penn State, NBS, Rockwell, AT&T, GM, Ford, USS, Argonne
Size/strain	IBM, Dow, Texas Instr., Union Carbide, AT&T, UCLA, UC Berkeley, Sandia, Los Alamos, Oak Ridge
Crystallinity	GE, GM, Ford, Rockwell, Dow
Texture	Du Pont, Exxon, Brimrose, Los Alamos, North Carolina State U.
Crystal structure	Searle, Mobil, DuPont, Dow, AT&T
Amorphous structures	Dow Corning, IBM, UCLA, UC Berkeley
High/low temperature	IBM, Dow, U. of Denver
Microbeam/microcamera	GE, Rockwell, Rigaku
Synchrotron radiations	Stanford, Brookhaven, Cornell, IBM, NBS
Search/match	Philips, Siemens, Rigaku, JCPDS, NBS, IBM, GE, Dow, Penn State, U. of Florida

7. CHEMICAL ANALYSIS

By far the most popular industrial application of x-ray diffraction is chemical analysis, such as phase identification, high/low temperature phases, solids solutions, and search/match programs. No other techniques can analyze polymorphs, solid solutions, salts, and hydrates as positively as XRD.

The chemical analyses in industrial laboratories are very demanding: the analyses are usually nonroutine, the samples are totally unknown, the requests are for quantitative data for multicomponents, the results for needed as soon as possible, and the costs should be as low as possible. The Matrix-Flushing technique of x-ray diffraction can adequately satisfy these demands.

The technique consists of three working equations relating intensity to concentration independent of absorption factors. Neither assumption nor approximation was made in the derivation of these equations. [11-17]

$$X_i = \left(\frac{k_i}{I_i} \sum_{i=1}^{n} \frac{I_i}{k_i} \right)^{-1} \tag{1}$$

where X_i = weight fraction, I_i = x-ray intensity, k_i = flush constant of component i, n = number of components, c =- flushing agent (internal standard, corundum), X_o = weight fraction of original specimen, < indicates the presence of amorphous materials in sample, = indicates all components in sample are crystalline, and > indicates wrong data.

$$X_i = \frac{X_c}{k_i} \cdot \frac{I_i}{I_c} \tag{2}$$

$$\sum_{i=1}^{n} \frac{I_i}{k_i} \genfrac{}{}{0pt}{}{<}{>} I_c \cdot \frac{X_o}{X_c} \tag{3}$$

Generally, all cases of quantitative chemical analysis by XRD can be classified into two categories. First, all components are crystalline and identified, in which case the percentage composition can be calculated using Eq. 1. Secondly, some components are amorphous and/or unidentified. Then the percentage composition of the identified components can be calculated using Eq. 2. The key points for accurate XRD analysis are to attain proper particle size (1-10 μm) and to avoid preferred orientations of needle or plate-like particles. The precision of the Matrix-Flushing technique is 8% relative or much better.

Normally, XRD provides little information about the amorphous components in a sample. An interesting feature of the Matrix-Hushing theory is that from its discriminant Eq. 3 one can determine whether a sample contains amorphous materials, and how much, if it does. The imbalance in intensity is due to the incoherent scattering of the amorphous materials. The amount of amorphous materials in a sample can be derived by mass balance.

8. POLYMER CRYSTALLINITY

Plastics, composites, and synthetic fibers need high performance over a wide range of temperatures. The high performance range of amorphous polymers is between T_g and T_g + 20°C where T_g indicates glass transition point. Below T_g, it is too brittle; above T_g + 20°C, it is too soft. Partial crystallization can broaden this range from T_g to T_m, as shown in *Table 5*, where T_m is the melting point of the crystalline domains. The crystalline domains act as a reinforcing network giving strength to the polymer similar to the wire mesh in concrete. However, too much crystallinity leads to brittleness measured as low impact strength and stress cracking (low tensile modulus). Obviously, crystallinity of polymers has to be controlled. Although there are five different techniques (XRD, NMR [nuclear magnetic resonance], DSC [differential scanning calorimetry], infrared, and density) to determine the crystallinity of polymers, only the XRD technique directly measuresthe polymer crystals in terms of the fundamental definition of crystal.[18-21]

Table 5
Crystallinity of polymers

Performance

	T_g	T_m	
PE	−85°C	137°C	High modulus over entire range
Nylon	56°C	260°C	Can subject to ironing
Dacron	50°C	270°C	Can retain crease

Measurement

	Density	Infrared	XRD	NMR	DSC
PET yarn	56%	81%	39%		
PET chips			52%	72%	
PET film			63%		57%
PET powder	65%		66%		

Crystallinity Scale

Alkyd rubber		Leather		PE		Rock metal
I____I	____	I____	____	I____	I____	I___
0	10	50		70	90	97

% Crystallinity

Any polymer can be considered a two-component system: crystalline component and noncrystalline component similar to the polymorphous systems anatase/rutile, calcite/aragonite, or quartz/cristobalite. The crystalline component scatters x-rays coherently, providing diffraction peaks, while the noncrystalline (amorphous) component scatters x-rays incoherently, giving a broad halo. Since the scattered intensity is proportional to the amount of scattering materials, the crystallinity X_c, defined as weight fraction of crystalline domains, can be obtained from Eq. 1 by putting $n = 2$ for two-component systems:

$$X_c = \frac{1}{1 + k\frac{I_a}{I_c}} \qquad (4)$$

where I_c indicates intensity of crystalline peak, and I_a indicates intensity of amorphous halo. The flush constant k has a physical meaning: it is an isotropic disorder factor closely related to lattice imperfection and thermal diffuse scattering. The flush constant $k = 1$ when I_c = the integrated intensity of all diffraction peaks, and I_a = the integrated intensity of the whole broad halo. If the intensities (integrated or peak height) of the strongest peak and the halo are to be used, the flush constant k can be determined by using a 100% amorphous polymer as flushing agent. This technique has been recently applied to the powder coating PET at Sherwin-Williams,[22] the aerospace composite PEEK at ICI,[23] and the commodity plastic polyethylene at Dow.[24] The procedures to quantitatively separate the intensity contributions from crystalline peaks, the amorphous halo, and the background can be found elsewhere.[22,25]

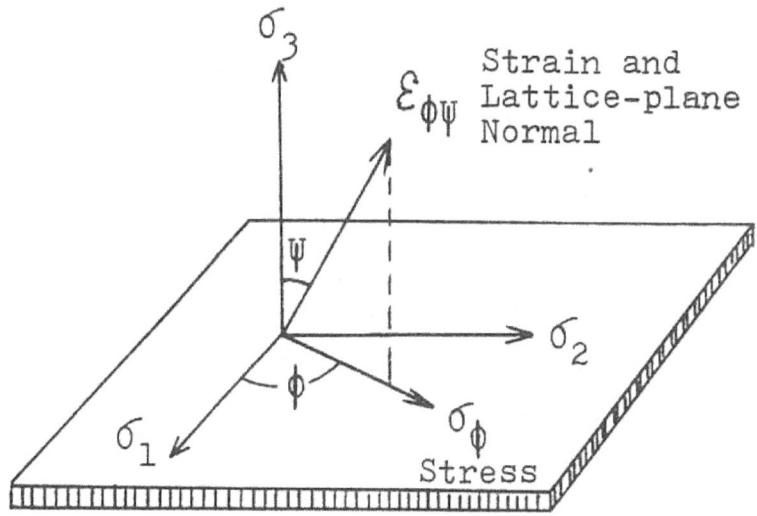

Figure 1: Directions of strain and stress.

Since the crystallinity of a polymer depends upon its thermal history, such as pellet extrusion/molding, Dow designed an instrument to simultaneously monitor crystallinity by XRD with a position-sensitive detector, and thermal history by differential scanning calorimetry, up to 600°C.[24]

9. RESIDUAL STRESS

The metal, semiconductor, and ceramic industries have a fundamental interest in residual stress, particularly in the top thin layers. Residual stress is the stress that remains in a material when no external force is applied. The residual stress can be introduced into materials, such as metals, by any mechanical (machining), chemical (plating/deposition), or thermal (welding/ quenching) process. The residual stress can be either detrimental or beneficial during industrial applications. For example, the initiation and propagation of cracks in fatigue loading, or in stress corrosion, can be impeded by compressive stress but greatly accelerated by tensile stress. The measurement of residual stress by XRD has been favored over other techniques (acoustic and magnetic) which are too sensitive to microstructures.

The basic principle of stress analysis by XRD is to measure the change in interplanar d-spacing due to lattice strain. Combining Bragg's law and the elasticity theory, the change in d or 2θ can be related to lattice strain, from which lattice stress can be calculated. Limited by the depth of penetration, the x-rays are diffracted from the surface thin layer (typically 10 μm). Under such conditions, it is plausible to assume a biaxial stress state where the stress normal to the specimen surface is zero *(Figure 1)* and the following equations hold:[26-31]

$$\sigma_1 + \sigma_2 = -\frac{E}{v}\left(\frac{d_n - d_o}{d_o}\right) \quad \text{(NIT)} \tag{5}$$

$$\sigma_\phi = \frac{E}{(1+v)\sin^2\psi}\left(\frac{d_i - d_n}{d_n}\right) \quad \text{(DET)} \tag{6}$$

$$\Delta 2\theta = 2\sigma_1 \frac{1+v}{E} \tan\theta \sin^2\psi + 2\frac{v}{E}(\sigma_1 + \sigma_2)\tan\theta \quad \text{(MIT)} \tag{7}$$

where E = Young's modulus,
  v = Poisson ratio,
  θ = Bragg angle,
  d_o = d-spacing in stress-free state,
  d_n = d-spacing from normal incidence,
  d_i = d-spacing from inclined incidence,
  ϕ = angle between principal stress direction and projection

of lattice-plane normal onto specimen surface (i.e., = 90),
ψ = angle between the normal of lattice planes to the normal of specimen surface,

τ_1, τ_2, τ_3 = principal stresses (τ_3 = 0 for biaxial stress), and

τ_ϕ = stress in any chosen direction ($\phi\tau$).

These equations represent three different ways to measure the residual stresses: (1) Normal incidence technique (NIT) from which the sum of principal stresses can be obtained. (2) Double incidence technique (DET) in which double exposures were made, one at normal incidence (ψ = 0), the other at inclined angle ψ to the surface normal (commonly ψ = 45° or 60°). The stress in any chosen direction can be calculated. (3) Multiple incidence technique (MIT) in which the principal stress can be obtained from the slope of a plot of α vs $\sin^2 \psi$[32-34].

When the plot of d vs $\sin^2 \psi$ is linear, the assumption of biaxial stress state is valid. When the plot of d vs $\sin^2 \psi$ is nonlinear, but curved, fluctuating, or splitting, then triaxial stress state prevails. The stress gradient of triaxial stress state can be explored with x-rays of different wavelength (hence different depth of penetration) from synchrotron or x-ray tubes of different targets. Recent developments in synchrotron x-ray sources, position sensitive detectors, and microprocessors have stimulated the activities in triaxial stress analysis, also made it possible to measure stress in the field as demonstrated in Table 6. The formulation and measurement of triaxial sresses are much more involved and still under development.[35-38]

Table 6

	Stress measurement	
	Data acquisition	Total analysis
Film/Guinier	4-8 hr	8 hr
Diffractometer	30 min	1-4 hr
PSPD/Guinier	3 sec-15 min	30 min
Portable gauges	USA:	TEC, PARS, Penn State, U. of Denver
	Germany:	Siemens
	Japan:	Rigaku, Strainflex
	France:	CETIM, Castex prototype
	Canada:	CANMET

10. THIN FILMS

At the heart of high technology are integrated circuits. The important steps in making integrated circuits are the deposition of epitaxial thin film(s) onto semiconductor crystals, and the implantation of dopant ions into semiconductors.[39,40] Thin-film composite systems are commonly used in electronic/magnetic/optical and other high-technology industries. A variety of materials (metals, alloys, oxides, nitrides, and carbides) are used to produce thin films of a few tens of Angstroms to a few microinches thickness. Their physical structures and chemical compositions greatly affect their performance, hence must be carefully characterized. X-ray diffraction is a powerful tool for studying thin films. Double crystal diffraction,[41-43] grazing incidence diffraction [44,45] and Buerger precession[46] techniques have been used for the study of thin films of all three general types: (1) the crystal orientation, lattice mismatch of epitaxial films (silicon wafers and supperlattice); (2) the phase identification, preferred orientation, stacking faults, residual strain of polycrystalline films (metals/ ceramics); and (3) the structure of amorphous thin films by small angle X-ray scattering.[47,48] Electron microscopy is often used to examine the defects in thin films; X-ray spectrometry is commonly used to measure the thickness and chemical composition of thin films; and the newer surface-sensitive techniques (ESCA, AES, SIMS [secondary ion mass spectrometry], and ISS [ion scattering spectrometry]) are widely used to probe the chemical composition of very thin surface layers (20 Å or less).[49]

11. PEAK PROFILE FITTING

In order to extract material information on atomic scale by XRD, such as crystal structure, polymer crystallinity, domain size, and residual stress, mathematical functions are necessary to represent the peak profile. For neutron diffraction peaks, the Gaussian function fits well. For x-ray emission peaks, the Lorentzian (Cauchy) function fits well. For x-ray diffraction peaks, neither fits well due to broadening caused by x-ray optics and crystal microstructures. Some functions useful for profile fitting are listed in Table 7.[50-53] It is rather easy to find a function to fit the intense part of XRD peaks but not so easy to fit the tails without significant deviations. Compared to experimental peaks, the tails of Gaussian function are too short, those of the Lorentzian function are too long, and those of Voigt and Pearson functions are closer to reality at the expense of computing time.[54-56] Fortunately, the bulk information of phase, crystallinity, size, and strain can be extracted from the intense part of the peak. Only the finer details of structure need to be derived from the tails. For example, both size and strain contribute to peak broadening, but size data are more sensitive to peak tails than strain data.

Both the Rietvelt method and Parrish method of profile fitting are able to resolve overlapping peaks and to observed peak profiles.[57-59] Although both methods are empirical, Rietvelt's refinement technique is constrained by lattice/structure/instrument parameters, and hence provides crystal structure information after refinement. Parrish's refinement technique is not constrained by structure parameters. It gives the position, intensity, and width of each peak, resolved or overlapping, which is useful for subsequent applications. No crystal structure information is offered after refinement by the Parrish method. However, the profile-fitted data can be further refined for crystal structure determination.[60]

Table 7

Functions for profile fitting

Gaussian	$y = A \cdot e^{-kx^2}$ or $A \cdot \exp(-kx^2)$	y = intensity
		x = peak position
Lorentzian (Cauchy)	$y = A(1+kx^2)^{-1}$	k = width parameter
		A = normalization factor
Modified Lorentzian	$y = A(1+kx^2)^{-2}$	m = decay rate of tails
Pearson VII	$y = A(1+kx^2)^{-m}$	$\begin{cases} m = 1 \text{ Lorentzian} \\ m = 2\text{-}4 \text{ Modified Lorentzian} \\ m = 20 \text{ or infinite, Gaussian} \end{cases}$
Hyperbolic	$y = A \operatorname{sech} x$	(Chung)
Trigonometric	$y = A \sin^2 x / x^2$	(Chung)
Voigt	$y =$ Lorentzian * Gaussian	(Langford)
Series of Lorentzian	3 to 7 terms	(Taupin/Parrish)
Series of Gaussian	A few terms of Edgeworth series	(Wilson/Young)

12. CONCLUSIONS

Although it is clearly impossible to give a comprehensive overview of this subject, an effort has been made to present a bird's-eye view of the X-ray diffraction field in U.S. industry.

In academia, independent research is a noble quality, but in industry, teamwork is important as well. Not only must the outcome of research be practical and relevant to company objectives, but a spirit of sharing, credit or blame must also be aroused and maintained among colleagues to whom one has to sell ideas and win support.

The stroke of luck greatly favors the prepared mind. A prepared mind can realize and accept challenges that are completely outside the realm of X-ray diffraction by venturing outside the confines of that field.

13. REFERENCES

Ahtee, M., et al. *J. Appl. Cryst.* 17, 352 (1984).

Alexander, L. E. *X-ray Diffraction Methods in Polymer Science* (John Wiley, New York, 1969).

Arndt, U. M. *J. Appl. Cryst.* 19, 145 (1986).

Bowen, D. K.; and S. T. Davies. *Nucl. Instrum. Methods 208*, 725 (1983).

Chung, E. X. *J. Appl. Cryst.* 8, 17 (1975).

Chung, F. H. *Adv. X-ray Anal.* 17, 106 (1974).

Chung, F. H. *J. Appl. Cryst.* 7, 519 (1974).

Chung, F. H., *J. Appl. Cryst.* 7, 526 (1974).

Chung, F. H. *J. Appl. Cryst.* 6, 225 (1973).

Chung, F. H. *Envir. Sci. & Tech.* 12, 1208 (1978).

Chung, F. H. *Envir. Sci. & Tech.* 16, 796 (1982).

Chung, F. H. "Imaging and analysis of airborne particulates," in *Instrumentation and Analysis*, P. N. Chermisinoff, Ed. (Ann Arbor Sci., Michigan, 1981), p. 89.

Cohen, J. B.; H. Dolle; and M. R. James. "Stress Analysis from Powder Diffraction Patterns," in *Accuracy in Powder Diffraction*, S. Block and C.R. Hubbard, Eds. (NBS Special Publication 567, 1980), p. 453.

Crowder, C. E., et al. *Adv. X-ray Anal.* 29, 315 (1986).

De Keijser, T. H. et al. *J. Appl. Cryst.* 15, 308 (1982).

Devine, T. J.; and J. B. Cohen. *Adv. X-ray Anal.* 29, 89 (1986).

Dolle, H., et al. *J. Strain Anal.* 12, 62 (1977).

Dolle, H.; and L. B. Cohen. *Metal. Trans.* 11A, 831 (1980).

Dolle, H., et al. *J. Strain Anal.* 12, 62 (1977).

Dragsdorf, R. D.; and C. P. Bhalla. *Adv. X-ray Anal.* 29, 381 (1986).

Ewald, P. P., Ed. *Fifty Years of X-ray Diffraction*. (International Union of Crystallography, Utrecht, Netherlands, 1962).

Fewster, P. F. *J. Appl. Cryst.* 18, 334 (1985).

Hasting, L. B.; W. Thomlinson; and D. E. Cox. *J. X-ray Cryst.* 17, 85 (1984).

Hauk, V. M. *Adv. X-ray Anal.* 27, 101 (1984).

Hauk, V. M. *Adv. X-ray Anal.* 29, 1 (1986).

Hill, M. J., et al. *J. Appl. Cryst.* 18, 446 (1985).

Hindeleh, A. M.; and D. J. Johnson. *Polymer* 19, 27 (1978).

Huang, T. C.; and W. Parrish. *Adv. X-ray Anal.* 29, 395

(1986).

Izumi, T.; T. Kurihama; and J. Nitta. *J. Appl. Cryst.* 17, 470 (1984).

James, M. R.; and D. P. Anderson. *Adv. X-ray Anal.* 29, 291 (1986).

Davis, B.; L. R. Johnson; and T. Mebrahtu. *Powder Diffraction* 1 (3), 244 (1986).

Kakudo, M.; and N. Kasai. *X-ray Diffraction by Polymers* (Elsevier, New York, 1972).

Klug, H. P.; and L. E. Alexander. *X-ray Diffraction Procedures* (John Wiley, New York, 1974).

Langford, J. I. *J. Appl. Cryst.* 11, 10 (1978).

Ledbetter, H. M.; and M. W. Austin. *Adv. X-ray Anal.* 29, 71 (1986).

Lonsdale, D. *J. Appl. Cryst.* 19, 300 (1986).

Marra, W. C.; P. Eisenberger; and A. Y. Cho. *J. Appl. Phys.* 50, 6927 (1979).

Matyi, R. J. *Adv. X-ray Anal.* 29, 375 (1986).

McSwain, R. H.; et al. *Adv. X-ray Anal.* 29, 49 (1986).

Naidu, S. V. M.; and C. R. Houska. *J. Appl. Cryst.* 15, 190 (1982).

Noyan, I. C. *Adv. X-ray Anal.* 28, 281 (1985).

Noyan, I. C.; and J. B. Cohen. *Adv. X-ray Anal.* 27, 129 (1983).

Noyan, I. C.; and J. B. Cohen. *Adv. X-ray Anal.* 27, 129 (1984).

Pardue, E. B. S.; et al. *Adv. X-ray Anal.* 29, 37 (1986).

Parrish, W.; and T. C. Huang. "Accuracy of the profile fitting method for x-ray polycrystalline diffractometry," in *Accuracy in Powder Diffraction*, S. Block and C.R. Hubbard, Eds. (NBS Special Publication 567, 1980), p. 95.

Prevey, P. S. *Adv. X-ray Anal.* 29, 103 (1986).

Rietveld, H. M. *J. Appl. Cryst.* 2, 65 (1969).

Ruud, C. O. *J. of Metals* 13 (6), 10 (1979).

Schafer, W.; et al. *J. Appl. Cryst.* 17, 159 (1984).

Segmuller, A. *Adv. X-ray Anal.* 29, 353 (1986).

Sotion, M.; A. Arniaud; and C. Rabourdin. *J. Appl. Polymer Sci.* 22, 2585 (1978).

Snoha, D. J.; C. O. Ruud; and H. D. Cassel. *Adv. X-ray Anal.* 29, 43 (1986).

Staudenmann, J. L.; R. D. Horning; and R. D. Knox. *J. Appl. Cryst.* 20, 210 (1987).

Tanner, B. K.; and M. J. Hill. *Adv. X-ray Anal.* 29, 337 (1986).

Thompson, P.; D. E. Cox; and J. B. Hasting. *J. Appl. Cryst.* 20, 79 (1987).

Will, G.; W. Parrish; and T. C. Huang. *J. Appl. Cryst.* 16, 611 (1983).

Wims, A. M.; and M. E. Myers. *Adv. X-ray Anal.* 29, 281 (1986).

Wolfel, E. R. *J. Appl. Cryst.* 16, 341 (1983).

Young, R. A. "Structural analysis from x-ray powder diffraction patterns with the Rietveld method," in *Accuracy in Powder Diffraction* (NBS Special Publication 567, 1980), p. 143.

Young, R. A.; and D. B. Wiles. *J. Appl. Cryst.* 15, 430 (1982).

11.

X-ray Diffraction Techniques and instrumentation

> **Happiness is the key to success. If you love what you are doing. You will be successful.**
> — Herman Cain

1. ABSTRACT

For research and development, for solving technical problems, or for making business decisions, one usually faces a question: what is the chemical composition of the sample(s) on hand. The best techniques to determine the quantitative composition of non-routine and totally unknown mixtures is X-ray diffraction (XRD) for chemical compounds and X-ray fluorescence (XRF) for chemical elements. This article presents a bird's-eye view of the XRD field. It summarizes the history, concept, X-ray sources, detectors, cameras, and instrumentation, followed by more details of qualitative and quantitative analyses, their shortcuts and traps.

2. INTRODUCTION

The stroke of luck greatly favors the prepared mind. Roentgen discovered X-rays in 1895[1]. It heralded the advent of modern physical sciences. Four branches of science have grown directly from X-rays: (1) X-ray crystallography for crystal structure determination; (2) X-ray spectroscopy for elemental analysis; (3) X-ray radiography for medical diagnosis or defect detection; and (4) X-ray therapy for treatment of disease, the extensions of which are radioisotope therapy and particle therapy.

Von Laue demonstrated the diffraction of X-rays by crystals in 1912. His experiment proved the wave nature of X-rays and the periodicity of atomic packing within a crystal. The success of X-ray diffraction led to the development of electron diffraction and neutron diffraction techniques for the study of matter. The theory and practice of these two techniques are nearly the same as that of X-ray diffraction. However, each has its own peculiarities in interacting with matter. Electron diffraction is useful for studying the structure of surfaces and films, while neutron diffraction is powerful for examining the magnetic structure and/or locating light atoms in crystals.

Laue's account of X-ray diffraction was read with great interest by two English physicists, W. H. Bragg and his son W. L. Bragg, who successfully interpreted Laue's data on zinc sulfide (ZnS) with the now well-known Bragg law and determined the crystal structure of the rocksalt series (NaCl, KCl, KBr and KI) the next year in 1913.

Before 1912, mineralogists and crystallographers had accumulated a great deal of knowledge on crystals concerning symmetry systems, group theory, morphological, chemical, physical and optical properties, but nothing about their internal structure. This knowledge we now call optical or classic crystallography. The use of X-ray diffraction to determine the internal structure of crystals pioneered by Laue and Bragg is called X-ray crystallography.

More than 95% of the solid materials on earth are crystalline. These crystalline materials are in the form of single crystal, powder or polycrystalline aggregate. Both powder and polycrystalline materials can be studied by powder diffraction techniques. Pauling's early work on the structure of minerals and Zachariasen's studies on thorides and actinides were done with powder diffraction data. However, except in rare cases, crystal structure determination is usually done by single-crystal techniques.

Three sets of experimental data can be obtained from X-ray diffraction of crystals: position, intensity and profile of the diffracted X-rays. These data carry the information of the specimen concerning its symmetry, atomic packing, crystallite size, lattice distortion, phase composition, stress and texture, etc. Various X-ray diffraction techniques and instrumentation are designed to extract this information from diffraction data. Among them, powder diffraction is the most frequently used technique in the realm of material science, and chemical analysis of unknown materials is the most popular application of powder diffraction in industry.

3. INSTRUMENTATION

All the X-ray diffraction instruments are of two types—camera or diffractometer. The first powder diffraction camera was designed by Debye and Scherrer in 1916. It was quickly accepted and improved on by many scientists. The major improvement is the use of electronic counters in place of the photographic film. Counter diffractometers appeared in the late 1950s using parafocusing optics. Although the camera techniques are still preferable in certain cases, powder diffractometer techniques are most popular in industry. The advantages of the diffractometer over the camera techniques are high intensity, linear response, better accuracy, simpler absorption effect and suitability for automation. There are in excess of 20,000 X-ray diffractometers in use in the world today[2]. These diffractometers have been manufactured by a number of companies during the past 30 years. Some companies offer complete X-ray diffraction systems. Others specialize in X-ray sources, detectors, cameras or other accessories. Among them are Philips, General Electric, Siemens, Rigaku, Diano, Supper, Rank, Syntex and many others. The International Union of Crystallography publishes and updates periodically an Index of Crystallographic Supplies, which lists various X-ray instruments, their manufacturers and suppliers. The many available techniques of using these instruments for structural or chemical analyses have been described by Bunn[3], Buerger[4], Klug and Alexander[5], Cullity[6], Azaroff[7] and Kaelble[6]. Almost all the publications pertaining to crystallography are available through Polycrystal Book Service, Pittsburgh, Pennsylvania.

3.1 X-Ray Sources

The wavelength of X-rays spans from 0.1 to 1000 Å (1 Å= 10^{-8} cm) in the electromagnetic spectrum. X-rays of about 1 Å wavelength (0.2-2.5 Å) are used for diffraction analysis because they have adequate penetration power yet are not destructive, and the interatomic distances in crystals have the same order of magnitude.

Radioisotopes emit X-rays in their decay process. For instance, the decay of Fe-55 to Mn-55 emits X-rays of 2.1 Å wavelength. Whenever matter is bombarded by high-energy particles, X-ray excitation takes place. The practical source of

X-rays for diffraction work is from an X-ray tube, where electrons are accelerated by an electric field to a high velocity in vacuum. These electrons then collide with a metal target and thus excite the characteristic X-rays of the target. The target metals for X-ray diffraction tubes and their applications are cited in Table 1. The X-ray tube and its electrical apparatus is called an X-ray generator and includes power supply, cooling system, safety device and control panel. Modern X-ray diffraction tubes have up to four windows, hence four pieces of X-ray apparatus can be set up simultaneously using a single X-ray source.

3.2 X-Ray Detectors

There are three ways to detect the presence of X-rays: fluorescent screen, photographic film and electronic counters. In X-ray diffraction work, a fluorescence screen is widely used to locate the position of the primary X-ray beam when adjusting the instrument. Photographic film is used for all camera techniques, and counters are used for all diffractometer techniques. The advantages of photographic film are simplicity, permanent record, and displaying all the diffracted beams with their relative disposition and intensity. Its major limitations are lower accuracy and less sensitivity. For single-crystal analysis, the advantages of films far outweigh their limitations, so that their use remains widespread. However, for chemical analysis, powder diffractometers are far superior to camera techniques.

Table I. Target Metals for X-Ray Diffraction Tubes

Target	Wavelength (Å)	Filter	Applications
Tungsten	0.211	—	Very intense white radiation, well suited for Laue exposure and microradiography.
Silver	0.561	Pd Rh	High penetration, permits transmission exposure for determination of texture in transformer sheets. The (100) plane of iron has an "easy" direction of magnetization [001] parallel to the magnetic field flux.
Molybdenum	0.711	Nb Zr	Short wavelength, high penetration and low resolution. Suited for inorganic compounds that usually have high absorption coefficient and/or small unit cell. Used for austenite determination and single-crystal analysis.
Copper	1.542	Ni	Best suited for most general diffraction applications. Standard tube for all kinds of powders, metals, ceramic and organic materials.
Nickel	1.659	Co	Measurement of stress in metal, such as the (331) plane in brass (cartridge) at $2\theta = 158°$.
Cobalt	1.790	Fe	Cobalt and iron X-rays do not excite iron fluorescence radiation. Suited for all kinds of steel analyses, especially by means of film techniques in which PHA discrimination is not possible.
Iron	1.937	Mn	Same as cobalt X-rays.
Chromium	2.291	V	Long wavelength, low penetration and high resolution. Suited for organic compounds that usually have low absorption and/or large unit cell. Chromium X-rays do not excite iron fluorescence radiation. Used for steel analyses and stress analysis such as the (211) plane in iron at $2\theta = 156°$.

X-ray counters used in diffractometry may be classified into three types: gas ionization counters[9,10], scintillation counters[11,12] and semiconductor counters[13,14]. Early Geiger-Muller gas ionization counters have been replaced by proportional gas ionization counters, which afford higher wavelength selectivity and much lower resolving time. Subsequently, scintillation counters now rival proportional counters for many applications and currently dominate the X-rays diffraction field. Semiconductor counters, originally developed for nuclear spectroscopy, were adapted for X-ray emission spectroscopy in the 1960s, and recently modified for X-ray diffractometry[15]. They offer exciting prospects for future applications such as fast data collection for dynamic systems[16]. Potentially, semiconductor counters could have superior resolving power due to efficacy of pulse-height discrimination and might require less stringent stability of the instrument[17]. Detectors of current use in diffractometers are all counters, the output of which is in digital form suitable for computer data processing. The features of these counters are compared in Table II.

Table II. Performance Data of X-Ray Counters

	Geiger-Muller	Proportional	Scintillation
Detection Process	Gas ionization	Gas ionization	Scintillation in NaI (Tl) crystal
Linear Count Rate (cps)	10^3	10^5	10^6
Background (cps)	1	0.5	10
Dead Time (μsec)	50~3000	<1	<1
Wavelength Region (Å)	0.3–4	Sealed 0.3–4 Flow 0.7–10	0.1–4
Energy Resolution	—	Excellent	Good
Pulse Size	Equal size	Proportional to X-ray energy	Proportional to X-ray energy
Pulse-Height Analysis	Cannot use PHA	Permits PHA	Permits PHA
Rating for Diffraction Use	Poor	Good	Good

3.3 X-Ray Cameras

X-ray diffraction instruments of any design are based on the Bragg law, which prescribes the geometrical conditions necessary to achieve X-ray diffraction. Bragg explained the X-ray diffraction effect in terms of reflection from a set of parallel and equispaced atomic planes. He conceived the following equation in 1912 now known as the Bragg law:

$$n\lambda = 2d\sin\theta \qquad (1)$$

where n = an integer denoting the order of reflection
λ = the wavelength of the incident X-rays
d = the interplanar spacing
θ = the angle between the incident beam and the crystal planes

Note that the phenomenon is not a surface reflection, as with visible light. X-ray reflection occurs only at the precise Bragg angle θ, which satisfies Equation 1, a condition of constructive interference. No X-ray reflection occurs at other angles due to destructive interference. Reflection of visible light by a polished surface occurs over a continuous angular range.

Perhaps the simplest way to obtain a diffraction pattern is to use the powder camera technique. When the sample is extremely small (a few micrograms), powder camera technique is the only way to obtain its diffraction pattern. Besides crystal structure, a powder photograph reveals other properties of the sample, including particle size, texture and crystallinity.

Note that the film in a camera intercepts simultaneously all the X-ray beams diffracted by a crystal, whereas the counter on a diffractometer detects only one diffracted beam at a time. Therefore, the power stability is not so critical for camera techniques as that for diffractometer techniques. Besides chemical analysis, various powder cameras can be used for structural analysis of polycrystalline materials including minerals, polymers, ceramics, metals and alloys. For structural analysis of single crystals, camera and diffractometer techniques are complementary to one another. Diverse kinds of X-ray cameras for powder or single-crystal analyses are summarized in Table III.

3.4 X-Ray Diffractometers

An improvement over the film technique is X-ray diffractometry. This approach is the one most widely used in modern analytical laboratories and provides for replacement of the film and camera by a goniometer-mounted X-ray detector and a strip chart readout system. In this application, the sample in solid or powder form is placed in a flat specimen holder in the goniometer. A fixed source of X-rays impinges on the sample, which diffracts the X-rays at discrete angles. By rotating the sample at half the angular velocity of the detector, diffraction maxima as a function of the Bragg angle are recorded. A full pattern recording angular positions and relative intensities of diffraction maxima may be obtained in about 30 minutes by this technique. The data obtained are more accurate and easier to interpret than the data produced by the film technique. A schematic representation of a powder diffractometer, which includes a curved crystal monochromator, is shown in Figure 1[18]. The crystal can be either LiF or graphite and is essential for elimination of undesired K_β radiation and excess background radiation. Prior to the utilization of the curved crystal monochromator, a nickel filter was used for this purpose. While adequate, the nickel filter does not provide the high signal-to-noise advantage of the monochromator (1.4/1 for Ni, 11.0/1 for LiF, ~25/1 for graphite).

Table III. X-Ray Cameras and X-Ray Diffractometers

Diffraction Instruments	X-Rays	Crystals
Debye-Scherrer Camera	Monochromatic	Powder
Parafocusing		
Seemann-Bohlin camera	Monochromatic	Powder
Guinier camera	Monochromatic	Powder
Guinier-deWolff camera	Monochromatic	Powder
Pinhole Camera	Monochromatic	Powder
Powder Diffractometer (Analog of Debye-Scherrer camera)	Monochromatic	Powder
Laue Cameras		
Front-reflection camera	White	Single-crystal
Back-reflection camera	White	Single-crystal
Weissenberg Camera (U.S. & Europe)	Monochromatic	Single-crystal
deJong & Bouman Camera (Europe) (i.e., Retigraph)	Monochromatic	Single-crystal
Buerger Precession Camera (U.S.)	Monochromatic	Single-crystal
Equi-inclination Diffractometer (Analog of Weissenberg camera)	Monochromatic	Single-crystal
Four-Circle Diffractometer (i.e., Eulerian geometry type)	Monochromatic	Single-crystal

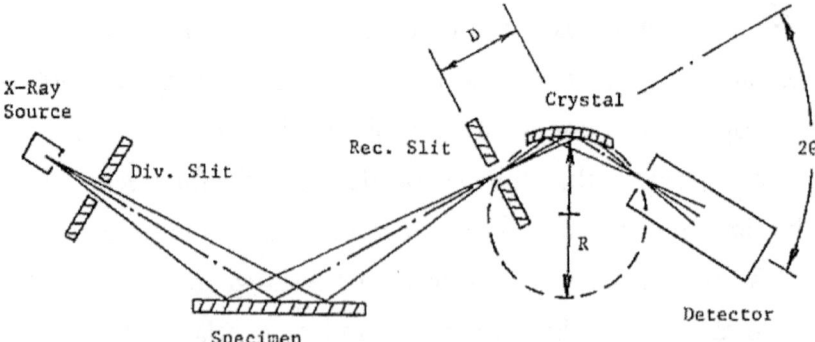

Figure 1. Schematic of an X-ray diffractometer for powder analysis. The crystal monochromator and the receiving slit travel with the detector in alignment.

Single-crystal diffractometers are of two basic designs: Equi-inclination geometry and Eulerian geometry. Equi-inclination geometry design is the analog of Weissenberg camera with film replaced by a counter. In the Eulerian geometry design, the detector and the primary beam always lie in a horizontal plane. The crystal is carried in an Eulerian cradle, which permits rotation about each of the Eulerian axes: omega (ω), chi (χ), and phi (ϕ). The intensity of the diffracted X-rays from a set of planes in the crystal can be measured by setting the ω, χ, ϕ and θ such that the Bragg condition is satisfied. A schematic diagram of the Eulerian geometry diffractometer is shown in Figure 2. The incident X-ray beam is fixed. The counter is constrained to move in the horizontal plane. The crystal is rotated until the normal to the set of diffracting planes lies in the horizontal plane. The angle between the incident beam and the set of diffracting planes in the crystal is made equal to $\theta = \sin^{-1}(\lambda/2d)$. The counter is then moved to 2θ position to measure the intensity of the diffracted beam.

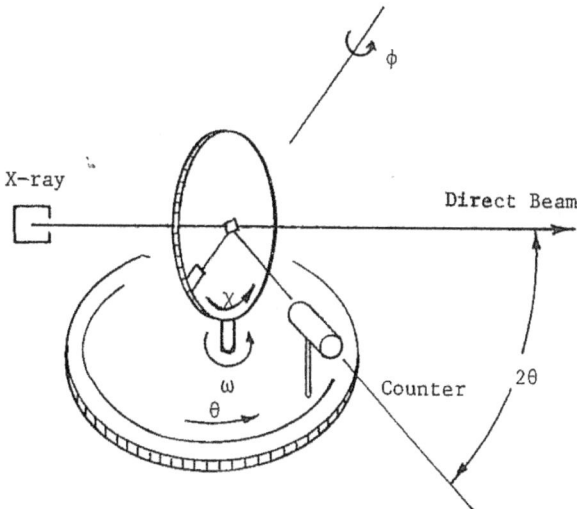

Figure 2.

3.5 Current Trend

Many advances have been made in the basic instrumentation used to obtain X-ray diffraction patterns. Computer-controlled powder diffractometers are now commercially available that enable the operator to search and match[19] patterns in the Powder Diffraction File[20], which is maintained by the Joint Committee on Powder Diffraction Standards (JCPDS). The completely automated data acquisition and search-match system depends on the quality of the X-ray diffraction pattern of the sample and the standard X-ray diffraction pattern in the File. An overall upgrading of the File is needed. One advantage of the new instrumentation will be in comparing known in-house materials with unknowns.

Advances in accessory equipment include a primary beam monochromator[21], which is inserted between the X-ray tube and the specimen. This monochromator, mounted at the tube stand, is used when radiation-sensitive specimens are examined. Also available is a low-and medium-temperature attachment[22], which will hold the specimen in a range of about $-180°t_i + 300°C$. This device is used for analysis of crystal structures, determination of temperature dependence of lattice parameters and examination of phase transitions.

New developments have been reported for three areas. The first area is in the use of synchrotron radiation[23], wherein the extreme energy intensity available and tunable wavelength feature provide an opportunity for performing experiments that were not possible in the past or, at best, were very difficult to carry out. This technique has been used to study high-temperature, high-pressure problems in geological systems[24]. The second area is in the use of energy-dispersive X-ray diffractometry[25], which is tied into the availability of the synchrotron source and the recent advances in electronics. This technique is especially suited to low- or high-temperature and/or high-pressure research. The third area is in the design or redesign of position-sensitive time-resolved detector systems. One such advance is the multiwire proportional counter system[26], which should speed up data gathering and has been used to obtain a powder pattern in one minute.

4. CHEMICAL ANALYSIS OF POWDERS
4.1 Qualitative Analysis
Each crystalline compound produces a characteristic X-ray diffraction pattern, whether in the pure state or as a component in a mixture. The identification of components in an unknown mixture is simply matching the diffraction pattern of the unknown to that of standards. Therefore, for chemical analysis by the X-ray diffraction method, a library of standard patterns is essential. To serve this purpose, the Joint Committee on Powder Diffraction Standards (JCPDS) publishes a Powder Diffraction File (PDF) annually. Over the years, the file has grown in size and improved in quality. Currently Year 2000, it contains about 193,000 standard patterns of crystalline organic and inorganic compounds.

4.2 Powder Diffraction File
The Powder Diffraction File is available on 3 x 5-inch cards. A single X-ray diffraction pattern is registered on each card. A typi-

cal PDF card, as shown in Figure 3, lists the interplanar spacing (d value) and the relative intensity (I/I_1) for every diffraction line of the compound. The intensity of the strongest line is assigned an arbitrary number of 100. The three most intense lines and the largest d value are sorted out and repeated on the upper left corner. Besides the X-ray diffraction pattern, the card contains the chemical name and formula of the compound, experimental conditions used for obtaining the pattern, optical and crystal data, source of the standard sample and references, when available. To effect a fast search-match for identification, three indexes accompany the Powder Diffraction File.

d	3.04	2.29	2.10	3.86	$CaCO_3$					
I/I_1	100	18	18	12	Calcium Carbonate		(Calcite)			

				d Å	I/I_1	hkl	d Å	I/I_1	hkl
Rad. CuKα	λ 1.5405	Filter Ni		3.86	12	102	1.297	2	218
Dia.	Cut off	Coll.		3.035	100	104	1.284	1	306
I/I_1 G.C. Diffractometer d corr. abs.?				2.845	3	006	1.247	1	220
Ref. Swanson and Fuyat, NBS Circular 539,				2.495	14	110	1.235	2	1.1.12
Vol. II, 51 (1953)				2.285	18	113	1.1795	3	2.1.10
Sys. Hexagonal S.G. D_{3d}^6 - $R\bar{3}c$				2.095	18	202	1.1538	3	314
a_0 4.989 b_0 c_0 17.062 A C 3.420				1.927	5	204	1.1425	1	226
α β γ Z 6				1.913	17	108	1.1244	<1	2.1.11
Ref. Ibid.				1.875	17	116	1.0613	1	2.0.14
$ε_a$ nωβ 1.659 $ε_γ$ 1.487 Sign -				1.626	4	211	1.0473	3	404
2V D_x 2.711mp Color				1.604	8	212	1.0447	4	138
Ref. Ibid.				1.587	2	1.0.10			(0.1.16
				1.525	5	214	1.0352	2	(1.1.15
Sample from Mallinckrodt Chem. Works Spect.				1.518	4	208	1.0234	<1	1,2.13
Anal.:<0.1% Sr;<0.01% Ba;<0.001% Al, B, Cs,				1.510	3	119	1.0118	2	3.0.12
Cu, K, Mg, Na, Si, Sn;<0.0001% Ag, Cr, Fe,				1.473	2	215	0.9895	<1	231
Li, Mn,				1.440	5	300	.9846	1	322
X-Ray Pattern at 26°C				1.422	3	0.0.12	.9782	1	1.0.17
				1.356	1	217	.9767	3	2.1.14
Replaces 1-0837, 2-0623, 2-0629, 3-0569, 3-0593, 3-0596, 3-0612, 4-0636, 4-0637				1.339	2	2.0.10	.9655	2	234

Figure 3. A data card on $CaCO_3$ from the Powder Diffraction File.

Davey (Alphabetical) Index. It lists alphabetically the names of compounds in the PDF. All entries are fully cross-indexed. For example, $CaCl_2 \cdot H_2O$ has three entries: calcium chloride hydrate; chloride hydrate, calcium; and hydrate, calcium chloride. Each compound name is followed by its chemical formula, its three most intense diffraction lines with their relative intensities indicated by subscripts (g for greater than 100, x for 100, 6 for 60, 2 for 20, etc.), then its card number and the microfiche number. The alphabetical index is useful when some information about the sample is available from elemental analysis, previous experience or educated guess.

Hanawalt (Numerical) Index. It lists the d values of the eight most intense lines of the pattern in order of decreasing intensities. Each d value has a subscript indicating its relative intensity. To cope with the possible uncertainty of relative intensities, each pattern is given three entries by permuting the three strongest lines. The patterns are divided into 45 groups according to the first d value of the entry. In each group the patterns are arranged according to the second d value in decreasing sequence. The corresponding compound name and PDF card number follow the d value listing. Some patterns are preceded with an "*" indicating high accuracy, "0" indicating less reliable and "c" indicating calculated pattern. The Hanawalt Index is used when there is little information about the sample.

Fink (Numerical) Index. It lists the d values of the eight most intense lines of the pattern in order of decreasing d values. Each pattern has four entries according to its four most intense lines. The format of the Fink Index is similar to that of the Hanawalt Index. It was known that both X-ray and electron diffraction patterns have the same groups of lines with high relative intensities although their intensity ranking may differ considerably. The Fink Index was developed mostly for the identification of electron diffraction patterns. It can be used for the identification of X-ray diffraction patterns where the intensity data are not reliable due to superimposed lines, preferred orientation or minor constituents.

4.3 Cases of Applications

Positive identification of the components in a mixture often leads to the solution of a technical problem, the mechanism of a reaction or the quality of a product. A few industrial cases are illustrated below:

> **Tin Plate.** A steel mill makes sheet metal for "tin" cans where the tin coating is applied by "hot-dipping." A thin film of tin adheres to the surface of the steel and protects it from corrosion. It was found that certain steel sheets could not be wetted by the molten tin[6]. X-ray diffraction analysis of these faulty steel sheets indicated that a finely divided graphite was deposited on the surface. The presence of graphite prevented the contact between steel and molten tin.
>
> **Martensite.** Many hardened steels consist of martensite and retained austenite. Austenite tends to transform and increase in volume, thereby leading to an increase in residual stress in the steel. Therefore, the presence of even a few percent retained austenite is undesirable in some ap-

plications. The presence and amount of austenite in hardened steel can be determined by microscopic examination when the austenite content is high. X-ray diffraction analysis must be used when the austenite content is below about 15%[27].

Fertilizer. After an application of composite fertilizer, the grass became withered instead of flourishing. Elemental analysis of the fertilizer could not find the cause. X-ray diffraction analysis indicated the presence of a potassium pyrosulfate, while a potassium sulfate was expected. Potassium pyrosulfate is highly acidic in water, hence killing the grass.

Paint. The sheet metal of an appliance was originally coated green. Its green color changed to blue after a brief period in service. X-ray diffraction analysis indicated that the pigment in the original paint was composed of lead chromate and iron blue; the exposed paint contained an extra component lead chloride. This finding led to the conclusion that the green color is a composite of yellow (lead chromate) and blue (iron blue). HCl fumes in that particular environment attacked the lead chromate forming lead chloride. The loss of a yellow component of the green color ended in a blue color[18].

Molybdate. Conventional corrosion-inhibiting pigments contain lead or chromium, which are toxic. A nontoxic pigment, Molywhite®, is being used for corrosion inhibitors. It is a basic zinc molybdate made from zinc oxide and molybdenum oxide[28]. The sequence of chemical reactions in its manufacturing process were followed by X-ray diffraction techniques. Therefore, the optimum reaction conditions for its production could be studied.

4.4 Limitations

In theory, the Hanawalt (Index) Search Manual should lead to the positive identification of any component in a mixture, and it does in most cases. However, it is not uncommon that a match of patterns may not be obtained for the following reasons:

1. The pattern of the unknown compound is not listed in the Powder Diffraction File.
2. The relative intensities of the diffraction lines changed drastically due to preferred orientation of the specimen.
3. There are superimposed lines from different components.
4. The intensities of the lines are too weak due to extreme absorption or very low concentration.
5. Some patterns have only one strong line, such as $CaCO_3$ (PDF 5-586) and β-SiO_2 (PDF 11-252).
6. The X-ray patterns of the unknown and the standard in the PDF were made with different instruments. For diffractometers, the absorption factor is independent of the Bragg angle, whereas in a Debye-Scherrer camera, the absorption is larger at lower Bragg angles. Therefore, low angle lines appear stronger relative to high angle lines on

a diffractometer chart than on a Debye-Scherrer photograph.
7. The pattern obtained nearly matches several patterns in the PDF due to isomorphism, such as the numerous spinet-type compounds[29].
8. Clays, minerals and other natural products contain random substitution and/or irregular stacking, thus defy exact match[30]. Specialized procedures are often necessary for their identification[31].
9. Organic compounds usually have complex patterns due to low symmetry and large unit cell of the crystals[32]. The pattern of an organic mixture usually has too many overlapped lines to unravel[33,34].

5. QUANTITATIVE ANALYSIS

In a powder mixture, each component produces its characteristic X-ray diffraction pattern independent of others. As already discussed, the identification of different components is achieved by unscrambling the superimposed patterns. Moreover, the intensity of the pattern of a component is proportional to the amount present. Unfortunately, the absorption effect of the total sample complicates the situation. The total absorption effect has been called the matrix effect. It depends on the nature of all other materials surrounding the component sought besides the component itself. To extract the weight fractions from intensity data, the matrix effect must be suppressed or circumvented. To this end, Klug and Alexander[5,35] developed an internal standard method in 1948. It involves the empirical construction of a calibration curve from standards, which is rather tedious, especially for multicomponent analysis. In 1974, Chung[36-38] conceived a Matrix-Flushing method, which avoided the calibration curve procedure, hence greatly simplifying the X-ray diffraction techniques for quantitative chemical analysis.

5.1 Matrix-Flushing Theory

The Bragg equation neatly prescribes the position of a diffraction line, but says nothing about its intensity. There are two theories dealing with the intensity of diffraction—Darwin's kinematical theory[39] and Ewald's dynamical theory[40,41]. Both theories lead to the same intensity expression for powder diffraction. For a single-phase powder with all crystallites randomly oriented, the intensity of diffracted X-rays in a diffractometer is given by Equation 2:

$$I(hk\ell) = \left(\frac{I_0 e^4 \lambda^3 d}{32 \pi m^2 c^4 r}\right)(N^2 p F^2)\left(\frac{1 + \cos^2 2\theta}{\sin^2 \theta \cos \theta}\right) TAV \qquad (2)$$

Where
    I = intensity of X-rays diffracted by (hk) plane,
    I_0 = intensity of primary X-rays,
    e, m = charge and mass of electron,
    λ = X-ray wavelength,
    d = slit width of detector,
    c = velocity of light,
    r = specimen-to-detector distance,
    N = number of unit cells per unit volume,
    p = multiplicity,

F = structure factor,
θ = Bragg angle,
T = temperature factor,
A = absorption factor, and
V = volume of powder in the beam.

For a multiphase powder, the first four factors in Equation 2 can be kept constant. A indicates the absorption factor of the total mixture, and V indicates the volume fraction of the component sought. Let μ_t and s be the linear absorption coefficient and thickness of the total sample, and let ρ_i and X_i be the density and weight fraction of component i. Since the absorption of X-rays, just like the absorption of visible light, follows the well-known exponential law, for infinite specimen thickness we have

$$A = \int_0^\infty \exp(-\mu_t s)\, ds = \frac{1}{\mu_t} \tag{3}$$

$$I_i = \frac{K_i}{\rho_i} \cdot \frac{X_i}{\mu_t} = k_i X_i \tag{4}$$

Note that K_i/ρ_i is a characteristic constant of component i, μ_t is a function of X_i, and k_i a factor containing the mass absorption coefficient of the total sample. At this moment, k_i is constant for a very small variation of X_i. Later on, it will be shown that k_i in the ratio k_i/k_j, is constant for any X_i. For the quantitative X-ray diffraction analysis of a mixture of n components, we have n unknowns (X_i, i = 1 to n), which must satisfy the following equation:

$$\begin{bmatrix} k_1 & 0 & 0 & ...0 \\ 0 & k_2 & 0 & ...0 \\ 0 & 0 & k_3 & ...0 \\ \multicolumn{4}{c}{............} \\ 0 & 0 & 0 & ...k_n \\ 1 & 1 & 1 & ...1 \end{bmatrix} \begin{bmatrix} X_1 \\ X_2 \\ X_3 \\ \cdot \\ \cdot \\ X_n \end{bmatrix} = \begin{bmatrix} I_1 \\ I_2 \\ I_3 \\ \cdot \\ \cdot \\ I_n \\ 1 \end{bmatrix}$$

(5)

The necessary and sufficient conditions for the existence of a unique solution of Equation 5 is that all its characteristic determinants vanish, which gives

$$X_j = \left(\frac{k_j}{I_j} \sum_{i=1}^{n} \frac{I_i}{k_i} \right)^{-1}$$

(6)

The unusual property of this unique solution is that the weight fraction of any component in a multiphase system is expressed in terms of ratios like I_i/I_j and k_i/k_j (i, j = 1, 2,...n). The use of an intensity ratio makes it immune to many sources of errors in intensity measurement. The use of k-ratio makes it free from matrix effect because all the absorption factors are exactly cancelled out. Furthermore, it can be shown[36] that the k_i's in this unique solution are the corresponding Reference Intensity Ratios defined and published in the Powder Diffraction File. Therefore, the X-ray diffraction patterns of mixtures can be interpreted quantitatively and directly, without using any internal standard.

When some components in the mixture are amorphous and/or unidentified, an internal standard, such as corundum α-Al_2O_3, must be added into the sample. Let the subscript c and o stand for corundum and original sample respectively. Then we have:

$$X_c + X_o = X_c + \sum_i^n X_i = 1 \qquad (7)$$

$$X_i = \left(\frac{X_c}{k_i}\right)\left(\frac{I_i}{I_c}\right) \qquad (8)$$

Note that the weight fractions in Equations 7 and 8 are referring to the composite mixture because X-rays faithfully reveal what is actually observed, but cannot differentiate an internal standard from an original component. Equation 8 prescribes the slope of a calibration curve for every component in the sample and thus greatly simplifies the internal standard method. It is the working equation for quantitative multicomponent X-ray diffraction analysis.

Usually X-ray diffraction gives little information about the amorphous component in sample. An interesting feature of the Matrix-Flushing theory is that it provides a way of detecting the presence of amorphous materials in a sample, and the amorphous content can be determined when necessary. The following discriminant Equation 9 can be derived from equations 7 and 8:

$$\sum_i^n \frac{I_i}{k_i} \gtreqless I_c \cdot \frac{X_o}{X_c} \qquad (9)$$

where
> indicates wrong experimental data
= indicates all components are crystalline
< indicates the presence of amorphous materials

Note that neither assumption nor approximation was made in the derivation of this theory. All absorption factors are exactly cancelled. Equations 6 and 8 are to be used for quantitative multiphase analysis. Equation 9 is to be used for testing the presence of amorphous components.

5.2 Applications

Practically all cases of quantitative chemical analysis by X-ray diffraction techniques can be classified into two categories. First, all components in a sample are crystalline and identified. Second, some components in a sample are amorphous and/or unidentified. The applications of Equations 6, 8, and 9 to these cases are illustrated below. The precision of routine analysis by this method is about 8% relative or better. Better precision is expected when both sample and standard have the same physical properties, such as particle size and lattice imperfection. The usual precautions are assumed such as homogeneity of samples, purity of standards, counting statistics, instrumental conditions and other sources of error. However, a set of Reference Intensity Ratios, k_j, has to be determined first.

Simultaneous k_i Determination. The JCPDS defined Reference Intensity Ratio, k_i, as the intensity ratio (I_i/I_c) of the most intense lines from a binary mixture made with a pure compound and synthetic corundum by one-to-one weight ratio. Before theoretically calculated k_i can be made more dependable[42,43], experimentally determined k_i should be used. A set of k_i of interest can be determined simultaneously by use of Equation 8, as shown in

Table IV, where the simultaneously determined k_i is compared with the individually determined k_i[44].

Table IV. Simultaneous k_i Determination

Component	Composition		Intensity		k_i	
	(g)	(wt %)	(hkl)	(cps)	Calculated	50/50 Mixture
ZnO	0.4183	16.29	101	1564	4.28	4.35
CdO	0.3562	13.87	111	2418	7.76	7.62
LiF	0.6302	24.54	200	729	1.32	1.32
CaF_2	0.5395	21.01	220	651	1.38	1.41
Al_2O_3	0.6242	24.30	113	546	1.00	1.00

Direct Interpretation. When all components in a sample are crystalline and identified, the X-ray diffraction pattern of the original sample can be interpreted quantitatively and directly, the weight fraction of every component in the sample can be easily calculated with Equation 6. When amorphous components are present, the results thus obtained represent the composition of the crystalline portion of the sample. The data of two examples are given in Table V, where all intensity data refer to the strongest reflections of corresponding components.

Quite often, only the relative amounts of concerned components in a sample are sought.

This information also can be obtained with Equation 6 by considering these concerned components only and ignoring all other crystalline or amorphous components.

Not that a special case in this category is binary systems, where n = 2. An unusual binary system is partially crystalline polymers[45], which consist of a crystalline component and a non-crystalline component. The slope of the straight line (k=2.83) in this reference verifies Equation 6 remarkably well.

Table V. Direct Interpretation of X-Ray Diffraction Patterns

Composition	Composition (g)	Intensity I_i (cps)	Reference Intensity k_i	% Composition Known	% Composition Found
ZnO	0.2236	610	4.35	9.87	9.2
NiO	0.5454	1412	3.81	24.06	24.5
CdO	0.6588	3303	7.62	29.07	28.7
KCl	0.8386	2207	3.87	37.00	37.7
ZnO	0.6759	2408	4.35	24.38	25.3
TiO_2	0.4317	931	2.62	15.57	16.2
$CaCO_3$	1.1309	2558	2.98	40.79	39.2
Al_2O_3	0.5341	420	1.00	19.26	19.2

Internal Standard. In many practical situations, the composition of a sample is only partially identified. The weight fractions of the identified components can be determined simultaneously with the addition of an internal standard such as corundum. The composite sample should be ground into a homogeneous fine powder before the X-ray diffraction analysis. The intensity and composition data of two examples are shown in Table VI. Equation 8 was used to calculate the weight fractions. The discriminant Equation 9 was used to indicate whether an amorphous component was present. The large intensity imbalance (1842 vs 2312) implies the presence of amorphous materials, which scatter X-rays incoherently, increasing the background. The nearly equal intensity balance (2721 vs 2736) implies that all components are crystalline where the use of Equation 6 should give the same results. Note that the weight fraction of total amorphous materials was calculated from material balance.

Table VI. X-Ray Diffraction Analysis with Internal Standard

Composition	(g)	Intensity I_i (cps)	Reference Intensity k_i	% Composition Known	% Composition Found	$\Sigma \dfrac{I_i}{k_i}$	$I_c \cdot \dfrac{X_0}{X_c}$
ZnO	1.8901	5968	4.35	41.49	41.1		
KCl	1.0128	2845	3.87	22.23	22.0	2721	2736
LiF	0.8348	810	1.32	18.32	18.4		
Al$_2$O$_3$	0.8181	599	1.00	17.96	—		
ZnO	0.9037	4661	4.35	34.43	36.4		
CaCO$_3$	0.7351	2298	2.98	28.00	26.2	1842	2312
SiO$_2$ Gel	0.4234	0	—	16.13	15.9		
Al$_2$O$_3$	0.5629	631	1.00	21.44	—		

When all the peaks in the X-ray diffraction pattern are accounted for, one is still not sure whether there is any amorphous material present. A normal scan of the last example in Table VI did not indicate the presence of amorphous materials, although the sample contained 16% silica gel. Therefore, it is safe to check with an internal standard for the following reasons: First, the true values of X_c and k_c of the internal standard are absolutely known. Secondly, the weight fraction, X_i, is dependent on I_i and k_i only, independent of any X_j, I_j and k_j, where $j \neq i$. Thirdly, the weight fraction of a component sought is independent of the presence or absence of any other component, crystalline or amorphous.

5.3 Factors Vital to Precision

The basic intensity Equation 2 was derived from an "ideally imperfect" mosaic crystal model. All mosaic blocks are small (0.1 ~ 1 μm size) and nonparallel (but with no more than 0.2 ~ 0.5° misalignment)[46]. All lattice planes are given every opportunity to diffract X-rays by rotating the crystal. For powder diffraction analysis, it implies that the crystallites have optimum particle size and proper amount of imperfection. The orientation of all crystallites is statistically random. These conditions are not always fulfilled. Besides, the sample preparation and the intensity measurement are equally important to attain reproducibility. These factors are discussed below.

Particle Size. The effective crystallite size of powder affects both the magnitude and the spread of intensity. To achieve 1% standard deviation, the crystallite size should not exceed 5 µm. If 2 ~ 3% standard deviation is sufficient, the crystallite size can be as large as 10 µm. For a spinning sample, the crystallite size can be 2-3 times larger. When the crystallite size falls below 0.1 µm, the peak height of the diffraction line decreases and peak width increases. The amount of line broadening provides a rather accurate measurement of particle size in the colloidal range (<0.1 µm). When the crystallite size is above 20 µm, the intensity of diffraction lines fluctuate sharply as a function of the orientation of the larger crystals. Powder of >50 µm size gives spotty lines in the Debye-Scherrer photograph. In essence, when the particle size of a powder is reduced to a few microns, say 1~10 µm, several sources of errors can be minimized, including line-broadening[5], preferred orientation[47], extinction[48] and microabsorption"[49] due to either the constant nature or trivial level of these effects. In many chemical precipitates and mineral clays, each particle may consist of a large number of crystallites, even though the apparent particle size is 50 µm.

Lattice Imperfection. Most crystals are neither perfect nor "ideally imperfect." Various imperfections such as dislocations, interstitials, stacking faults and random layers may exist. The profile of diffraction lines reveals the average of all sorts of lattice imperfection. Highly perfect crystals give lower intensity due to extinction. Highly imperfect crystals give lower intensity due to line broadening. For free-flowing powders of 1 to ~50-μm particle size, the extinction and line broadening are insignificantly small. Iodine, iodide and many organic compounds are very soft and malleable. Extended grinding may cause lattice distortion. In such cases, strain-free powder can be obtained by grinding the sample with liquid nitrogen or dry ice.

Preferred Orientation. Plate- or needle-like crystallites tend to set in certain preferred orientation. Preferred orientation due to particle shape is called shape texture. Talc and asbestos powders show extreme shape texture. Preferred orientation due to mode of creation is called orientation texture. Fiber and film of polymers, wire and sheet of metals show orientation texture. Shape texture is important in sample preparation for chemical analysis of powders. Orientation texture is important in structural analysis of polycrystalline aggregates. Because of preferred orientations, some lines of the diffraction pattern are intensified and others weakened. When serious discrepancies arise in relative intensity between the pattern obtained and the pattern in the PDF, preferred orientation should be suspected and some correction steps be taken. Size reduction by grinding, dilution with amorphous powder, embedding in collodion and roughening the top surface can reduce the preferred orientation and make the intensity data more reproducible.

Sample Preparation. For any meaningful chemical analysis, the sample must be representative and homogeneous. It is extremely difficult to attain a truly homogeneous powder mixture. Mixing and grinding are usually done by use of an automatic mortar-and-pestle grinder, a ballmill or a Wig-L-Bug. The grinding time ranges from a few minutes to several hours, depending on nature and amount of sample. Sometimes wet grinding in a nonsolvent liquid can speed up the dispersion and size reduction. The powder sample should be free flowing into the sample holder from a side opening to avoid preferred orientation and induced packing[50].

Intensity Measurement. Integrated intensity rather than peak height should be used for serious quantitative X-ray diffraction analysis, since the peak height varies due to particle size and lattice distortion. Observe counting statistics and accumulate enough counts, N, such that the standard deviation $N^{-1/2}$ is within expected limits. The integrated intensity may be measured from the area under the line profile or by recording the total counts, while the counter scans the line profile. It is essential that a normal scan of the total sample be made to observe intensity level, background and possible interferences.

6. STRURAL ANALYSIS OF SINGLE CRYSTALS

Many volumes have been written on the determination of single-crystal structures by X-ray diffraction techniques[3,4,46,51]. The short description here presents a concise concept for those with casual interest in this field about its goal, work and problems.

Single crystals are a type of laboratory curiosity. However, much of our understanding of the properties of polycrystalline materials such as metals, plastics and minerals has been gained by studies of isolated single crystals. Single-crystal structure determination is quite an involved science. The primary goal of this particular science, X-ray crystallography, is to obtain a set of coordinates of various atoms in the repeating unit (Unit cell) of the crystal. Thereupon, a detailed three-dimensional picture of atomic positions and molecular packings within the crystal can be visualized as through a super-power microscope.

The size and symmetry of the unit cell in the crystal are deduced from the positions of the diffraction lines, and the coordinates of atoms in the unit cell are calculated from the intensities of the diffraction lines. The main work for single-crystal structure analysis is twofold: (1) experimental, which includes the selection of a single crystal of 0.2 to ~ 3-mm size, the determination of its density, the indexing of reflections and the collection of a set of intensity data from diffraction experiments by use of Weissenberg/Precession camera or diffractometers; and (2) computation, which includes the derivation and establishment of a set of most probable coordinates of atom's, thermal ellipsoids, bond lengths and bond angles in the crystal based on the set of intensity data collected from diffraction experiments.

The computation is far from straightforward. Given the position of atoms in the unit cell, the direction and intensity of diffracted X-rays can be calculated precisely. However, given a set of direction and intensity data, the position of atoms in the unit cell cannot be calculated straightforwardly because the phase angle (out of alignment of wavelets) in the Fourier series is unknown. Hence, the central task of structure analysis is to deduce the missing phase angles and to derive a trial structure. The Patterson map and the Direct Method of Karle-Hauptman are two widely used approaches to the phase problem. But neither method will always result in a trial structure. Chemical information, broad knowledge, experience and imagination all play important roles.

A trial structure must be chemically plausible and give good agreement between observed and calculated intensities. This trial structure is then refined by least-squares process or Fourier synthesis to obtain a set of most probable coordinates of atoms in the crystal.

The discrepancy index, R, denotes how well the derived structure fits the observed intensity data. The lower the R, the greater the confidence in the derived structure. $R = 0.2 \sim 0.3$ indicates good trial structures. $R = 0.03 \sim 0.08$ indicates most reliably determined structures.

7. CONCLUSIONS

X-ray diffraction techniques are widely used in industry for chemical analysis of powders, and elegantly handled in academia for structure analysis of single crystals. Powder analysis is rather simple and routine, while structure analysis is filled with puzzles and excitement. Both are well established and yield quantitative results. Other applications include the analysis of particle size and lattice imperfection from the line profiles, the crystallinity and orientation of polymers from the total diffraction pattern, the analysis of stress in metal or alloy from the total diffraction pattern, the analysis of stress in metal or alloy from small changes in d-spacing, the texture of metal or alloy from pole figures, and the phase diagram from the composition of quenched specimens. Various models and mathematical expressions have been developed for each case. However, both expertise and diligence are required to attain reliable data. The use of simplified approaches gives only simplified pictures. Nevertheless, the derived results are practically useful and academically interesting.

An X-ray powder diffraction pattern is not necessarily made from a dab of finely pulverized solid particles. It can be made from any polycrystalline aggregate, such as metal wire or sheet, polymeric fiber or film, plastic plate or block, etc. The term particle size may indicate the size of crystallites (small single-crystals) in powders, the size of grains in metals, the size of crystalline domains in polymers, or the size of micelles in organic fibers. With X-ray diffraction techniques, the size, perfection, orientation and composition of crystallites in a specimen can be studied. Their effects on properties and performance can be correlated. Therefore, X-ray diffraction has been an important tool for research in materials science.

Besides the composition and performance of materials, physicists and physical chemists are seriously concerned about the electronic configuration of atoms, the lattice and atomic vibrations, the structure of biopolymers, and the mechanism of enzyme actions, etc. They need either the most accurate data or a huge collection of data to solve these problems. With the advent of modern electronics and computers, X-ray diffraction instrumentation has progressed from simple manual cameras to monstrous automated diffractometers handling all data collection and processing. However, the effectiveness of this automated instrumentation is yet to be fully realized. The conventional X-ray sources and detectors have prevented automated systems from a real breakthrough. New developments[52,53] in this direction include synchrotron X-ray sources, laser plasma X-ray sources, position-sensitive detectors, energy-dispersive detectors, holographic image processing and three-wavelength method of phase determination. A real breakthrough could be expected at the turn of the century.

8. REFERENCES

Abrahams, S. C.; and J. B. Cohen. *Phys. Today* 29: 34 (1976).

Akimoto, S.; and Y. Takenchi. *Nippon Kessho Gakkaishi* 18: 163 (1976).

Alexander, L. E. *X-Ray Diffraction Methods in Polymer Science* (New York: John Wiley & Sons, Inc., 1969).

Alexander, L.; and H. P. King. *Anal. Chem.* 20: 886 (1948).

Arndt, U. W.; and B. T. M. Willis. *Single Crystal Diffractometry*, Chapter 4 (New York: Cambridge University Press, 1966).

Averbach, B. L.; and M. Cohen. *Trans. AIME* 176: 401 (1948).

Azaroff, L. V.; and M. J. Buerger. *The Powder Method* (New York: McGraw-Hill Book Co., 1958).

Bindley, G. W. *Phil. Mag.* 36: 347 (1945).

Birks, J. B. *The Theory and Practice of Scintillation Counting* (London: Pergamon Press, Inc., 1964).

Brown, G., Ed. *X-Ray Identification and Crystal Structures of Clay Minerals*, 2nd ed. (London: Mineralogy Society, 1961).

Buerger, M. J. *Crystal Structure Analysis* (New York: John Wiley & Sons, Inc., 1960).

Bunn, C. W. *Chemical Crystallography* (New York: Oxford University Press, 1961).

Chung, F. H. *Adv. X-Ray Anal.* 17: 106 (1974).

Chung, F. H. *J. Appl. Crystallog.* 7: 519 (1974).

Chung, F. H. *J. Appl. Crystallog.* 7: 526 (1974).

Chung, F. H. *J. Appl. Cryst.* 8: 17 (1975).

Chung, F. H.; and R. W. Scott. *J. Appl. Cryst.* 6: 225 (1973).

Cole, H., Ed. "Instrumentation for Tomorrow's Crystallography," *ACA Trans.* 12 (1976).

Cullity, B. D. *Elements of X-Ray Diffraction* (Reading, MA: Addison-Wesley Publishing Co., Inc., 1956).

Darwin, C. G. *Phil. Mag.* 27: 315, 675 (1914).

Darwin, C. G. *Phil. Mag.* 43: 800 (1922).

Denne, W. A. *J. Appl. Crystallog.* 9: 510 (1976).

Ewald, P. P. *Acta Crystallog.* 11: 888 (1958).

Ewald, P. P. *Ann. Phys.* 49: 1, 117 (1916).

Ewald, P. P., Ed. "Fifty Years of X-Ray Diffraction," International Union of Crystallography, Utrecht, Netherlands (1962).

Frankel, R. S.; and D. W. Aitken. *Appl. Spectrosc.* 24: 557 (1970).

Frevel, L. K. *Anal. Chem.* 44: 1850 (1972).

Garska, K. *J. Appl. Spectros.* 30: 204 (1976).

Giessen, B. C.; and G. E. Gordon. *Norelco Rept.* 17(2): 17 (1970).

Hashizume, H.; Y. Memiya; K. Kohra; T. Izumi; and K. Mase. *J. Appl. Phys.* 15: 2211 (1976).

Hubbard, C. R.; E. H. Evans; and D. K. Smith. *J. Appl. Crystallog.* 9: 169 (1976).

Hubbard, C. R.; and D. K. Smith. *Adv. X-Ray Anal.* 20: 27 (1977).

JCPDS. "Powder Diffraction File," Swarthmore, PA (1979).

Jenkins, R.; Y. Hahm; S. Pearlman; and W. N. Schreiner. *Norlco Rept.* 26(1): 1 (1979).

Johnson, G. G. *Comput. Chem. Instr.* 5: 45 (1977). Kaelble, E. F. *Handbook of X-Rays* (New York: McGraw-Hill Book Co., 1967).

Kirkpatrick, T.; and J. J. Nilles. U.S. Patent 3,677,783 (1972).

Klug, H. P.; and L. E. Alexander. *X-Ray Diffraction Procedures* (New York: John Wiley & Sons, Inc., 1974).

Koch, E. E; C. Kunz; and E. W. Weiner. *Optik (Stuttgart)* 45: 395 (1976).

Lang, A. R. *J. Sci. Instrum.* 33: 96 (1956).

Mantler, M. *Mikrochim. Acta Suppl.* 7: 555 (1977).

Matthews, F. W.; and J. H. Michell. *Ind. Eng. Chem. Anal. Ed.* 18: 662 (1946).

Muller, R. H. Anal. Chem. 38: 155 (1966).

Nuffield, E. W. *X-Ray Diffraction Methods*, Part 2 (New York: John Wiley & Sons, Inc., 1966).

Parrish, W.; and T. R. Kohler. *Rev. Sci. Instr.* 27: 795 (1956).

Rudman, R. *Low Temperature X-Ray Diffraction* (New York: Plenum Publishing Corp., 1976).

Scott, R. W. *Treatise on Coatings*, Vol. 2, Part II (New York: Marcel Dekker, Inc., 1976), p. 597.

Standard X-Ray Diffraction Powder Patterns, NBS Monograph 25, U.S. Government Printing Office, Washington, D.C. (1971).

Stout, G. H.; and L. H. Jensen. *X-Ray Structure Determination* (New York: Macmillan Publishing Co., Inc., 1968).

Sturm, E.; and W. Lodding. *Acta Crystallog.* A24: 650 (1968).

Tanner, B. K. *Prog. Crystallog. Growth Charact.* 1: 23 (1977).

Walker, G. F. *Nature* 164:577 (1949).

Williams, P. P. *Anal. Chem.* 31: 140 (1959).

12.

Progress and Potential of X-ray Diffraction

Like all brilliant original ideas, it seems so obvious when pointed out.

—W L. Bragg

Education is what remains after all school learning is forgotten. It is less the factual knowledge that is important, than the facility one requires for scientific reasoning.

—M. V Laue

1. ABSTRACT

Smart materials and complex bio-systems are the two major frontiers of contemporary research. The XRD data link structures to properties. The brilliant ideas link properties to applications. Recent progress includes (1) New methods for solving the structures of large molecules such as protein and enzymes. (2) Matrix Flushing techniques unify and simplify quantitative XRD & XRF analyses. (3) Synchrotron radation provies high intensity, high resolution, small divergence X-ray beam. (4) Structure of new materials including fullerenes (cage-like C_{60}), high temperature superconductors, thinfilms, and multi-layers. (5) New concepts on drug-receptor interactions, molecular wires, molecular switches, molecular memories, and others. To predict future is risky. The future potentials could be (1) New techniques and applications, such as imperfection and microstructure studies, etc. (2) High-tech materials include molecular electronics, and (3) Emerging topics such as computational chemistry, holographic atomic imaging, and so forth.

2. INTRODUCTION

X-ray diffraction is a quantitative and mathematical method for crystal structure analysis. It has been an indispensable research technique in industrial, academic, and government labs. Government research is usually mission oriented. Its purpose is to find more effective ways to defend (defense), to cure (health), or to protect (environment). Academic research is publication oriented. Professors select topics of research with reasonable prospect of success, so that the findings will be acceptable for publication and suitable for students' theses. Industrial research is profit oriented. Its attention is focused on the "bottom line" in terms of new and improved products. X-ray diffraction is one of the most important analytical tools to accomplish these research goals. The evolution, progress, and potentials of the X-ray diffraction techniques are briefly overviewed here with minimum attention to finer points or structural details that have been covered by the authoritative presentation under specific topics.

3. EVOLUTION

A burst of discoveries marked the start of the 20th century. Roentgen discovered X-rays in 1895. Planck introduced the quantum theory in 1900. Einstein published the relatively theory in 1905. Laue demonstrated X-ray diffraction and Bragg interpreted it in 1912. Bohr's atomic theory appeared in 1913. Debye and Scherrer designed the powder diffraction method in 1916. The entire structure of science and philosophy was reconsidered and rebuilt. Now near the end of the 20th century, it is probably an opportune time to review what has been accomplished and what are the future prospects for X-ray diffraction, in particular or relevant science in general.

3.1 Roots

Crystallography, the science of crystals, studies the symmetry, growth, structure, and properties of crystals. X-ray diffraction of crystals provides the exact molecular geometry and mode of packing. The derived hard parameters (crystal symmetry, unit cell dimensions, bond lengths, and bond angles) and soft parameters (hydrogen bonds, interatomic separations in complexes) are the primary data for studying structure-property relationships. For evolution reasons, in Germany and many other countries, crystallography is mainly taught as part of *mineralogy*, where the habits and birefringence of crystals are studied. In England, it is taught as part of *physics* because the Braggs and many other pioneers were physicists. In America, it is considered as part of *chemistry and mineralogy*, because under the leadership of Linus Palling (twice Nobel laureate, in chemistry and in peace), Caltech attained a dominant position as a center of crystal research. Today, the boundary is disappearing. The study of crystal structures has permeated into the field of geology, petrology, metallurgy, materials science, biology, pharmacy, medicine, and mathematics, simply because structures prescribe properties, which in turn govern applications/functions of all materials, including natural products and living matters.

3.2 Three Generations

The evolution of crystallography constitutes one of the cornerstones of contemporary science. It becomes an important factor of progress in technology. It is convenient to divide the evolution of crystallography into three periods according to the tool used and the information extracted:

> **Morphological crystallography (before 1912):** This includes all the descriptive, optical, and morphological information, the group theory, and the tensor calculus about crystals mustered before the discovery of X-ray diffraction in 1912.
> **X-ray crystallography (1912 to 1962):** The theory and practice for solving crystal structures by X-ray diffraction were essentially developed in this first 50 years. It was a glorious and exciting period, filled with many discoveries and enormous advances. The activities in this period are detailed by the book *Fifty Years of X-Ray Diffraction* (Ewald 1962) published by the International Union of Crystallography It contains reviews and memoirs by numerous pioneers who grew up together with the subject.
> **Modern crystallography (1962 onward):** The use of synchrotron radiations started in the 1960s. Fast advances derived from the newly found light source and from the

blossoming computer science allow smaller and smaller single crystals, faster and faster experiments, as well as shorter and shorter time-resolved X-ray diffraction. Crystallography is no longer just structure determination per se, nor just for fundamental knowledge. Crystal structure determination is done with applications in mind. It provides answers to specific problems, leads to new products, or opens new perspectives.

3.3 Structural Repertoire

The essential aim of research in chemistry is the synthesis and characterization of new materials. The goal of characterization is to learn the structure-function relationships. The most basic characterization is the molecular and crystal structures. The repertoire for structure determination consists of three tools: (1) *crystallography*, mainly by X-ray diffraction, complemented by neutron diffraction and electron diffraction; (2) *spectroscopy*; mainly by infrared (IR) and nuclear magnetic resonance (NMR); and (3) *molecular modeling*, mainly by quantum mechanics and/or molecular mechanics calculations. Only X-ray diffraction provides the information that allows the determination of the molecular conformation and crystal structure directly and exactly Single-crystal techniques are most likely used for biological/medical materials. Powder diffraction is frequently used for engineering materials. Both spectroscopy and molecular modeling are supplementary. However, spectroscopy provides the information for determining the molecular conformation in solids as well as in solution. The value of molecular modeling lies in extensions to experimental investigation when crystallography is not amenable. Note that quantum mechanical calculations can also give the electron-density distribution in a molecule of known structure, which is subtle in many chemical/biological reactions.

Computational chemistry gives the most stable structure of a single molecule at standstill, which means an isolated molecule in vacuum at absolute zero temperature. In order to assess the capability of theoretical models, they must first be judged against data provided from crystallography In most cases the calculated structure with the global minimum energy does correspond to the stable conformation observed in crystals, whether of ab initio, semi-empirical, or molecular mechanics variety. However, when strong directional forces are present such as temperature, pressure, solvent, gravitational, electromagnetic field, etc., the structure of the grown crystal may be different from that of theoretical model, because the external (field) force overcomes the internal (molecular) force. In such cases, certainly the fact (by X-ray diffraction) overrules the theory (by calculation).

3.4 Frontier Research

Crystal structural determination of macromolecules is complex and involves multidisciplines. Molecular structure and conformation is at the heart of chemistry. As a bridge, chemical structure links mineralogy, geology, and materials sciences in the high-tech field. It also links biological, pharmaceutical, and medical sciences and materials in the health field. It is the basis for structure-function studies in all sciences.

Smart *materials* and complex *biosystems* are the two major frontiers of contemporary research. Both are areas of pivotal importance in the quality of our daily lives (equipment, health, and environment). Both the materials and the biosystems will be central concerns of research into next century, Besides, there is always the big science of national labs: nuclear programs to search the innermost quarks and space programs to explore the outermost solar systems and beyond. More than ever, collaboration is a reality and a necessity in frontier research. Interactions and cooperations between labs and among scientists produce synergy; and hence contribute more and accomplish bigger. Future research teams will be a synergistic integration of experts in different fields, including even business and law In Max von Lane's words, "Science would soon come to a standstill if all scientists were of the same intellectual type. Science needs scholars with many different talents."

4. PROGRESS

Progress in science and technology depends essentially on the quantitative relationships between structure (steric, hydrophobic, and electronic) and activity (stability, reactivity, functional property, drug efficiency, and biological effects). Great strides forward have been made after the first 50 years of X-ray diffraction (1912 to 1962). The developments and advances in X-ray diffraction after 1962 are briefly summarized here.

4.1 New Horizon in Single-Crystal Diffraction

The only basic obstacle in single-crystal structure determination is the *phase problem*. In recent decades, a full range of techniques were developed for solving the phase problem. Traditionally, there are three most extensively used methods: the Patterson map with heavy atom, the direct method, and dynamic modeling. These methods can be made routine for small molecules of up to 100 nonhydrogen atoms. Theoretical and experimental advances in techniques for solving the phase problem have led to three newer methods for large molecules such as proteins and enzymes: (1) single or multiple isomorphous replacement (SIR or MIR), (2) anomalous scattering (SIR/AS), and (3) multiple-wavelength anomalous dispersion with synchrotron radiation (MAD/SR). The details of these newer methods are treated in recent books (Woolfson 1997, Rao 1997, Powell 1996, Woolfson and Fan 1995, Vainshtein, Fridkin, and Indenbom 1994, Ladd and Palmer 1994, Glusker, Lewis, and Rossi 1994). Ever since Crick and Watson's double spiral structure of nucleic acids, many biomolecular structures, such as insulin and glycogen phosphorylase, have been elucidated and provide clues to fundamental understanding of biological structures and functions, including new drugs by molecular design, effective actions of viruses, and mechanisms of genetic processes.

A suite of computer programs is available for automatic instrument control, data collection, data processing, structural parameter refinement, and color graphic display. These devices are coupled with ever-increasing computer power: the fiendish rate of doubling the power of computer chips in 18 months (Moore's Law). Because of a recent breakthrough in "multilevel cell" flash memory, the time now required to double the memory of computer chips will be even shorter. Taking advantage of the fast growth of affordable CD-ROM storage and the prevalent PC systems, six structure databases were set up for search and research: the Cambridge Structure Database (CSD), Inorganic Crystal Structure Database (ICSD), NRCC Metal Crystallographic Data File (CRYSMET), Protein Data Bank (PDB), Powder Diffraction File (PDF), and NIST Crystal Databases (Organic, Inorganic, Biological, Macromolecule, and Neutron Diffraction). The gross sizes of these growing databases indicate an impressive wealth of structural data (Gorter and Smith 1995, Vainshtein, Fridkin, and Indenbom 1994). More details about crystallographic databases may be found in Chapter 2 on the practice of diffraction analysis.

4.2 Renaissance of Powder Diffraction

Powder diffraction has two major areas of applications: (1) structure and microstructure (imperfections) analysis, and (2) chemical (phase) analysis. There are two basic problems in powder diffraction analyses: the resolution of overlapping peaks for structural analysis, and the correction for matrix effects for chemical (phase) analysis. These two problems were essentially solved by two series of significant work in the 1970s. Subsequent developments have greatly expanded or perfected its applications in these two areas. Langford and Louer (1996) published a comprehensive and authoritative review on powder diffraction.

It is a heartfelt joy to watch powder diffraction grow from an ugly duckling into a beautiful swan. The flourishing European Conference (EPDIC), the sustained Denver Conference, the stature of the *Journal of Applied Crystallography*, the firmly established journal *Powder Diffraction*, and the frequent special topical symposia or workshops all indicate a renaissance of powder diffraction techniques (Jenkins and Snyder 1995, Proceedings of EPDIC 1996, Chung 1989).

In 1967, Rietveld first introduced his powerful method for refining the crystal structure from powder data. It has been subsequently improved and extensively used since the 1970s, especially when proper single crystals are not available. It has been applied successfully for structural investigations of high-temperature superconductors, fullerenes and fullerides, zeolite catalysts, intercalates, phase transitions, magnetic moments, and many others. Note that the *Rietveld method* can also be applied to neutron diffraction data of powders. The advent of the Rietveld method has made powder diffraction one of the most widely used techniques for structure and microstructure analyses of materials (Rietvelt 1967, 1969, Young 1993).

The success of structure determination from powder data has greatly enhanced the capability of the diffraction techniques. The powder diffraction pattern is the projection of a three-dimensional diffraction space onto a single dimension linear representation. Consequently, the pattern consists of many overlapping peaks, partially or totally superimposed. How to resolve these scrambled peaks is still a challenge to solving crystal structure by powder data. To date, the maximum entropy technique, the simulated annealing, the anomalous scattering, and the ab initio approach, have all had some success, but a total solution is still pending.

As shown by Klug and Alexander (1974), in the powder diffraction pattern of a mixture, the integrated intensity of reflections for each component is related to the concentration of that component present in the mixture. However, the intensity-concentration relationship is plagued with matrix effects. Zevin and Kimmel (1995) reviewed the various methods to retrieve the composition data by minimizing the matrix effects for quantitative X-ray diffraction analysis.

In industrial scenarios, such as factory malfunctions, defective products, or customer complaints, people frequently request fast response (the same day or the next day) for the quantitative composition of nonroutine, totally unknown samples. Most methods can not deliver the "results of analysis" so rapidly to solve urgent problems. Chung (1974a, 1974b, 1974c, 1975) published a series of papers on the *Matrix-Flushing method* (now known as the *RIR method*) for quantitative chemical (phase) analysis of powder mixtures. It provides three powerful equations relating intensity to concentration (Chung 1981). Numerical examples of these three equations can be found in Chapter 20 on X-ray diffraction analysis in the paint and pigment industry. The Matrix-Flushing method is simple, exact, and free from matrix effect, and hence quite popular in industrial labs. Under favorable conditions, the weight percentage of every component can be determined from a single scan of the original sample without any internal standards or calibration curves. Subsequently, Hill and Howard (1987) Bish and Howard (1988), Hill (1991), Snyder (1992), Bish and Post (1993), Taylor and Aldridge (1993), Gorter and Smith (1995), and others developed the Rietveld equivalent of the Matrix-Flushing method using integrated intensity of all reflections in the powder pattern. It has the benefits of minimizing the effects of preferred orientation and extinction at the expense of more time for data collection and processing (Davis 1986, Davis and Smith 1988, Smith 1992).

4.3 Synchrotron Radiation

In the X-ray diffraction field, probably no technical advances since Laue and Bragg's experiments has had greater impact than the advent of synchrotron radiation (SR) sources. Electron/positron storage rings were first developed as "colliders" for high-energy physics. They were later successfully used as sources of very-high-intensity X-ray beams known as synchrotron radiation. In 1960s, the first generation of SR was used "parasitically" from storage rings built for high-energy experiments. The second-generation, "dedicated" SR sources were built in the 1970s and 1980s when storage rings were optimized as source of SR. In 1990s, the third generation of SR provides greater brilliance X-rays of higher intensities from high-performance storage rings. The high-performance was derived from two developments: (1) the optimized *magnetic lattice* design (the array of magnets which constrain the electron beam) that increases brilliance, and (2) the insertion devices (wigglers and undulators placed between the bending magnets) that increase flux density. Currently, some 60 such facilities worldwide are operational or under construction. A fourth generation of SR is being planned beyond the year 2000, striving to achieve still higher brightness, better spatial coherence, and shorter pulses compared to existing sources (Terminello et al. 1996, Johnson et al. 1996).

Synchrotron radiation offers high intensity, high resolution, small divergence, small beam size, tunable wavelength, polarization, and pulsed time structure. The combined effect of the first four features signifies *high brightness*. High brightness improves spatial resolution, allows time resolution, imparts the ability to study tiny sample volumes (surfaces and microcrystals) or weakly scattering objects (magnetic scattering, plasmon scattering, and nuclear resonance scattering) and gives faster data collection times. High brightness also implies high spatial coherence, which allows applications in holography, interferonaetry, and time-correlation spectroscopy. High brightness is the main improvement of the latest "third generation" of SR sources (Lovesey and Collins 1996, Kenway et al. 1992).

4.4 New Materials

New materials including synthetic crystals rapidly invade modern technology such as radio-technics, electronics, semiconductors, quantum electronics, nonlinear optics, acoustics, and others. Major efforts of contemporary science and technology are searching for crystals with valuable properties, studying their structures, and developing new techniques for their syntheses. Some topics have created great excitement in the scientific world, such as fullerenes in the 1990s, and high-T_c superconductors in the 1980s. The progress of modern crystallography (new X-ray sources, new position/area detectors, new diffractomer design, and new mathematical data analysis) has contributed extensively to comprehending new materials or shaping new concepts.

Fullerenes and Fullerides

Besides diamond (a perfect electric insulator but good thermal conductor) and graphite (a conductor), a third allotropic form of carbon, fullerene, was discovered in 1985. The structure of fullerenes has been thoroughly studied, mostly by X-ray diffraction (Blank et al. 1997, Vainshtein et al. 1994). The cluster of 60 carbon atoms of fullerene, C_{60}, forms a cage-like network of bonds with icosahedral symmetry. The pseudospherical structure is built in the same way as the icosahedral soccer ball. It is made of 20 hexagons (like graphite and benzene rings) and 12 pentagons (like cyclopentadiene). The bucky ball has a diameter of about 7.1 Å. The carbon atoms forms clusters C_n, n=24, 28, 32, 50, 60, 70, 76, 78, 84, or even up to 960. The C_{60} and C_{70} are most stable. The number of pentagons is always 12; the rest are hexagons in all these fullerenes. The C_{60} molecule is like an onion made of concentric shells, with the distance between shells 3.4 Å. Their crystal structures were studied by X-ray and neutron powder diffraction. Note that spherical viruses also have icosahedral symmetry like C_{60}.

The excitement in fullerene research has been enhanced by the discovery of *superconducting fullerene* compounds. Fullerides or fullerites are A_nC_{60} crystals in which the atoms A are mainly alkali metals placed in the interstices of the close packing. Fullerides exhibit a wide range of fascinating properties such as high T_c superconductivity, molecular ferromagnets, molecular traps, molecular bearings, catalysts, and electroconductive properties (from insulators, to semiconductors, conductors and superconductors). For n <4, A_nC_{60} compounds are metals. The compound TDAE-C_{60} show ferromagnetic properties, where TDAE is tetrakis (dimethylamino) ethylene,

High Temperature Superconductors

Superconductivity was discovered in 1911 by Kamerling-Onnes. He found that at $T<4.15\ K$ the electric resistance of mercury dropped down to zero. Before 1985, superconductivity was found in a number of pure metals, alloys, and intermetallic compounds such as Nb_3Ge with $T_c = 23.2$ K. After 1985, a series of high-temperature superconductors was discovered; the first was $(La,Ba)_2CuO_4$ with $T_c = 36$ K. Note that La_2CuO_4 is not a superconductor. It is necessary to replace some La atoms by Ba or Sr to create some La deficiency or oxygen redundancy to attain superconductivity (Bednorz and Muller 1986). A giant step forward was the discovery of $YBa_2Cu_3O_7$ ($T_c = 93$ K) by M. K. Wu et al. (1987). X-ray diffraction analysis indicates that all copper-containing high-T_c, superconductors are formed by planar CuO_2 layers alternating with alkaline earth metals. To date, superconductivity is also found in other ceramics fullerides, doped semiconductors, organics, and even polymers. The number of superconducting compounds now exceeds 600, and is increasing. All ceramic superconductors have hole conductivity The T_c of the compound $Tl_2Ca_3Ba_2Cu_3O_{10}$ reached 125 K in 1991. Most recently Chu et al. (1997) reported a *nontoxic superconductor* of $T_c = 126$ K, which contains no volatile toxic elements Hg or Tl.

The structures of high-T_c superconductors were determined by both single-crystal and powder diffraction. Practically all the high-T_c superconducting materials are nonstoichiometric phases, and contain at least one element with variable valency. All have perovskite-like layer structures that belong to the cuprate class. All contain an element capable of forming peroxide compounds, including Ca, Sr, Ba, and K. The compound (Ba,K)BiO$_3$ has perovskite structure, T_c=30 K, but no copper atoms are present. This prevents the interpretation of superconductivity in terms of the two-dimensional CuO$_2$ nets. The syntheses, structure, properties, and applications of superconductors are currently hot research fields.

Thin Films and Multilayers

A thin film is up to several 10 Å thick (monolayer to a few atomic layers) in surface science, but could be several hundreds of micrometers for protective coatings. The concept of a thin film depends on its structural features and material applications.

Thin films, multilayers, and superlattices found applications in modern microelectronics. nonlinear optics, laser technology magneto-optical recording, random access memories, quantum devices. biosensors, biological membranes, conducting polymers, wear-resistant coatings, and molecular electronics (Fewster 1996). Their performances depend on their structures. For example, the performance of semiconductor materials rely heavily upon three factors: (1) the layer-by-layer thin-film structure grown by molecular-beam epitaxy, vapor-phase epitaxy or chemical-beam epitaxy, (2) the level of impurity doping and energy-band structure modification, and (3) the sizes, shapes, and positions of defects in the crystal lattice. X-ray diffraction is good tool, often the only tool, for measuring/controlling the thickness, composition, and microstructure of the layers (Ulman 1991).

Langmuir-Blodgett (LB) films are artificial mono- and multimolecular layers consisting of the amphiphilic surfactant molecules. *Superlattices* are LB films containing two or more kinds of molecules. LB films may open the way to create artificial biomembranes, monochromators, optical lenses, molecular lubricants (for optical and magnetic disks), high-capacity condensers, electron/light lithography, and so forth (Vainshtein et al. 1994).

4.5 New Concepts

The drive of science has always been to imitate the life systems. The computer simulates the brain, polymers mimic the tough tissues, and composites resemble the biominerals (teeth, bones, and shells). The living body gives sensitive, selective, fast, and intense responses to all sorts of external stimulations. Molecular biology shows that the responses are results of specific properties of organic molecules. Molecular structures have three levels of complexity: minerals, organics, and biomolecules. To date, most of the materials for electronic devices are minerals (metals and semiconductors). The rich possibilities of organic chemistry are just starting to be mature enough to open promising avenues leading into molecular electronics. The bionics age will begin when we know how to make and use biomolecules. Then, robot cops and bionic women will be everywhere.

Modular Structures

Active developments in theoretical and experimental crystallography produced numerous new data on crystal structures. The number of determined structures has reached some 180,000.

Most of them have been solved since 1980. Statistical analyses of these data indicate that many large superstructures could be considered as being composed of some simple prototype *building modules* differing in their combination, distribution, and stacking. Building modules can be recognized by common determinative features such as blocks (polyhedra islands, rings), layers (planar nets, sheets), and rods (strings, chains, ribbons). The building modules present in different structures need not be exactly identical. They may vary in size, shape, orientation, and even composition, but some close relationships exist, usually recognizable as some common parts (fragments) in different structures such as the many varieties of zeolites and silicates (Vainshtein et al. 1994).

Phyllosilicates (mica, kaolinites, chlorites, etc.) are composed of octahedral and tetrahedral sheets that stack in random or ordered fashion. Modular structures of crystal substances bear important properties. It has been shown that the building module of SnSe and SnS consists of closely packed octahedral layers. The electronic band gap is wider for rhombohedral than for hexagonal polytypes. Both of them change from dielectrics to semiconductors (decreasing electronic band gaps) with increasing layer repeat. Under different doping atoms, either p- or n-conductivity is realized. Different polytypes may be combined to form *solar batteries* of up to 25% efficiency (Rao 1997). Similar modular structures of high-T_c superconductors are being observed. Modular structures can be single-module or mixed-module, ordered or disordered layers. The intermodular space may be intercalated with variable amounts of water or organic compounds. The property-structure dependence is affected by the module type, intermodular bonds, periodicity of module stacking, stacking faults, and distortion inside modules.

Functions and Structures of Biomolecules

Protein and nucleic acids are the most important materials in all lifeforms. Proteins are the basic building blocks of living organisms. Nucleic acids provide information storage and transfer. Over 2000 biomolecular structures have been investigated to date, including various ligand modifications. As more protein structures have been solved, more synthetic chemists have become involved in rational drug design based on *receptor geometry* and *charge distribution*. One can also make deliberate changes in structure by replacing certain amino acid residues in a protein, thus improving its enzymatic, regulatory, or some other functions. The structures of biological molecules have wide applications in medical research (such as drug design) and genetic engineering (such as vaccines and cloning).

The structures of small molecules (up to 100 atoms) have been solved routinely. mostly by the direct method. Given a raw data set from a diffraction experiment, it takes only a few minutes on a PC to obtain an approximate structure, and a few more minutes to run full-matrix least-square refinements. Computer graphics put the molecule in color-coded stereo display. At present, the medium-size (100 to 500 atoms) structures and macromolecules (larger than 500 atoms) defy the direct method. They are solved generally by an isomorphous replacement method (SIR, MIR, MAD and MASC), or by a molecular replacement technique when a similar structure is already known. A majority of macromolecular structures were solved by molecular replacement. The isomorphous replacement method, such as the MAD technique, is an extremely demanding experiment. Popular tools for protein structure investigations include tunable synchrotron radiations, area detectors (TV type, multiwire proportional chamber, charge-coupled device, or phosphor image plate), and four-circle diffractometer. Low-temperature data collection by cold N_2 or He gas stream is commonplace for greater accuracy and precision with the least radiation damage to the crystal sample.

Macromolecules do not diffract well. The diffraction limit of macromolecules, in term of angstrom resolution, is larger than the interatomic bonding distances. The electron density map is thought of as "globs" of electron density of a size appropriate to an amino acid residue. To clearly resolve the atoms in the structure needs a resolution better than 1.2 Å. A resolution of 2 to 2.5 Å allows one to solve problems regarding their tertiary structure and protein function.

Major advances in structures down to atomic level have revealed the functionality of biological molecules, drug-protein interactions, information transmission, metabolism, and many others (Nogrady 1988). The structure features versus biological functions of many biomolecules have been investigated (Horn and De Ranter 1984), for example, *antibiotics* (gramicidins S and A, transmembrane channels for ion transport), *enzymes* (glycogen phosphorylase, regulate metabolism), *ribonuclease* (enzymes decompose RNA), *ribosomes* (interactions between RNA and various protein molecules), *photosynthesizing bacteria* (mechanism of photochemical reactions), and *viruses* (plant, animal, insect, and bacterial viruses, bind to the host cells, result in body infection).

Drug-Receptor Interactions

Molecular pharmacology the study of drug actions at the molecular level, is not quite an exact science due to the complexity of the biological systems. However, enormous advances were made in the past 25 years, mainly due to the better understanding of the ligand-receptor interactions. Receptors are binding sites in or on animal cells (cell membrane, cell nucleus, and cytosol). They interact with and mediate the effects of guest molecules endogenous to the body (hormones, neurotransmitters, autacoids). They also interact with xenobiotics (herb medicines, synthetic drugs, and toxins), which are foreign to the body. When the drug binds the target receptor sites as designed, the disease gets cured, but when it binds to similar receptors in other part of the body, it causes side effects. Receptors perform two vital functions: The first is to recognize and discriminate, and the second is to transduce a signal that initiates a cascade of biochemical reactions that lead to a characteristic physiological or pharmacologic response. Incidentally, *acupuncture*, an ancient Chinese therapy, has been recently endorsed by the National Institutes of Health (NIH). Acupuncture physically, instead of chemically, stimulates the receptors that transduce a signal for releasing pain-suppressing endorphins. Actual isolations of various receptors such as certain steroid hormones and acetylcholine are milestones for this discipline. The binding of a drug to the receptor is stereospecific. Consequently, great emphasis has been laid on the elucidation of stereostructural properties of drugs by using X-ray diffraction for solids, NMR for liquids or solutions, and quantum or molecular mechanics calculations for initial screening (Lipkowitz and Boyd 1997).

Receptor structure is critical in the design of drugs that conform to the receptor site and correct unwanted behavior of cells. Molecular biology and genetic engineering provide great impetus to the molecular design of drugs. The drug-receptor concept greatly helped to develop new *biotech drugs* (large molecules) and to improve the efficiency of conventional drugs (small molecules). For example, Merck's Mevacor blocks the production of cholesterol, which can clog the arteries that deliver blood to the heart and cause heart attack. Squibb's Capoten regulates blood pressure, by inhibiting overproduction of angiotensin II and hence stopping the rise in blood pressure. Eli Lilly's growth hormone treats dwarfism and obesity control. Ulcers are generated by overproduction of gastric acid in the stomach triggered by histamine; SB's Tagamet, an antiulcer drug, blocks the acid-secreting cells by inhibiting action of the histamine receptor (Kourounakis and Rekka 1994).

Advances have been made in the knowledge of many cell receptors in different human organs; some of their structures and functions were investigated by X-ray diffraction. However, the structural studies of biological molecules are difficult. The biological molecules are very large, they do not generally crystallize well if they crystallize at all, and they diffract weakly. An emerging field is called supramolecular crystallography. It bridges between small-molecule crystallography and protein crystallography. One of its major thrusts is to build large assemblies that will exhibit features of biomolecules. Supramolecules are large enough to be relevant to biological systems but small enough to reveal detailed structure by X-ray diffraction. Supramolecules are synthetic receptor molecules to which functional groups can be attached, such as calix-and-arenes, resorcinarenes, porphyrins, and cyclodextrins (Vogtle 1991, Tsouearis et al. 1995, Echegoyen and Kaifer 1996).

Molecular Electronics

Molecular crystals ushered in broad progress in high-tech industries. It inspired the concept of molecular electronics. Molecular electronics are chemical systems that are structurally organized and functionally integrated to perform highly selective operations. They can be activated by means of photon, electric, magnetic, thermal, ionic, or chemical signals.

> **Molecular Crystals:** Recently the optical properties of molecular crystals have found broad applications in high-tech industries. These range from bar codes to fiber optics. The electro-optic integrated circuits require the electronic control of light. They incorporate guided wave modulators and switches using molecular crystals. The origin of the electro-optic coefficient on molecular crystals is its electron distribution, which can be measured by X-ray diffraction, or calculated by quantum mechanics. The emerging electro-optic technology requires molecular crystals with an optimum electro- optic coefficient, low

dielectric constant, high optical quality, and processibility (Wright 1995, Chemia and Zyss 1987). Some molecular crystals studied and/or used in high-tech industries are cited below:

WORM memory: The write-once-read-many-times (WORM) memory systems provide very high-density information storage. Thioaryl phthalocyanines are commonly used as the dye layer, protected with a polyacrylate overlayer.

Color microfilters: These materials are used for liquid crystal color displays (LCD). Such filters must provide clearly defined red, green, and blue arrays of subpixel dimensions. Suitable dyes are copper phthalocyanine for blue, octaphenyl copper phthalocyanine for green, and peryleneimide dye for red.

Electrophotography: Photocopiers and laser printers both work on the same principle and frequently use molecular crystals for improved performance, lower cost, and lower toxicity than corresponding inorganic alternatives. The drum consists of a metal-coated base, a thin charge-generating layer (photoconductor), and a charge-transport layer (optically transparent hole-conducting). Organic materials are chosen for the charge-generating layer, because their spectral response can be tailored. Photocopiers use perylene. Laser printers with 820-nm diodes use a range of phthalocyanines. The charge-transport layers are commonly triphenylmethanes, hydrazones, or oxadiazoles.

Solar photoelectric cells: Hundreds of organic solar cells have been investigated. The principal motivation is that organic materials can be cheap, have large area, be flexible, be lightweight, and have light absorption matching the solar spectrum. The most promising are based on combinations of p-type phthalocyanines with n-type perylene biscarboximides.

Molecular wires: The so-called *organic metals*, polyenes, are crystalline compounds composed of two partners, an electron donor and an electron acceptor, which are often arranged in stacks yielding a one- or two-dimensional conductivity. Their conduction is achieved through overlap of the pi orbitals of neighboring molecules, such as a radical-anion salt (Cu, Ag, Li, etc.) of 2,5-dimethyl dicyanoquinone diimine (DCNQI). The conductivity of the copper salt at low temperature is as high as that of copper at room temperature, and it increases monotonically with decreasing temperature, as is typical of metals. The behavior of other salts is similar to that of a metallic semiconductor.

Molecular rectifiers: Another hoped-for molecular electronics is the molecular rectifier or *diode*, a chain molecule with a -CN group at one end as acceptor, and an amino group -NH_2 at the other end as donor. There are two problems to conquer here: irreversible changes in parts of the molecule through oxidation and reduction, and the attachment of the functional unit to a metallic or a semiconducting interface.

Molecular switches: These switches are activated by light. Some, such as thiophene fulgide, which is photochromic, can be reversibly switched by opening and closing a ring and its conjugated chain of pi bonds; for example, polyene closes by ultraviolet and opens by visible light.

Molecular memories: For *molecular storage* and retrieval, one approach is to use the phenomenon of photochemical hole-burning by laser light such as for porphyrin in an organic or inorganic glass matrix. The molecule bacteriorhodopsin has been demonstrated for stable information storage and for optical holography.

Optoelectronic devices: Electroluminescent devices have practical applications as light-emitting diodes (LED). They are the most important component in integrated optoelectronic devices, where they serve to transform electrical signals into optical ones. Organic molecules and polymers can also be used for this purpose. The polymer poly (p-phenylene vinylene) (PPV) is an interesting candidate.

5. POTENTIALS

The future potentials of the industrial applications of X-ray diffraction can be guestimated from its past progress. The easy veins have already exhausted, so greater endeavor is necessary to make a strike. Certainly, there will be unsuspected discoveries to generate a surge of interest or open a field of newfound land. When gazing at such a broad field in feeble twilight, certainly there are many blind spots. Nevertheless, here is one set of opinions about the challenges, the opportunities, the promising directions, and the potential advances.

5.1 New and Improved Techniques
Direct Method for Macromolecules

The modern computer programs of the direct method are so quick and so successful in solving the structure in the majority of the cases. It becomes a routine tool in the analysis of new products. However, it can be used only for molecules of 100 non-hydrogen atoms or less. Major efforts should be extending it to more complex molecules (100 to 500 atoms), even to macromolecules (larger than 500 atoms, such as proteins). Karle (1995) and Hauptmann (1995) pointed out that the normalized structure factor $/E/$ and the scaling term ϕ are interdependent. Given one, obtain the other. Given both, obtain higher resolution. Other mathematical formulations could exist pending discovery (Powell 1996).

Imperfection and Microstructure

The powder diffraction pattern is a fingerprint of the sample material. The position and intensity of its lines (peaks) have been successfully used for structure and phase analyses. Its line shape reveals the details of microstructural imperfections in the phase giving rise to the particular reflection. However, the current methods to extract this information from line profile (shape) are less than perfect. The parameters obtained normally bear no immediate relation with the parameters used in solid-state physics. Mathematical analyses are needed to establish the connection such as crystallite size, shape, composition, mosaic block molecular alignment, stress, porosity, defects, and amorphous content. Note that imperfections are inherent and affect the properties of all real materials in our daily life (Proceedings of EPDIC 1993, 1996, Kurzydlowski and Ralph 1995). The molecular/crystal structure of a compound *(single-crystal analysis)* determines its *bulk properties*. Microstructural imperfections *(powder diffraction analysis)* in the single-crystal or poly-crystalline material control its *performance*.

Flash Laue Diffraction

Flash Laue diffraction provides snapshots in real time of dynamic processes during chemical reactions or phase transitions with milliseconds or nanoseconds resolution time. For unstable samples and very small single crystals (less than 0.1 mm diameter), it is necessary to collect data rapidly before the crystals decompose and cease to diffract. Virus crystals, for instance, are often fragile and sensitive to radiation damage. The Laue method using synchrotron radiation with an image plate detector appears to be feasible, although it still needs technological improvement in resolution. Flash Laue diffraction takes only a fraction of a second for data collection. (Wulff et al. 1996).

Instrumentation with Synchrotron Radiation

In the foreseeable future, there will be three types of X-ray sources: X-ray tubes, synchrotron radiation, and free-electron lasers (FEL). In the X-ray tube, typically only 0.1% of the electron beam power is extracted. The other 99.9% of power is dumped as heat. Cooling of the anode limits the power of the X-ray beam to a few watts, with rather large angular divergence, resulting, in poor brilliance. Of course, the advantages of X-ray tubes are low cost, small size, and that they are readily built in-house. Free-electron lasers are still in the planning stage (Johnson et al. 1996).

The features of synchrotron radiation have allowed dramatic improvements in many well-established fields including crystallography, X-ray fluorescence, topography, absorption spectroscopy (both XANES and EXAFS), powder diffraction, surface diffraction, protein structure, magnetic scattering, small-angle scattering, and tomography. In order to take advantages of the sharper and more brilliant X-rays, parallel developments in instrumentation are crucial, such as diffractometer goniometry, detectors, and counters. For example, for the diffraction experiment on a single grain of a few micrometers size, the tolerances on the stability, accuracy, and reproducibility of the diffractometer system must be at a matching level. For fast time-resolved or real-time on-line experiments, the demands on instrumentation are even more stringent. Potential improvements of synchrotron X-ray sources include the use of larger storage rings, refinement of insertion device technology, and optimization of the lattice functions at the source point.

5.2 High-Tech Materials
Smart Materials
Smart or intelligent materials refers to molecular systems with properties like biological systems in the living world. They have

the abilities of self-organization, self-reproduction, correction of defects and errors, and adaptation to external conditions. These far-off goals are interesting and challenging to materials scientists as well as solid-state physicists (Parker 1994, Cheetham and Day 1992, Proceeding of Ray Smallman Symposium 1995). Certainly, their research needs the structure data obtained by X-ray diffraction analysis.

Many fields of applied research need structurally based information for theoretical analysis and quality assurance, fields such as advanced ceramics, superconductors, optoelectronics, and molecular electronics. In molecular engineering, the tailoring of electronic and optical properties of organic and polymeric materials for specific applications certainly needs quality assurance in various stages. X-ray diffraction is a major technique-often the only technique-for these purposes.

For example, it is difficult to grow large, high-quality single crystals with the required processibility, mechanical integrity, and ambient stability. One alternative is by host-guest alignment. In order to make organic photoconductors, a highly specialized molecular species such as N,N-dimethylamine nitrostilbene (DANS) are doped (around 1%) into polymers such as polymethyl methacrylate (PMMA). Heat the polymer to its glass transition temperature T_g; then apply an electromagnetic field to align polar orientation and cool below T_g to freeze the alignment, thus imparting the ability to transport photogenerated charges (Fox 1997, Gambardella 1995, Koller 1994). X-ray diffraction will be used for molecule/crystal structure analysis and quality control.

Thin Films and Superlattices

Thin-film and multilayer structures will continue to grow because they give a greater range of mechanical, electrical, magnetic, optical, and structural properties relative to those in the bulk. Superlattices are multicomponent lamellar devices with alternating physical properties in the nanometer range. Many applications in molecular and biomolecular electronic devices are in sight. Increasingly powerful X-ray sources, advanced diffraction theory, and extensive computer power could solve the complex problems in extracting the structural parameters in thin films, multilayers, and superlattices (Vainshtein et al. 1994, Fewster 1996, Ito et al. 1994).

Interfacial Structures

The atomic structure of interfaces has received less study than either the surface structure or the bulk structure, because it is difficult experimentally and theoretically to extract information that is localized to the *buried interfaces*. In recent years, there have been many papers that examined the incommensurate structures in buried interfaces such as grain boundaries, encapsulated surfaces, and inter-diffusions by using various techniques including X-ray diffraction. Interface structures are very important for adhesion (Chung 1991), diffusion bonding, multi-layers, super-lattices, microelectronic devices, hetero-epitaxial oxide hetero-junctions, etc. The performance of many optoelectronic devices is sensitive to deviation from perfection, orientation, surface topography, and interface details (Howe 1997, Bringans et al. 1990). This field is fertile for applied research through ingenious design of X-ray diffraction experiments.

Structure-Property-Application Trinity

The study of structure-property relationships is basic research, the study of property-application relationships is applied research, and the study of structure-application relationships is theoretical research. There is a clear need to establish the relationships among the structure, property, and applications for fundamental understanding, theoretical formulation, and technological advances. Note that application is the real goal of materials research in industry. The characteristics of many optical or electronic devices are sensitive to structural imperfections, molecular orientation, surface topography, and interface details. The structural parameters accessible by X-ray techniques include dislocation generation, dopant segregation, layer thickness fluctuation, layer tilting, lattice distortion, stress level, strain profile, strain relaxation, porosity, topography, surface and interface microstructures, molecular alignment, twisting, local chemistry, long-range order, and so forth (Gambardella 1995, Koller 1994).

5.3 Molecular Devices: Molecular electronics is only a goal set for the not-too-distant future. It serves as a stimulus for combined efforts of solid-state physicists and molecular-design chemists. The molecular functional units can serve as switches, conductors, logic elements, and memory/storage devices. The attraction of these ideas is the hope of better performance and further miniaturization. X-ray diffraction is to play an important role in their structure-property studies and quality control. Some concepts and promising candidates are cited here (Smith 1996, Haken and Wolf 1995):

5.4 Emerging Topics
Computational Chemistry

Advanced mathematical operations and ever-increasing computer power may bring about a breakthrough in computational chemistry. The ab initio methods calculate the three-dimensional structure and fundamental properties of an isolated molecule at rest, that is, a single molecule at absolute zero temperature in vacuum. Their behaviors in solution are being approached by using the Monte Carlo methods. Given a large computer, these methods are easier and faster than crystal structure analysis, because they require neither the crystal (compound synthesis) nor the lab work (diffraction experiments). Nevertheless, computational chemistry reduces but never replaces experimental work. Theoretical results always need experimental data to prove or disprove their validity (Lipkowitz and Boyd 1997, Gans et al. 1996).

Holographic Atomic Imaging

Gabor proposed using holography to record the *amplitude* and *phase* of scattered wavefronts, then to derive the crystal structure at atomic resolution from such holograms (interference pattern). Recent reports indicate that the photoelectron or the fluorescent X-ray wavefronts do form an *electron hologram* or *X-ray hologram*. To improve their quality they need an X-ray source of highly coherent radiation at very short (subangstrom) wavelength and beam size. The fourth generation of synchrotron radiations can likely provide such an X-ray source. The reconstruction algorithms also need further improvement (Len et al. 1996).

Free-Electron Lasers

Free-electron lasers (FEL) are powerful sources of coherent electromagnetic radiation, operating in the microwave, infrared, and ultraviolet regions. Recent progress in FEL physics and technology makes it possible to extend into the X-ray region with peak power and brightness many orders of magnitude larger than those obtainable from other sources. There is much activity directed toward the future "fourth generation light sources," including FEL. In principle, the self-amplified spontaneous emission (SASE) mode of operation of FEL is expected to give brilliance similar to SR but much higher peak brilliance, because of the much shorter electron beam. The SASE-FEL technique promises very powerful coherent X-rays, but has yet to be demonstrated experimentally in the next century (Walker 1996).

Spallation Neutron Beam

Most of the molecular structures in solids have come from X-ray diffraction on single crystals. But neutron diffraction is particularly advantageous for hydrogen locations and magnetic structures. However, continuous neutron beams from nuclear reactor sources have the disadvantages of high cost and limited availability. Even those of "high flux" are of lower effective intensity than beams from X-ray tubes, partly because much of the beam is wasted in wavelength selection. The generation of a pulsed neutron beam by spallation of accelerated protons from a Li or Ta target enables much more efficient use of neutron diffraction (David 1987).

6. CONCLUSIONS

X-ray diffraction is a major tool of research. Over a dozen Nobel Prizes were awarded for its brilliant applications. It will continue to play a dominant role in the advancement of science and technology. Particularly in the frontier research of biomedical and engineering materials, new insights are essential for breakthroughs in rational drug/material designs. Molecular/crystal structure data will be indispensable to develop these crucial insights. The competing techniques for molecular/crystal structures are no match of the X-ray diffraction analysis, and hence there are plenty of opportunities as well as challenges ahead.

Because the X-ray diffraction techniques are becoming more and more sophisticated, and the arrays of modern instrumentation are awesome and mindboggling, one can easily be preoccupied and overwhelmed. As an ending note, a few thoughts and observations are presented here that might enhance our capability, broaden our career, or elevate our profession:

> **Teamwork**: In academia, independent research is a noble quality. In industry, however, teamwork takes precedence. A team of varied expertise produces synergy. A joint program is more versatile and likely more creative.
>
> **Open-minded**: X-ray diffraction or any other technique cannot achieve big by standing alone. Be open-minded and take advantage of the many complementary techniques to solve problems, find truths, create products, or open new avenues.
>
> **Big Picture:** Industry emphasizes applied research. Ap-

plied research is just basic research plus applications. Always think about applications, the fit into the big picture, no matter whether it is a technique, a theory, a material, or a product.

Extra mile: A bright banker is still a banker. A banker with unique knowledge of petroleum exploration is a bright star. Walk the extra mile. Learn the extra trick. Mind other people's business. The stroke of luck greatly favors the prepared mind.

7. REFERENCES

Bednorz, J. G.; and K. A. (1986). Possible high T_c superconductivity in the Ba-LA-Cu-O system. *Z. Physik, B64,* 189.

Bish, D. L.; and S. A. Howard. (1988). Quantitative mineralogical analysis using the Rietveld full-pattern fitting method. *J. Appl. Cryst.* 21, 86.

Bish, D. L.; and J. E. Post. Quantitative mineralogical analysis using the Rietveld full-pattern fitting method. (1993). *Am. Miner.,* 78, 932.

Blank, V. D.; et al. (1997). Physical properties of superhard and ultrahard fullerites, *Applied Physics,* Special Issue on Polymeric Fullerenes, *A64,* 247.

Bringans, C. D.; R. M. Feenstra; and J. M. Gibson, Eds. (1990). *Atomic Scale Structure of Interfaces,* Materials Research Society, Pittsburgh, PA.

Cheetham, A. K.; and P. Day, Eds. (1992). *Solid State Chem-*

istry Clarendon Press, Oxford. Chemia, D. S., and Zyss, J. (1987). *Nonlinear Optical Properties of Organic Molecules and Crystals*, Academic Press, New York.

Chu, C. W.; et al. (1997). Superconductivity up to 126 Kelvin in interstitially doped BaCaCuO, *Science*, 277, 1081.

Chung, F. H. (1974a). Quantitative interpretation of X-ray diffraction patterns of mixtures: I. Matrix flushing method, *J. Appl. Cryst.*, 7, 519.

Chung, F. H. (1974b). Quantitative interpretation of X-ray diffraction patterns of mixtures: II. Adiabatic principle, *J. Appl. Cryst.*, 7, 526.

Chung, F. H. (1974c). A new X-ray diffraction method for quantitative multi-component analysis, *Adv. X-Ray Anal.*, 17, 106.

Chung, F. H. (1975). Quantitative interpretation of X-ray diffraction patterns of mixtures: III. Simultaneous determination of a set of reference intensities, *J Appl. Cryst.*, 8, 17.

Chung, F. H. (1981). X-ray diffraction techniques and instrumentation, p. 151 in *Analytical Measurements and Instrumentation for Process and Pollution Control*, Eds. P. N. Chereminsinoff and H.J. Perlis Ann Arbor Science, Ann Arbor, MI.

Chung, F. H. (1989). Industrial applications of X-ray diffraction, *Am. Lab. 21(2)*, 144,

Chung, F. H. (1991). Unified theory and guidelines on adhesion, *J. Appl. Polymer Sci*, 42, 1319.

Conference Proceedings 389, AIP Press, New York.

Davis, B. L. (1986). *Reference Intensity Method of Quantitative X-ray Diffraction Analysis*, South Dakota School of Mines and Technology, Rapid City SD.

Davis, B. L.; and Smith, D. K. (1988). Powder Diffract. 3, 205.

David, W. I. F. (1987). The scope and possibilities of crystallography with pulsed neutrons, in *Chemical Crystallography with Pulsed Neutrons and Synchrotron X-rays*, eds. M. A. Carrondo and G. A. Jeffrey, Reidel, Boston.

Echegoyen, L.; and A. E. Kaifer, Eds. (1996). *Physical Supramolectilar Chemistry*, NATO ASI Series, Kluwer Academic, London.

Ewald, P. P., Ed. (1962). *Fifty Years of X-Ray Diffraction*, Int. Union of Cryst., Utrecht, Netherlands.

Fewster, P. F. (1996). X-ray analysis of thin films and multilayers, Rep. Prog. Phys., 59, 1339.

Fox, M. A. (1997). Beyond the traditional confines: Physical chemistry in the 21st century, Pure and Applied Chemistry *IUPAC*, 69, 235,

Gambardella, A. (1995). *Science and Innovation*, Cambridge University Press, Cambridge.

Gans, W.; A. Amann; and J. A. Boeyens, Eds. (1996). *Fundamental Principles of Molecular Modeling*, Plenum Press, New York.

Glusker, J. P.; M. Lewis; and M. Rossi (1994). *Crystal Structure Analysis for Chemists and Biologists*, VCH, New York.

Gorter, S.; and D. K. Smith (1995). *World Directory of Powder Programs*, Release 2.2, Leiden University, CPD.

Haken, H.; and H. C. Wolf (1995). *Molecular Physics and Elements of Quantum Chemistry*, Springer, New York.

Hauptmann, H. (1995). *Acta Crystallogr.*, Sect. B, 51. 416.

Hill, R. J. (1991). *Powder Diffract.*, 6, 74.

Hill, R. J.; and C. J. Howard (1987). *J. Appl. Cryst.* 20, 467.

Horn, A. S.; and C. J. De Ranter (1984). *X-Ray Crystallography and Drug Action*, Clarendon Press, Oxford.

Howe, J. M. (1997). *Interfaces in Materials.* John Wiley, New York.

Ito, H.; S. Tagawa; and K. Horie, Eds. (1994). *Polymeric Materials for Microelectronics Applications*, ACS, Washington, D.C.

Jenkins, R.; and R. L. Snyder (1995). *Introduction to X-Ray Powder Difflactometry* Wiley & Sons, New York.

Johnson, R. L.; H. S. Backing; and B. F. Sonntag (1996). *X-Ray and Inner-Shell Processes*, AIP Conference Proceedings 389, AIP Press, New York.

Karle, J. (1995). *Acta Crystallogr.*, Sect. B, 51.

Kenway, P. K.; et al., Eds. (1992). *X-Ray Optics and Microanalysis*, Inst. of Physics, Conference Series Number 130, Philadelphia.

Klug, H. P.; and L. E. Alexander (1974). *X-ray Diffraction Procedures for Polycrystalline and Amorphous Materials*, 2nd ed., Wiley, New York.

Koller, A. (1994). *Structure and Properties of Ceramics*, Elsevier, New York.

Kourounakis, P. N.; and E. Rekka (1994). *Advanced Drug Design and Development*, Ellis Horwood, New York.

Kurzydlowski, K. J.; and B. Ralph (1995). *The Quantitative Description of Microstructure of Materials*, CRC Press, New York.

Ladd, M. F. C.; and R. A. Palmer (1994). *Structure Determination by X-Ray Crystallography*, 3rd ed., Plenum Press, New York.

Langford, J. I.; and D. Louer (1996). Powder Diffraction, *Rep. Prog. Phys.*, 59, 131.

Len, P. M.; C. S. Fadley; and G. Materlik (1996). Atomic holography with election and X-rays, ATP Conference Proceedings 389 on X-Ray and Inner-Shell Process, New York.

Lipkowitz, K. B.; and D. B. Boyd (1997). *Reviews in Computational Chem. 10*, VCH, New York.

Lovesey, S. W.; and S. P. Collins (1996). *X-Ray Scattering and Absorption by Magnetic Materials*, Clarendon Press, Oxford.

Nogrady, T. (1988). *Medicinial Chemistry: A Biochemical Approach*, 2nd ed., Oxford University Press.

Parker, G. (1994). *Introductory Semiconductor Device Physics*, Prentice Hall, New York.

Powell, H. R. (1996). X-Ray analysis of single crystals, in *Annual Reports on the Progress of Chemistry* Vol. 92, Section C,

Royal Soc. of Chem.

Proceedings of EPDIC III (1993). *Mat. Sci. Forum.*, *Vol. 3*, Preface.

Proceedings of EPDIC IV (1996). *Mat. Sci. Forum.*, *Vol. 4*. p. 325.

Proceedings of Ray Smallman Symposium (1995). *Towards the Millennium: A Materials Perspective*, *Institute of Materials.*, University of Birmingham, Birmingham.

Rao, C. N. R. (1997). *New Directions in Solid State Chemistry* 2nd ed., Cambridge University Press, Cambridge.

Rietveld, H. M. (1967). *Acta Cryst.*, 22, 151.

Rietveld, H. M. (1969). *J. Appl. Cryst*, 2, 65.

Smith, C. G. (1996). Low-dimensional quantum devices, *Rep. Prog. Phys.*, 59, 235.

Smith, D. K. (1992). Particle statistics and whole-pattern methods in quantitative X-ray powder diffraction analysis, *Adv. X-ray Anal.*, 35, 1.

Snyder, R. L. (1992). *Powder Diffract*, 7, 186.

Taylor, J. C.; and L. P. Aldridge (1993). *Powder Diffract.* 8, 138.

Terminello, L. J.; S. M. Mini; H. Ade; and D. L. Perry, Eds. (1996). *Applications of Synchrotron Radiation Techniques to Materials Science III*. MRS Proceeding 437, Pittsburgh, PA.

Tsoucaris, G.; J. L. Atwood; and J. Lipkowski, Eds. (1995*). Crystallography of Supramolecular Compounds*, NATO ASI Series, Kluwer Academic, London.

Ulman, A. (1991). *Ultrathin Organic Films*, Academic Press,

New York.

Vainshtein, B. K.; V. M. Fridkin; and V. L. Indenbom (1994). *Structure of Crystals*, 2nd ed., Springer-Verlag, New York.

Vogtle, F. (1991). *Supramolecular Chemistry*, John Wiley & Sons, New York.

Walker, R. P. (1996). Synchrotron radiation and free-electron laser X-ray Sources, AIP Conference Proceedings 389 on X-Ray and Inner-Shell Process, New York.

Woolfson, M. M. (1997). *An Introduction to X-ray crystallography*, 2nd ed., Cambridge University Press, Cambridge.

Woolfson, M. M.; and H. F. Fan (1995). *Physical and Non-Physical Methods of Solving Crystal Structures*, Cambridge University Press, Cambridge.

Wright, J. D. (1995). *Molecular Crystals*, 2d ed., Cambridge University Press, Cambridge.

Wu, M. K.; et al. (1987). *Phys. Rev. Lett.*, 58, 908.

Wulff, M.; et al. (1996). Single-pulse Laue diffraction, AIP Conference Proceedings 389 on X-Ray and Inner-Shell Process, New York.

Young, R. A., Ed. (1993). *The Rietveld Method;* Int. Union of Cryst., Oxford University Press, Chester, UK.

Zevin, L. S.; and G. Kimmel (1995). *Quantitative X-Ray Diffratometry*, Springer, New York.

13.

The Principles of Diffraction Analysis

> The richest man is not he has the most, but who needs the least.
> — Unkown

1. ABSTRACT

Roentgen discovered X-rays in 1895. Laue proofed the wave nature of X-rays and designed the first XRD experiment in 1912. The application of XRD developed slowly over the ensuing 20 years, first on the structures of common crystals. Powder diffraction for chemical analysis started in the middle 1930s. The first set of Powder Diffraction File was published in 1941. Single crystal structure analysis retains the three dimensional spatial relationship of the diffracted X-rays. The position and intensity data determine the space group, unit cell dimensions, density, and coordinates of atoms in unit cell. In powder diffraction analysis, the three dimensional spatial imformation is compressed to one dimensional powder pattern crowded with overlapping peaks. The peak position and intensity are used for qualitative and quantitative chemical analyses. The peak profile refinement reveals micro properties including particle size, preferred orientation, stress, and others. In general, single crystal structure analysis is elaborate, hence esoteric in academia, while powder mixture phase analysis is simpler, hence popular in industry. The history and fundamentals of XRD are briefly summarized.

2. HISTORY OF POWDER DIFFRACTION ANALYSIS

Following the discovery of X-rays by Roentgen in 1895 and the proof that X-rays do have wave properties and diffract from a periodic atomic array by von Laue and his students in 1912-1913 analytical applications of X-ray diffraction developed slowly over the next 20 years. Most of the earliest efforts were directed at the solution of crystal structures of common phases. Debye and Scherrer (1916) and Hull (1917) suggested that powder diffraction patterns could be used for identification and quantification of crystalline compounds, and the basic Debye-Scherrer camera was perfected (Debye and Scherrer, 1917). However, because most of the early developments were directed toward solving single-crystal structures, it was really the middle 1930s before the powder diffraction method began to attract followers with the publication of the procedure of Hanawalt and Rinn (1936) and the database of patterns by Hanawalt, Rinn, and Frevel (1938). Following these publications and the conversion of the data sets into the first set of the Powder Diffraction File in 1941, phase identification applications expanded and the modern counter diffractometer was developed by Parrish, Hamacher, and Lowitzsch (1954). Since this time, powder diffraction methods have evolved rapidly to include many aspects of materials characterization from determining the periodicity and symmetry, solving the crystal structure, phase quantification, and measuring the physical properties of crystals, either unconstrained or without a product.

It is really remarkable that an analytical method with its roots in the 1930s and earlier is still one of the most useful of all methods in use today, but it is true. Perhaps the only methods that can provide equivalent important information about crystalline materials are elemental chemical analysis and the optical microscope. Of all the spectroscopies, which include infrared, visible, Raman, ultraviolet, etc., X-ray diffraction (XRD) is the only one where all the members of the data set may be related to every other member through fundamental equations. The relationships allow the use of the data to prove many points with considerable confidence including phase purity and the properties of the crystal structure. Not only is the XRD method still in use, but XRD is still advancing rapidly in the types of problems it can attack and the accuracies that may be achieved.

Phase identification was one of the first applications to grow to a useful level, Hanawalt and Rinn (1936) published the first systematic method for numerical pattern recognition for matching experimental and reference X-ray powder diffraction patterns, Hanawalt, Rinn, and Frevel (1938) published a set of 1000 reference powder diffraction patterns with the approval and support of the Dow Chemical Company. At this same time, a Joint Committee on Chemical Analysis by X-ray Diffraction Methods was formed under the auspices of the National Research Council, Canada (NRCC) and Committee E-4 of the American Society for Testing and Materials (ASTM). Subsequently, the Institute of Physics (Britain) and the American Society of X-ray and Electron Diffraction (ASXRED; now the American Crystallographic Association) also participated. This group, which has evolved into the International

Centre for Diffraction Data (ICDD), was responsible for the first publication of the Powder Diffraction File (PDF). The 1000 diffraction patterns, previously published by the Dow Chemical Company (Hanawalt, Rinn, and Frevel, 1938) were republished in 1941 as Set 1 of the PDF. The Institute of Physics sponsored the addition of 500 patterns primarily from the British Museum as Set 2 in 1945, and another set of 1000 patterns supplied from several sources, including the General Electric Company, became Set 3 in 1949. Because of the multiplicity of data sets with different data for the same material in these early sets, an Associateship was sponsored at the then National Bureau of Standards (now the Nation Institutes for Science and Technology, NIST). This group reran many of the phases under well-defined, controlled conditions using then modern equipment, and supplied new data sets for the PDF. The first of these patterns appeared in Set 5 in 1955. Since 1955, a new set of the PDF has been issued each year, and the PDF now contains over 110,000 data sets. For a history of the PDF, see Hanawalt (1983) and Smith and Jenkins (1989).

With the ability to measure diffraction data more accurately using the counter diffractometer, quantitative applications of X-ray diffraction, primarily starting from Alexander and Klug (1948), have also advanced to a sound theoretical and practical level. Their theory, based on converting volume fractions of phases in a mixture into weight fractions using attenuation coefficients, has led the development of this application into a useful, almost routine, procedure with modern computer-controlled equipment and computer-based data processing. An excellent text on this topic has been prepared by Zevin and Kimmel (1995).

Other major applications of diffraction analysis include following phase changes under nonambient conditions and atmospheres. Both low- and high-temperature and low-pressure studies are possible with minor modifications of the standard diffractometer or Guinier camera. Many specialized attachments have been developed and marketed for operating at extreme ranges including temperatures up to 3100 K and down to 2 K and vacuums to 10^{-8} torr. Pressure studies to a few atmospheres can be done in diffractometer attachments, but high-pressure studies in the GPa range require very specialized devices and are usually used only with very-high-intensity sources such as the synchrotron.

With the introduction of computer-controlled data-collecting devices, the diffraction traces can now be obtained entirely in digitized form. Mathematical methods may be applied directly for many applications, including profile fitting to obtain peak positions and resolve overlapped peaks; full-pattern fitting to yield accurate cell data, structure refinements, and even crystal structure solutions; and extraction of profile parameters to determine crystallite properties such as crystallite size and strain. New techniques still are being developed. Part of the role of this book is to illustrate many of these new methods and to direct the reader to sources of additional information.

3. FUNDAMENTALS OF X-RAY DIFFRACTION

Diffraction of a beam of X-rays by a crystalline material is a process of scattering of the beam by the electrons associated with the atoms in any crystal, and interference of these scattered X-rays because of the periodic arrangement of the atoms in the crystal and its symmetry. This process leads to the enhancement of the intensity of the scattered radiation in certain directions due to constructive interference controlled by the periodicity and symmetry and destructive interference in all other directions. The measurement of the directions of the constructive beams allows one to determine the fundamental properties of the crystalline state, the magnitude of the fundamental unit cell of the crystal and its symmetry.

The directions of the diffracted X-rays from a single crystal may be measured in several ways that lead to a full three-dimensional description of the directionality and direct determination of the information necessary to decipher the unit cell data. In the powder method, the three-dimensional information is compressed to one dimension of the linear powder pattern, making it somewhat more difficult to decipher the unit cell data. However, with the very good data now obtainable on modern powder diffractometers, Guinier cameras, or the high-resolution synchrotron devices and the availability of several computer programs to analyze or evaluate the data, the same success is possible with the linear powder data. These methods are now incorporated in every data analysis package supplied with the new computer-controlled diffractometers used in industry as well as other laboratories.

The intensities of the diffracted rays are controlled by the types of atoms and their arrangement in the unit cell of the crystal. Atoms with more electrons scatter more strongly than low-Z atoms. This difference in scattering power leads to marked differences in the intensities of the various diffracted rays. Measuring the intensities of the diffracted rays provides data on the atom.

arrangements in the crystal structure. Single-crystal methods have been used since the early days of X-ray diffraction studies to determine the crystal structures of many materials; however, many of the simpler structures (e.g., Bragg, 1913) and several materials unobtainable as single crystals (Zacharisen, 1949) were also solved by powder methods. More recently, methods and computer programs have been developed to solve more complex structures from powder data. These methods are the subject of one chapter by Louer. These new methods have made powder diffraction a very powerful method for analyzing new materials where no structure data are available on the material or any other material that may be isostructural or otherwise partially related. Once a structure has been revealed by these methods, the structure may be further refined by the Rietveld technique to reveal even more information on the crystal rather than just the atomic coordinates (Young, 1993). Crystal structure determination is now available to industry on a relative routine basis.

With the introduction of the computer-controlled diffractometer, the diffraction pattern is digitized. This digitization has allowed even the profile shapes of the diffraction peaks in a powder diffraction pattern to be analyzed. Analytical profiles have been improved from the early Gaussian and Cauchy (Lorentzian) models to models that use weighted combinations of these shapes along with asymmetry parameters, full-width at half maximum (FWHM) functions for each shape, and other analytical factors. These new models allow more direct interpretations of the resulting data to be related to properties of the crystal such as crystallite size and strain. This field of accurate profile fitting of the diffraction data has opened up many new directions for diffraction analysis important to the industrial community. More detail on these three types of data in the diffraction pattern and their uses are described later in this chapter.

Powder diffraction has become one of the most important material characterization techniques available to industry. It is relatively simple to obtain very good experimental data, and the computer programs are readily available to do the desired analysis right in the laboratory. For very special experiments requiring more sophisticated equipment, users can arrange to use the synchrotron and neutron sources by following procedures for submitting proposals and carrying out the experiments at the many sources available around the world. Powder diffraction is still an advancing science in the field of new materials and their applications. Industry is where these new materials are developed into products, and the data produced by diffraction methods provides the necessary information to improve these products.

4. SINGLE CRYSTAL VERSUS POWDR DIFFRACTION

The first diffraction experiments were actually done on single crystals. Knipping and Friedrich used a very unlikely candidate, $CuSO_4 \cdot 5H_2O$, for the first experimental tests to prove von Laue's theory of diffraction. The following year Bragg examined a powdered sample of NaCl and used the diffraction data from a primitive diffractometer to solve its structure, establishing X-ray diffraction as a real experimental science for understanding the structure of matter. The theories of crystallography from the 1800s predicted the periodic nature of crystal structures, and these experiments provided the confirmation needed for the field of X-ray crystallography to take its place as an important field of materials science and as an effective analytical technique. The early experimental X-ray diffraction applications emphasized examining single crystals, with only a few workers realizing its value for the examination of powders. This early history is described in Ewald (1962).

Single-crystal methods are primarily directed toward determining the crystal periodicity and symmetry of the crystal and solving the arrangement of atoms in the material, because this information is difficult to obtain from powder experiments. The advantage of the single-crystal experiments for collecting diffraction information is that they retain the three-dimensional spatial relationships of the diffracted rays so important in the determination of the diffraction symmetry and resolving intensity data necessary for the solution of the crystal structure. Single-crystal cameras, which are quite different from the powder devices, evolved in the 1930s to the 1950s to provide the resolution and collection of the diffracted intensities from all possible diffracted rays provided by a moving single crystal in a collimated X-ray beam. More recently, recording devices have diversified to record in reasonable data collection times the large amounts of data necessary to determine the structure of large molecules and proteins. Developments in the theory of single-crystal diffraction allowed the use of the set of diffracted intensity measurement to be analyzed directly to yield the crystal structure. The advent of computers as number processors in the 1950s allowed the large amount of data in a set of intensities to be reduced to vector maps, electron-density maps, and refinement of models.

It is not the purpose of this book to consider the single-crystal methods in any detail. Rapid employment of computers for the collection and analysis of the data from the 1960s onward along with developments in the theory of direct methods analysis have now made the solution of many crystal structures routine. There are several companies that specialize in structure solutions, and these companies may be contracted for such studies at prices that are considerably less than those at which most in-house diffraction laboratories can do the same studies. Several modern textbooks on the single-crystal method exist that cover the subject thoroughly and are to be recommended, including Glusker and Trueblood (1985) and Stout and Jensen (1989).

Single-crystal diffraction studies are not limited to crystal structure analysis. Diffraction topography is a large field that has provided much information on the perfection of crystals used in industry as integral parts of devices. Examples include crystals used in sensing and control devices, substrates for electronic components, tools and dies, turbine blades, and many other applications. Diffraction topography has allowed the direct observation of dislocations and other defects and the modification of these defects under changes in conditions such as heating and stresses. This important field also will not be covered in this book, but it is covered well in Tanner (1976).

4.1 Types of Information in Powder Diffraction Data

In a powder diffraction pattern, the three-dimensional diffraction information is condensed to a single direction in diffraction space. In such a pattern there is a superposition of information that makes extraction of the information more complex than for the single-crystal experiment. Regardless, modern computer-controlled diffractometers and high-intensity, narrow-profile sources are allowing this information to be collected and reduced to usable data by routine procedures. There are three types of information in any diffraction pattern: geometrical, structural, and physical state. These types are reflected in the diffraction data, respectively, as the position of the diffracted ray in space, the intensity of the diffracted ray, and the angular deviation of the intensity from the ideal diffraction angle. In a powder diffraction pattern, these three types reduce to the angular position of the diffraction peak, the integrated intensity of the diffraction peak, and the profile of the peak. The position of the peak is controlled by the Bragg equation. The intensity is controlled by the powder pattern power equation. The profile is a convolution of all the effects that influence the angular intensity distribution including the sample function. Tables 1 to 3 illustrate the three types of information, with their source, functionality, and applications. Powder diffraction techniques are designed to measure one or more of these types of information to solve specific problems of industry and research.

Table 1 Information in a Diffraction Pattern: Crystal Geometry, Position of Diffraction Peaks

Controlled by:
  Wavelength of X-rays
  Instrument aberrations
  Periodicity of crystal structure
  Functional dependence:
  $\lambda = 2d_{hkl} \sin \theta$
Applications:
  Characterization of lattice properties—indexing of data sets
  Determination of crystal symmetry
  Chemical analysis in solid solutions
  Phase identification (Fink method)
  Proof of phase purity
  Thermal expansion
  Compressibility
  Phase changes

Table 2 Information in a Diffraction Pattern: Crystal Structure, Intensities of Diffraction Peaks

Controlled by:
  Source intensity
  Instrument aberrations
Functional dependence:
  $I = KLPTA|F|^2$
  $F = \sum f_n e^{2\pi(hx_n + ky_n + lz_n)}$
Applications:
  Phase identification (Hanawalt method)
  Crystal structure determination
  Quantitative phase analysis
  Percent crystallinity
  Element substitutions
  Order–disorder
  Texture

Table 3 Information in a Diffraction Pattern: Crystallite Perfection, Shapes of Diffraction Peaks

---

Controlled by:
  Source intensity distribution
  Instrument aberrations
  Sample effects
Functional dependence:
  $F(x) = \int G(y) H(y - x)\, dy$
Applications:
  Crystallite size
  Crystallite strain
  Other crystallinity parameters
  Chemical inhomogeneity
  Stacking faults and their arrangement
  Concentrations of other defects

The positions of diffraction peaks are now easy to measure to accuracies of $0.01°2\theta$, which allow lattice parameters to be determined to 1 part in 5×10^4. Good instrument calibration coupled with the use of an internal d-spacing standard now make such measurement routine with modern computer-controlled diffractometers. The intensities of diffraction peaks are very dependent on the nature of the specimen used, but accuracies of $\pm 10\%$ can usually be achieved with a little effort and $\pm 1\%$ with considerable care during specimen preparation (Buhrke, Jenkins, and Smith, 1998). Most of the techniques of powder diffraction use either the **d**-value information or the **d**-value information combined with the intensity information to solve many different problems of industry and research. Many examples are found in the chapters of this book.

Diffraction profiles may now be recorded digitally, which allows computer analysis to fit different analytical functions, aiding in their interpretation. The complication of profile analysis is that it is the sample contribution to the profile that is required, but the experimental profile is a convolution of the contributions from the source and instrument as well. Advanced mathematical techniques are required to extract the sample contribution to the profile, but these techniques are being improved rapidly. Coupled with the use of high-energy, narrow source profile radiation and parallel-beam instruments, the profile information is becoming easier to obtain and very useful, as illustrated in several of the chapters in this book.

4.2 Crystal Structure Analysis

Crystal structure analysis is the use of the position and intensity information from a diffraction experiment to solve for the periodicity, symmetry, and arrangement of the atoms in crystals of a material. It is the arrangement of atoms that controls the intensities observed in any diffraction experiment, and the mathematical dependence is well known. It is easy to calculate the intensities once the crystal structure is known, but it is not easy to go in the reverse direction from a set of collected intensities to the crystal structure. The problem is that the structure is related to the amplitude of scattering, and the data collected reflect the intensity, which is the square of the amplitude. Although the intensity is real and always positive, the amplitude may be positive, negative, or even complex imaginary, so it is not routinely possible to use the square root of the measured intensity to calculate the crystal structure. Many elaborate procedures and advanced theories called direct procedures have been developed for the single-crystal community to solve this problem and are beyond the scope of this book. However, these methods are now being extended into the realm of powder diffraction, and at least one chapter in this book is devoted to the use of powder data to solve even moderately complex structures today.

One procedure that is extensively used with powder data is Rietveld refinement (Rietveld, 1967, 1969), where a known crystal structure may be improved using the experimental data compared with the calculated diffraction trace to adjust parameters to yield a better data fit. This method has advanced rapidly to an almost routine procedure for an experienced diffractionist to obtain many different types of information about the specimen not just the refined crystal structure (Young, 1993). If the calculations are accomplished in a logical sequence, the refinement can produce structural parameters that may be more meaningful with respect to the structural state of the material that produced the diffraction data. It is not unreasonable that some aspects of materials may be different in the powdered state compared to the single crystal state. The Rietveld method starts with a model, usually based on a reported single-crystal structure, and refines the structural parameters along with pattern and profile parameters to provide a more accurate description of the material in the powdered state. Several chapters in this book illustrate applications of the Rietveld method.

5. TYPES OF RADIATION USED FOR DIFFRACTION ANALYSIS

This book is primarily directed toward the use of conventional X-rays in the diffraction experiments; however, it is not restricted to X-ray applications. The main requirement for radiation to be useful for diffraction experiments is that the wavelength be in a range such that the diffraction effects can be recorded by the equipment in use. Useful ranges are around 0.03Å for electron diffraction and 0.3 to 6Å for X-ray and neutron diffraction. The most common values for neutron diffraction are around 1.0Å for neutron beams and 1.5406Å for $CuK_{\alpha 1}$ X-rays. The choice of radiation for an experiment depends on many factors, including availability of the alternate choices, economics of the experiment, type of material under study, and problem to be solved. Electron diffraction is available at any material laboratory with an electron microscope. Many neutron sources are available through the world where arrangements may be made to run samples. Pulsed neutron sources are more common now than thermal sources because of the higher intensity that can be generated. In addition, high-energy beams in the X-ray wavelength range are available at synchrotron facilities around the world. Neutron and synchrotron X-ray wavelengths can be tunable when nonochromators are used to select the wavelengths. Several chapters in this book discuss the use of neutron diffraction and electron diffraction in industrial settings.

6. SAFETY AND REGULATIONS IN THE DIFFRACTION LABORATORY

Because the radiation used for diffraction experiments is highly energetic and can cause considerable damage to human tissues, it is extremely important to design equipment and experiments with safety in mind. The primary consideration is to confine the radiation to a volume where operators cannot encroach during active experimentation. Federal and state regulations usually dictate stringent limitations on the amount of radiation that may be detected outside the enclosure confining the beam. Individual state regulations are usually more restrictive, so any industry establishing a diffraction laboratory will have to work with both the suppliers of the equipment and the state regulators to guarantee a safe environment for personnel. The requirements usually include radiation safety training for all personnel using the equipment and radiation monitoring of personnel and equipment. Safety is a necessary consideration in all diffraction studies, whether they are being done in an industrial, university, or government laboratory, and is a significant cost in all diffraction analyses.

One of the effects of safely regulations is the limitation on the use of older equipment, particularly generators of radiation. Regulations over the years have become more restrictive, and older equipment rarely meets the modern criteria without considerable modification. These regulations usually restrict industrial laboratories to purchasing new equipment; however, it is possible for older equipment to be retrofitted with the necessary safety features for companies on a low budget. The major modification is usually a total enclosure of the radiation chamber with doors that are interlocked so that accidental opening will deactivate the generation of the X-ray beam.

7. REFERENCES

Alexander, L. E.; and H. P. Klug (1948). X-ray diffraction analysis of crystalline dusts. *Anal. Chem.* 20:886-894.

Barrett, C. S.; and T. B. Massalski (1992). *Structure of Metals* (3rd ed.). Pergamon, New York. Bragg, W. H. (1913). Crystal structures of NaCl, KO, KBr, and KI. *Royal Society Proceedings*, June.

Buhrke, V. E.; R. Jenkins; and D. K. Smith (1998). *A Practical Guide for the Preparation of Specimens for X-Ray Fluorescence and X-Ray Diffraction Analysis*. Wiley-VCH, New York.

Cullity, B. D. (1978). *Elements of X-ray Diffraction*. Addison-Wesley, Reading, MA.

Debye, P.; and P. Scherrer (1916). Interferenzen an regellos orientierten teilchen im rontgenlight [Interference of irregularly oriented particles in X-rays]. *Phys. Ziet*, 17:277-283.

Debye, P.; and P. Scherrer (1917). X-ray interference produced by irregularly oriented particles (III). Constitution of graphite and amorphous *C. Phys. Zeit.* 18:291-301.

Ewald, P. P., Ed. (1962). *Fifty Years of X-ray Diffraction*. International Union of Crystallography, N.V. A. Oosthoek's Uitgeversmaatschappij, Utrecht.

Glusker, J. P.; and K. Trueblood (1985). *X-Ray Structure Analysis*: A Primer. Oxford University Press, New York.

Hanawalt, T. D.; and H. W. Rinn (1936). Identification of crystalline materials. *Ind. Eng. Chem. Anal*, Ed. 8:244-247.

Hanawalt, J. D.; H. W. Rinn; and L. K. Frevel (1938). Chemical analysis by X-ray diffraction-Classification and use of X-ray diffraction patterns. *Ind. Eng. Chem. Anal. Ed.* 10:457-512.

Hanawalt, J. D. (1983). History of the powder diffraction file (PDF). In *Crystallography in North America*, J. P. Glusker and D. McLachlan, Jr., Eds., pp. 215-219. American Crystallography Association, Buffalo, NY

Hull, A. W. (1917). A new method of X-ray crystal analysis. *Phys. Rev.* 10:661-696.

Jenkins, R.; and R. L. Snyder (1995). *Introduction to X-Ray Powder Diffractometry*. J. Wiley & Sons, New York.

Klug, H. P.; and L. E. Alexander (1974). *X-Ray Diffraction Procedures for Polycrystalline and Amorphous Materials*. J. Wiley & Sons, New York.

Parrish, W.; E. A. Hamacher; and K. Lowitzsch (1954). The Norelco X-ray diffractometer. *Philips Tech. Rev*: 16:123-133.

Rietveld, H. (1967). Line profiles of neutron powder diffraction peaks for structure refinement. *Acta Crystallogr.* 22:151-152.

Rietveld, H. (1969). A profile refinement procedure for nuclear and magnetic structures. *J. Appl. Crystallog* 2:65-71.

Smith, D. K.; and R. Jenkins (1989). The Powder Diffraction File: Past, present, and future. *Rigaku J.* 6:3-14.

Stout, G. H.; and L. H. Jensen (1989). *X-Ray Structure Determination*. Wiley Interscience, New York.

Tanner, B. K. (1976). *X-Ray Diffraction Topography*: Pergamon Press, Oxford, UK.

Zacharisen, W. H. (1949). Crystal chemical studies of the 5f-series of elements. VII. The crystal structure of Ce_2O_2S, La_2O_2S and Pu_2O_2S. Acta Crystallogr, 2:60-62.

Zevin, L. S.; and G. Kimmel (1995). *Quantitative X-Ray Diffractometry* Springer, New York.

14.

The Practice of Diffraction Analysis

> Formal education will make you a living. Self education will make you a fortune.
> — Jim Ramsey

1. ABSTRACT

Any radiations with wavelength in the order of atomic distance can be used to study crystal structure by diffraction experiments. There are three types of radiation used in diffraction studies: X-ray, Neutron, and electron. Neutron diffraction enhances the scattering contribution of lighter atoms: hydrogen (as deuterium), carbon, nitrogen, and oxygen, while X-ray radiation favors atoms heahier than oxygen. Electron beam has high absorption coefficient, it must be used in vacuum for surface structure studies. XRD is least expensive, hence most popular. As to quantitative XRD analysis of mixtures, the Internal Standard method was developed in 1936 by Clark and Reynolds. The Matrix Flushing method was developed in 1974 by Frank Chung. The Rietveld Refinement method for mixtures was developed in the 1980s by Bish, Howard, Hill, Smith et al. The Matrix Flushing method is simple, fast, and least expensive, hence most popular in industry worldwide. The sources of radiation, instrumentation, databases, and lab practice are reviewed.

2. INTRODUCTION

The purpose of this chapter is not to be a text on the principles and practices of diffraction analysis. Rather its goal is to illustrate the many techniques that are available to the industrial diffractionist and to point the reader to sources of more detailed information on each topic. Further information on many topics can be found in the chapters in this book. The literature on diffraction analysis is voluminous, and any effort to be even partially comprehensive is beset with failure.

There are several key textbooks that cover the applications of powder diffraction quite adequately. However, there are also a few main sources of concentrated information other than textbooks. They include the *Journal of Applied Crystallography*; *Powder Diffraction*, the proceedings of the Dever Analytical X-Ray Conference: *Advances in X-Ray Analysis*; the proceedings of the European Powder Diffraction International Conference (EPDIC); and the proceedings of the Australian X-Ray Analytical Association (AXAA).

3. SOURCES OF RADIATION

There are three major types of radiation used in diffraction studies: X-ray, neutron, and electron. Although the aim of this book is to concentrate on the use of X-rays, some chapters are devoted to neutron diffraction applications and one is devoted to electron diffraction. It is important to consider all three sources, as they augment each other rather than duplicate.

X-ray sources are easily setup in the laboratory, but most sources are of limited power and selected wavelength. The power level is usually less than 5 kW. This relatively low power level is adequate for most studies and serves the laboratory facility quite well. Modern instrumentation, including focusing diffracted beam monochromators and devices that take advantage of focusing in parallel beam optics, has improved the efficiency of intensity measurement. Primary beam monochromators are used to pass only the Kα_1 component of the primary beam on some diffractometers, minimizing fluorescent effects from the specmen and resulting in patterns with very low background and improved resolution. Rotating-anode high-intensity laboratory sources are available with power levels over 100 kW, but systems above 35 kW are usually limited by high maintenance problems and short continuous use times. In spite of the maintenance difficulties, these high-intensity laboratory-based sources have proved very useful for experiments that were impossible without them.

Higher power beams that have the advantage of wavelength tunability in the X-ray wave-length range may be obtained on a synchrotron. The disadvantage of synchrotron sources is that for most users the experiment must be moved from the laboratory to the source installation, and the experiments require scheduling at arranged times suitable to the operations of the source. Synchrotron sources are located at national laboratories and high-energy physics sites associated with universities throughout the world. Because these sites are supported by government funding, access to experiment time is obtained either by supporting instrumentation on a dedicated beam line (which has already been done by many industries) or requesting time through an established procedure for each facility. Regardless of the extra work to run an experiment, the effort is usually worth the trouble. The higher intensity beam allows experiments on extremely tiny specimens, specimens that only yield weak intensity responses for the critical part of the information, and experiments where rapid time resolution or ultra-high spatial resolution is important. All of these conditions are ones that limit applications of powder diffraction in the laboratory on conventional equipment.

Neutron sources are becoming readily accessible now with beam intensities sufficient for almost any diffraction experiment. Sources are located at national laboratories throughout the world and are accessible through application to an oversight committee that reviews the proposed experiment and schedules the time similar to the synchrotron facilities. There are several major differences for neutrons compared to X-rays for diffraction experiments that provide information that augments the information obtained by the X-ray experiment. The neutron scattering mechanism is quite different compared to X-rays, such that elements adjacent in the periodic table may have very different scattering power, and low-Z elements scatter with the same general magnitude as the high-Z elements. Neutrons are sensitive to the magnetic properties of the atoms. In addition, neutrons are not significantly attenuated by the specimen, so relatively large specimens are used, allowing more particles in the beam and a gain in particle statistics as well as intensity. Although the early diffraction devices used with neutron sources employed constant-wavelength beams obtained by monochromating the reactor output and measuring angle-dependent diffraction, modern facilities use very high pulsed sources and measure the energy differences (time-of-flight) in the diffracted beam to obtain the desired information. This topic is thoroughly discussed in the chapter on neutron diffraction.

Diffraction by electron beams is available usually as an option on transmission electron microscopes (TEM). Compared to X-ray diffraction, it is more rarely applied for routine problems, but it does have some special applications. Electron beams are usually in the very short wavelength region compared to X-rays. Also they are quickly attenuated by any specimen, so very thin specimens are required that are stable in a high vacuum. It is usually the specimen preparation that limits the application of electron diffraction. Actually, the TEM can take advantage of diffraction effects not only to generate classical diffraction patterns, both spot and ring patterns, but also to image the diffracting ray, revealing the portions of the crystal contributing to the beam and dynamical diffraction effects. This imaging is similar to diffraction topography using an X-ray beam. Dynamical diffraction is much more prevalent in electron diffraction because the efficiency of diffraction is very high compared to X-ray diffraction, where it is less than 0.1%. Applications of electron diffraction are covered in another chapter.

4. DIFFRACTION INSTRMENTATION

Diffraction instrumentation for powder applications has evolved considerably from the days of the Debye-Scherrer camera and the Parrish diffractometer. Although these instruments are still in use in their essentially original configuration, many new modifications have been added to enhance the applications of each device. The Guinier camera was perfected in Europe in the 1940s and has become the dominant laboratory device in many laboratories in that part of the world. More recently, the diffractometer has also become common in Europe. Diffractometers have evolved from the Bragg-Brentano fixed-source focusing geometry to:

- Theta-theta fixed-specimen Bragg-Brentano moving-source focusing geometry, useful for liquids, hot-stage studies, special environmental attachments, and general diffractometry.
- Fixed-source Seeman-Bohlin geometry for grazing incidence devices, used principally for thin-film specimens.
- Incident-beam focusing optics following Guinier geometry, for obtaining excellent resolution.
- Incident-beam defocusing optics, producing parallel beams in Debye-Scherrer geometry and also producing excellent resolution.

In addition, the Debye-Scherrer camera has been improved with the moving-specimen Gandolfi attachment that allows randomization of very small specimens.

Single-crystal cameras have also evolved considerably over the years, although the major innovations occurred in the 1950s and 1960s when the Weissenberg and Buerger cameras were perfected. Film was an essential component of these cameras, and the intensities were read visually by comparison with intensity strips or by a densitometer. Cameras allowed the recording of not only the spot intensities but also the diffraction space between the spots for sections in reciprocal space. This record often revealed unusual diffraction effects other than Bragg diffraction. In the late 1960s, computer-controlled single-crystal diffractometers were developed that measured one peak at a time and took many days to collect a full data set for crystal structure analysis. These devices only sampled the peaks and missed any effects that appeared in other parts of reciprocal space. In the 1970s, area detection devices were added to single-crystal cameras and allowed the collection of large numbers of intensities at one time and shortened the data collection time. Even the massive data sets of intensities from large protein crystals were accessible with these cameras. Textbooks such as Glusker and Trueblood (1985) and Stout and Jensen (1989) contain details on the use of these cameras for single-crystal structure analysis.

Single-crystal cameras are not all used for obtaining intensities for structure analysis. One major use of early cameras was to determine the symmetry and unit cell to allow the unequivocal indexing of powder diffraction data. Both the Weissenberg and Buerger (precession) camera reveal the symmetry by direct inspection, and the unit cell was easy to measure with some accuracy. Buerger (1960, 1964) discusses the interpretation of the data obtained on these cameras in considerable detail. Laue cameras were used to orient large crystals for cutting and property measurements. They can also be used for revealing gross defects by employing parallel beams and imaging individual diffracted images. Amoros, Buerger, and Canut de Amoros (1975) described the use of the Laue camera. Topographic cameras such as the Lang and Berg-Barrett cameras are used to examine linear and planar defect distributions in nearly perfect single crystal wafers such as silicon. Lang cameras can now examine wafers up to,30 cm in diameter. Although these studies are important in industrial applications, the subject of single-crystal diffraction is outside the scope of this book. Tanner (1976) described the topographic cameras and their uses.

Detector systems have also evolved considerably in the last 50 years. The first diffraction images were made on film, and film was the most important detector into the 1970s. Film is still used in some single-crystal cameras and in the Debye-Scherrer/Gandolfi camera; however, appropriate film is rather difficult to obtain today. The Geiger-Müller detector was used on the early diffractometers (Parrish, et al., 1954), but modern Bragg-Brentano diffractometers use either a scintillation counter, a gas-filled proportional counter, or a solid-state Li-drifted silicon detector. These detectors are employed to measure a single diffracted beam that exists the receiving slit (and monochromator) assembly. Position-sensitive detectors (PSD), based on the proportional counter concept with the sensor wire curved to fit the diffractometer circle, are used on some Seeman-Bolin diffractometers and on Guinier devices. PSDs are used often on diffractometers modified for measurements under non-ambient conditions. The wire concept is employed also in the area detector with a two-dimensional grid coupled with multichannel output. Jenkins and Snyder (1995) give a comprehensive discussion of detectors used for diffraction applications.

5. CRYSTALLOGRAPHIC DATABASES

There are many databases located throughout the world that contain crystallographic information that is very useful to the diffractionist. Some of these databases are private collections of data maintained in individual laboratories for specific purposes. However, there are several major collections that are maintained for general usage. Most of these databases have been supported by government and other grants to get them started, and their present income from sales partially supports the continued operations to keep them up to date and useful. One exception to this statement is the Powder Diffraction File (PDF), which was sponsored by industrial grants when it was initiated under the Joint Committee for Chemical Analysis by X-Ray Diffraction Methods. The PDF, as published by ICDD, is totally self-supporting through its sales. As government support diminishes in the 1990s, other databases are restructuring to operate with income from sales.

Except for the PDF and the Electron Diffraction Database (EDD), most of the information included is based single-crystal parameters, However, with the availability of appropriate computer programs, many of these databases may be used to produce information for comparisons with experimental powder diffraction data. An overall discussion of the databases may be found in Allen, Bergerhoff, and Sievers (1987). This publication provides considerable more detail on the databases listed in Table 1.

These databases contain crystallographic data on many thousands of phases that allow the user to make comparisons with experimental measurements. Many of the databases provide basic crystal structure descriptions from which diffraction effects may be modeled using modern computer programs that simulate the measurements and allow subtle structure interpretations to be made. The PDF contains direct diffraction pattern information as the reduced d–I sets primarily for phase identification, but it also contains crystal data, references, and supporting descriptions. The CDIF contains the unit-cell descriptions of most of the phases in the other databases along with the reference information and provides an excellent first source on the world literature on crystals that have been studied. All diffractionists need to be aware of these databases and their contents. The most pertinent of the databases are discussed in some detail next.

Table 1 Databases Containing Crystallographic Information

Title of database	Acronym	Source
Powder Diffraction File	PDF	ICDD
Inorganic Crystal Structure Database	ICSD	FIZ
Cambridge Structural Database	CSD	Cambridge
Metals Structure Database	CRYSTMET	TOTH
Electron Diffraction Database	EDD	ICDD
Crystal Data Identification File	CDIF	NIST
Protein Data Bank	PDB	Brookhaven
Database of OD (Order–Disorder) Structures	ODDB	ZPC

5.1 The Powder Diffraction File

The Powder Diffraction File (PDF; Jenkins and Smith, 1987) is maintained by the International Centre for Diffraction Data (ICDD), located in Newtown Square, PA. The primary source of information is the scientific literature, but the ICDD does maintain a large grants-in-aid program to promote the collection and submission of good data on materials of research and technological interest to augment the PDF. New data are thoroughly edited by established checking procedures and a team of experienced scientists, and 2500 new data sets are added each year. The 1998 statistics for the PDF are given in Table 2.

In 1997, the ICDD began a project to use the single-crystal structure databases as a source of information to calculate the theoretical powder diffraction patterns and add these patterns to the PDF. The first agreement has been made with the Inorganic Crystal Structure Database (ICSD), and around 38,000 calculated data sets were added in 1998. As agreements are finalized with the other pertinent databases, other data will also be added.

The PDF is a primary source for powder diffraction patterns for phase characterization and identification. It is also a source for the data collection parameters; the crystallographic parameters, including the crystal system, symmetry, and lattice parameters; physical property information; and literature source reference. Many efforts have been made to provide many entries to the data, including indexes, crystallochemical codes, and subfile classifications to assist the user in locating the desired phase.

Many search/match programs are available to assist the user in addressing the PDF. Some of these programs are in public domain and a list may be obtained by addressing the Gorter and Smith Powder Diffraction Program Listing on the World Wide Web page of the International Union of Crystallography. Many other search/match programs are instrument specific and are supplied by the instrument manufacturer as part of the software package.

Table 2
The Powder Diffraction File

Inorganic phases	97,133
Organic phases	21,981
Minerals	13,184
Metals and alloys	22,721
FIZ/ICSD	37,833

Table 3 The Crystal Data Identification File

Inorganic phases	149,140
Organic phases	158,531

5.2. Crystal Data

Crystal data (Mighell, Stalick, and Himes,1987) are maintained at the National Institutes of Standards and Technology in Gaithersburg, MD, and distributed by the ICDD and others in computer-readable form as the Crystal Data Information File (CDIF). The primary information is the lattice parameters of the crystallographic unit cell. The file also contains other crystallographic information including the reduced cell, the crystal system, space group if available, and the literature reference, It also indicates the type of data on which the information is based, powder or single-crystal. The entries are primarily obtained from all the other databases, so it contains very comprehensive coverage of both inorganic and organic materials that have been examined throughout the world. The statistics are given in Table 3.

The CDIF contains unit cell information along with crystal symmetry, chemistry, density, experimental type, functional groups, literature reference, and other data. This database is exceptionally useful for locating information in the literature on a large number of phases. In addition, it can be used as an identification file if the unit cell parameters can be determined on the unknown phase. This cell data may be obtained from a powder diffraction pattern by measuring accurate peak positions and using one or more of several computer-based indexing programs to index the pattern. Once the cell is determined, the data can be matched against the data in the CDIF using lattice matching software (Himes and Mighell, 1987). This method is a very powerful alternative to pattern matching with the PDF, and is a necessary step for phases not in the PDF. Once the match is located, the other information on the phase becomes useful to provide further tests for confirmation of the match.

Software to search the CDIF is distributed by ICDD, and a few manufacturers have included programs in the package associated with single-crystal diffractometers. The first step in any crystal structure analysis is to locate the unit cell in diffractometer space, and the user should immediately check the CDIF to confirm that the structure has not already been reported in the literature. It is also important for the user to consider possible isostructural compounds when searching the CDIF; as many new phases are related to phases that have already been studied. Some surprises are encountered in lattice matching, such as for BeH_2 and OH_2 (ice-III) (G. S. Smith et al. 1988).

5.3 Inorganic Crystal Structure Database

The Inorganic Crystal Structure Database is maintained at the Fachsinformationzentrum (FIZ) in Karlsruhe, Germany. This database is a primary source for crystal structure information for more than 40,000 inorganic phases other than metals. It contains the complete description of the structural arrangement of the atoms including the unit cell, symmetry, atom coordinates and temperature factors where they were available in the source literature. It also contains the R-factor reported by the original authors, the original reference, and remarks based on the editing of the data where appropriate. The ICSD is supplied in several computer-readable forms including CD-ROM disks along with software for accessing the data. The statistics of this database are given in Table 4.

Table 4 The Inorganic Crystal Structure Database (Release 97-2)

Entries with structure parameters	43,275
Minerals	6,957

The value of the ICSD to powder diffraction users is great. Besides being a source of general structural information, the structure description is complete for most entries and forms the basis for the calculation of the theoretical powder diffraction pattern. If a best-fit analytical function can be determined for the profile shape obtained from a specific diffractometer, it is possible to calculate the angle-dependent diffraction trace that should be obtained from any given crystal structure. There are several computer programs that perform this calculation; one of the best known is POWD12 (Smith and Johnson, 1997). All Rietveld refinement packages also can provide a diffraction trace based on the structure model. In fact, the ICSD (or CRYSTMET) is the starting place for most structure descriptions used for Rietveld studies.

It is very advisable to examine the ICSD for any structure related to material under study, for isostructural compounds as well as the specific compound, and calculate the diffraction pattern to be expected from the material. Differences between the experimental and calculated traces allow the user to interpret the data better than simply relying on a table of t's and I's. If the profile function is well known for the instrument in use, then subtle differences in the experimental profiles can be interpreted in terms of chemical differences or physical effects of the crystal perfection. When calculating a powder diffraction pattern from any structure description, it is important to confirm the correctness of all the input descriptive data by also calculating empirical formulas, densities, and interatomic distances and comparing the results to reality.

5.4 NRCC Metals Crystallographic Data File (CRYSTMET)

The CRYSTMET Database has been maintained until recently at the National Research Council, Canada (NRCC), in Ottawa, Ontario. Currently, it is maintained by Toth, Inc., Gloucester, Ontario, Canada. It contains structure descriptions for metals, alloys, intermetallics and some corrosion products in a format similar to the ICSD. In fact, the same software will access both files. Uses of this database are the same as for the ICSD. The 1993 statistics for CRYSTMET are given in Table 5.

Table 5 The CRYSTMET Metals Crystallographic Data File

Entries with structure information (1993)	52,538

Table 6 The Cambridge Structural Database

Entries with structure information (1997)	163,057
New Entries for 1998	16,000+

5.5 Cambridge Structural Database

The Cambridge Structural Database (CSD; Allen and Kennard, 1987) contains the crystal structure descriptions for organic compounds. It is based at Cambridge University in the United Kingdom, and it is maintained by the Cambridge Crystallographic Data Centre. It contains the same type of crystallographic information as the ICSD except it does not report the temperature factors and it emphasizes the molecular structure in the asymmetric unit of the structure. It does provide additional information of interest to the organic chemist such as the connectivity in the molecular unit. It is useful to the powder diffraction community both as a source of the crystal description and literature source and as a basis for the calculation of the theoretical powder diffraction pattern. Usage of powder methods in the organic chemistry field is less than for inorganic applications, but they are still important. Although there are many more organic compounds in the world compared to inorganic ones, the PDF contains only a very limited number. Consequently, alternate methods of study are often used where diffraction methods would provide the most useful data. The ability to calculate the diffraction pattern opens up the applications of powder diffraction methods in the field. The CSD provides that information. The statistics on the CSD are given in Table 6.

5.6 Other Crystallographic Databases

In addition to the databases already listed that provide data for simulating powder diffraction patterns, there are several others that are more specific in their coverage. The Protein Databank (Abola, et al., 1987) provides structure descriptions on proteins that would rarely be studied by powder methods. The Biological Macromolecule Crystallization Database (Gilliland, 1987) contains the unit cell crystallographic information but insufficient other data for pattern calculations. It is a source primarily for how to make specific crystals. The Database of OD (Order-Disorder)

Structures (ODDB; Backhaus, Krell, and Fitchner, 1987) emphasizes structures that show polytypism and stacking fault disorder. It contains the structural description of the layer unit and the stacking possibilities, but the user will have to complete the specific structure description to use the data to calculate the diffraction pattern.

6. EXTRACTION OF INFORMATION FROM DIFFRACTION PATTERNS

The early years of powder diffraction required the reading of film images and graphical strip charts to extract the peak positions and peak intensities. Measuring integrated intensities required calibrated intensity strips for film data, area measuring devices for strip charts, counting squares on the strip chart, or cutting out and weighing the peaks. A few ingenious mechanical devices were created for area measurement with the diffractometer that were successful but required long times to obtain useful data. With the use of the computer to control data collection, store the digitized diffraction pattern, and reduce the pattern to position and intensity values, data reduction is both routine and more accurate. There are several levels at which the raw digital data may be used: individual peak information, the complete set of d and I information, and as the whole trace. Individual peak information and the d-I set may be obtained from the trace by utilizing various peak-finding algorithms or by peak profile fitting using programs such as PRO-FIT (Toraya, 1986, 1993) and SHADOW (Howard and Preston, 1989). The full pattern may be analyzed for only the peak positions by programs such as WPPF, which fits the unit cell parameters for one or more phases to the trace. Programs such as that of LeBail, Duroy, and Fourquet (1988) may accomplish the extraction of the intensities for structure analysis. The full structure refinement may be carried out by many Rietveld programs (Young, 1993). It is also possible to graphically decompose patterns by superposing a reference pattern from a full-pattern database without processing the raw data in any way (Smith, Hoyle, and Johnson 1993).

It should be mentioned that there are two ways to extract information from a diffraction peak: decomposition and deconvolution. These terms are commonly misused, especially deconvolution. Decomposition is the separation of overlapped peaks by using a selected analytical profile and optimizing the fit of the cluster by adjusting the profile parameters. Deconvolution is the extraction of the sample contribution in a profile from the source and instrument contribution by elaborate mathematical methods that may involve some approximations. The profile-fitting procedures used for extracting position and intensity information involve decomposition. Separation of size and strain information by procedures such as the Warren-Averbach method involves deconvolution.

7. ANALYTICAL METHODS

Powder diffraction techniques may be classified into those that yield a qualitative answer and those that yield a quantitative answer. Qualitative answers include confirmation or identification of the phases present in a specimen and crude estimates of how much is present. Quantitative analysis provides numbers with estimated accuracy for crystallographic parameters including lattice dimensions, phase abundances, crystal structure coordinates, and other physical property data.

The data in the PDF are most commonly used for phase identification. It must be remembered, when using the PDF, that all the d-I tables are based on peak information. That is, the d's and I's are reported for the peak positions from the experimental trace, and the peaks in a cluster are not decomposed into individual components. When making comparisons of diffraction data with the PDF, one must use peak heights versus peak heights not peak areas versus peak heights (i.e., apples versus apples, not apples versus oranges).

7.1 Qualitative Phase Analysis

The identification of phases based on their powder diffraction pattern dates back primarily to Hanawalt and Rinn (1936) and became a common procedure when the Powder Diffraction File was published and contained sufficient data sets to yield snatches. As the coverage of the PDF improves, phase identification also improves. It is impossible to locate a match for an experimental data set if the data for the phase in question are not included in the reference file.

Where the user only wants to confirm the presence of a phase in a sample, it is a simple matter to locate an appropriate reference powder diffraction pattern and collect the experimental data for comparison. The problem arises when the user has little or no information from which to start. Fortunately, the powder diffraction pattern can be a sufficient fingerprint to allow identification of phases even when no initial information is available. Conflicting results in such situations may require further testing by other methods, but the user now has some hint as to what tests may be appropriate.

There are several approaches to numerical phase identification with powder diffraction data. There are two stages in the identification of phases by diffraction data: the *search* step and the *match* step. The search step locates potential candidate phases by any of several means. Once the candidates are found, the data sets are matched to confirm which candidate is the best identification. Very commonly, the user has some idea of the potential phases that might be present in the sample, and appropriate reference data sets can be compared directly with the experimental data to confirm the match.

Where the user has no ideas on possible phases, numerical search procedures must be employed. Numerical methods are based on the *d-I* listings derived from the experimental data. Initially, the search procedures were all "manual," generally known as the Hanawalt method and the Fink method. When the number of phases was small (≤ 3), success was possible in a reasonable time frame. But numerical procedures are amenable to computer analysis, and there are now many computer-based techniques. In fact, all instrumentation manufacturers include software for phase identification in their automated powder diffractometer package. A listing of public domain search/match programs may be found under the International Union of Crystallography on the World Wide Web as the Garter Powder Diffraction Program List.

More recently, search procedures are based on pattern matching techniques using the experimental full-trace pattern rather than just the derived d's and I's (Nusinovic and Winter,

1994). Procedures are under way to develop full-trace databases (Smith, Johnson, and Jenkins, 1995) and programs to do full-pattern matching (Smith, Hoyle, and Johnson, 1993) for phase identification.

The basis of phase identification is to locate the pattern of a known phase, from the reference database, in the experimental pattern of the specimen that may be a mixture. This procedure can be very straightforward if the experimental data are of high quality and free of preferred orientation, and the reference patterns are of equal quality. As the quality of either or both categories of pattern degrades, the identification procedure becomes more difficult. The PDF is the primary database for phase identification, and the ICDD attempts to maintain the file with the most accurate available data for all the phases. Unfortunately, much of the data is obtained from the literature where the reported information was not always collected under optimal experimental control. These data are editorially evaluated and graded to assist the user in the identification match step. The ICDD does maintain a grants-in-aid program to provide pattern data that are collected under conditions leading to the best possible diffraction information. It is the users' responsibility to see that the experimental pattern is also of the highest possible quality to allow successful matching.

Even though computer phase identification is very successful, many times the computer only provides a graded list of candidates that must be evaluated by the user. Sometimes the best identification is a structurally related phase that is chemically different from the true phase, so search procedures should be run both with and without chemical restraints. A simple X-ray fluorescence (XRF) scan can help provide the information that ultimately leads to positive identification. Isostructural series, either complete or partial, are very common in inorganic systems, and the user can usually discern the correct identification by applying a little knowledge of crystal chemistry. Single-phase problems are usually straightforward, but mixtures even of simple phases can be very difficult to unravel. The computer is generally unnecessary for the single-phase problem. It is important in the mixture problem especially when three or more phases are present. Modern search/match programs are very successful when the data are good even with as many as eight phases in the mixture.

7.2. Quantitative Phase Analysis

Quantitative phase analysis by X-ray diffraction (QXRD) is the only analytical technique that is truly phase sensitive. Alexander and Krug (1948) developed the theoretical basis for the technique, and a recent text by Zevin and Kimmel (1995) reviewed the whole topic. The fundamental equation for the ith peak of the Jth phase is:

$$I_{iJ} = K_{iJ} x_J / \mu_M \qquad (1)$$

where K_{ij} is the phase constant for the ith (hkl) peak, x_j is the weight fraction of the Jth phase, and μ_M is the linear attenuation coefficient of the mixture. This coefficient could be calculated only if the abundances of all the phases in the mixture are known a priori, which is true only in very special situations, such as when all phases have exactly the same chemistry. Thus,

μ_M must be measured or its effect must be eliminated by some compensatory technique.

The limitations of the QXRD method have always been experimental. Regardless of the technique used, the diffraction intensities must be measured with high accuracy, and such accuracy is difficult to attain. The effect of the attenuation of the X-ray beam is only one of the limiting factors. The specimen and counting statistics are the other problems. Probably the most serious problems are reduction of the particle (crystallite) size and achieving a truly random orientation of the particles in the specimen. Elton and Salt (1996) have reviewed the previous analyses of the particle statistics problem and confirmed that the size estimates of de Wolff (1950) and Smith (1993) were correct. In order to achieve intensity measurements with accuracies on the order of 1%, the typical specimen must have particles (crystallites) on the order of 1 μ in maximum dimension. Comminuting a sample to this small dimension is very difficult and limits the accuracy of the analysis, even if all other requirements are satisfactory.

There are three principal techniques of QXRD. They are the absorption-diffraction method, the internal-standard method, and the external-standard method. In the absorption-diffraction method, the attenuation coefficient of each sample is measured, usually by a transmission technique. In the internal-standard method, a reference material in known abundance is added to the mixture, and all intensities are ratioed to the intensities of this reference material. A single specimen may be used if all the phases are analyzed, or several spiked specimens may be used to analyze for just one phase in the mixture. In the external-standard method (also called the reference-intensity-ratio [RIR], or Chung [1974b] method), prior measurements of the intensity ratios of each phase with respect to some reference must be made. So-called "standardless" methods fall into this latter category, as they require precalibration.

(1) The Absorption-Diffraction Method

The direct use of Eq. 1 requires the knowledge of the absorption coefficient of the specimen. Unless the sample is composed only of phases that have identical absorption coefficients (e.g., TiO_2), it is not possible to know this information a priori, although it might be estimated from a bulk chemical analysis, so it must be measured directly on each specimen examined. Such measurements are not easy because of the particle nature of the sample, the unknown porosity of the packing, and the general X-ray opaqueness of the specimen. If the specimen is thin, the attenuation can be determined by placing the specimen in the diffracted beam in a diffractometer and comparing the attenuation with the nonattenuated intensity and the specimen thickness. Unfortunately, thin specimens rarely have sufficient particles to satisfy the conditions of crystallite statistics, thus limiting the potential accuracy of the quantification to 5 or 10%. Davis (1992) described good examples of the use of the absorption-correction method.

(2) The Internal-Standard Method

Alexander and King (1948) also developed the theory for the internal-standard method of quantitative X-ray powder diffraction. Considering intensity ratios rather than individual intensities as the measurement quantities can eliminate the effect of the absorption coefficient. Equation 2 shows that a constant, k, and the ratios of peak intensities relate individual weight fractions to each other. With the addition of a known amount of a reference material (often Al_2O_3, corundum, or similar binary oxide or halide), the weight fractions of individual phases may be determined. The basic equation for this method is

$$x_\alpha = k \frac{I_{i\alpha}}{I_{is}} x_s \qquad (2)$$

where x_α and x_s are the weight fractions of the analyte and the standard respectively, $I_{i\alpha}$ and I_{is} are the measured peak intensities, and k_i is the proportionality constant. This relation applies to all the phases in the sample. If all the crystalline phases in a sample are determined in an analysis and the sum is less than J, the remainder can be interpreted as the amorphous component. Chung (1974b) proved that when Al_2O_3 is chosen as the internal standard, the proportional constant k_i is equal to the reciprocal of the analyte RIR as listed in the PDF. All RIRs in the PDF are based on I/I (corundum) as proposed by de Wolff and Visser (1964).

(3) The Method of Standard Additions (Spiking)

This method for the determination of the abundance of a phase α in any sample is an extension of the internal-standard method where the problem of the attenuation coefficient is overcome by preparing several samples in which known amounts of phase β have been added. This method, commonly known as **spiking**, was first proposed by Leroux, Lennox, and Kay (1953). If there is a second phase α in the sample with a resolved strong peak, the intensity ratios of a characteristic peak of phase to the peak of phase β are plotted against the known addition, and the best-fit line is extrapolated to zero addition. The intensity ratio for each sample may be expressed as

$$\frac{I_{i\alpha}}{I_{i\beta}} = K(x_\alpha + y_\alpha). \tag{3}$$

The quantities x_α and y_α are the weight fractions of α in the unspiked sample and the amount of added phase β.

(4) The External-Standard Method (Matrix Flushing)

The external-standard method is also known as the reference-intensity-ratio (RIR) method or the Chung **"matrix-flushing"** method (1974a). In this method, the K of the internal-standard method is determined from reference-intensity ratios measured prior to the experiment for each phase in the study. The RIR was first suggested by de Wolff and Visser (1968) as a method to place diffraction patterns on a relative-absolute intensity scale. The RIR was suggested as the ratio of the strongest peak of a phase to the strongest peak of α-Al_2O_3, corundum, in a 1:1 weight ratio sample. Other phases are also used including SiO_2, TiO_2, CeO_2 Cr_2O_3 and ZnO, to take advantage of ubiquitous quartz or absorption matching. The equation for the RIR method is

$$x_\gamma = \left(\frac{I_{i\gamma}}{I_{is}}\right)\left(\frac{I_{is}^{rel}}{I_{i\gamma}^{rel}}\right)\left(\frac{K_s}{K_\gamma}\right) x_s \tag{4}$$

where the k in the internal-standard equation is replaced by

$$k_{\gamma s} = \left(\frac{I_{is}^{rel}}{I_{i\gamma}^{rel}}\right)\left(\frac{k_s}{k_\gamma}\right) \tag{5}$$

Some numerical examples may be found in Chapter 20 on X-ray diffraction in the paint and pigment industry.

(5) Pattern Fitting for Quantification

With the availability of digitized diffraction patterns, the collection of the intensity data for quantitative analysis is more routine than it was when the theory and early applications were developed. Because it is the integrated intensity, not the peak intensity, that is related to the abundance of a phase, elaborate devices were invented in the 1950s and 1960s for stepping across peaks to determine the area. These devices are not necessary today, as the digitized pattern not only can yield the areas of individual peak profiles, but they can also be analyzed as the whole trace. In the 1980s and 1990s, profile-fitting and pattern-fitting procedures became the normal methods for extracting the pertinent intensity information. Although one can still use individual peaks for phase quantification, it is the practice today to use multiple peaks or the whole pattern to represent each phase, and the basic techniques mentioned have been modified to use these data. Bish and Howard (1988) and Hill and Howard (1990) used the Rietveld algorithm, and Smith et al. (1987) developed the whole-pattern matching method specifically for quantification. Both Rietveld and whole-pattern fitting methods have the potential to approach the theoretical accuracy limit of ± 1%.

7.3 Lattice Characterization

Lattice characterization includes both determination of the correct lattice, crystal system, and lattice type, and the refinement of the data to achieve the accurate magnitudes of the parameters.

Lattice dimensions are extremely sensitive to very subtle structural effects, and accurate measurements often reveal interesting features of the crystal that might not be suspected initially. Very low concentrations of impurities and defects can be detected by accurate lattice parameter differences. Phases can be identified by lattice matching even for cubic materials.

Indexing

Fundamentally, indexing is a very simple concept that is easily solved with very accurate geometric data, that is, complete resolution of all peaks and accurate peak position measurements. The simplest form of the indexing relationship is expressed as the linear function:

$$\frac{1}{d_{hkl}^2} = h^2A + k^2B + l^2C + 2hkD + 2hlE + 2klF$$

(6)

where d is the experimental d-spacing, h, k, l are the Miller indices, and A, B, C, D, E, and F are functions of each of the cell parameters that are irrelevant until this equation is solved for all measured peaks. The goal of indexing is to find a set of A–F values that allows hkl solutions for all observed d-spacings. This relationship is classified as a Diophantine problem because all h, k, and l values must be integers. It is worth noting that the equation is a function of reciprocal d [$(1/d)=2\sin\theta)/\lambda$] and is sometimes expressed as a function of the measurement angle θ. Most relationships in diffraction are in their most simple form when written in this manner, the so-called "reciprocal" geometric space. It is analogous to the physicist's momentum space.

There are several ways to solve this equation for a set of measured d-spacings, from analytical to trial-and-error techniques. The initial theory on its analytical solution was presented by Runge (1917). The Runge approach first assumed that the data set was triclinic. The method examined the data for subsets where one of the Miller indices is zero, a technique known as looking for "zones." Once a series or zones is recognized, the zones are combined by intersecting common rows, and the three-dimensional lattice is tested against the experimental data. When a solution is determined, the lattice is further tested for metric relations that suggest the symmetry is higher than triclinic. This method is better known as the Ito (1949) method and was promoted by de Wolff (1962) and Visser (1969). The well-known VISSER/ITO program uses this method of analysis to produce results on most good data sets.

Indexing is a problem that is natural for computer analysis, where a trial-and-error approach may be used because the computer does not get tired. Trial-and-error methods are viable because of the Diophantine condition and the fact that many phases have high symmetry that simplifies the number of peaks that have to be identified. Shirley (1980) classifies this approach as "exhaustive" because it examines all possibilities in crystal space for a solution. The approach assumes a crystal system, (cubic {A only}, tetragonal and hexagonal {A and C only}, orthorhombic {A, B, and C only}, etc.) and uses the low-angle, high d-spacings to provide trial values for A, B, …that are tested against the other data in the set employing all the integral multipliers (Miller indices) allowed by the symmetry of the crystal system. An early attempt to employ the computer for this approach was proposed by Goebel and Wilson (1965). The exhaustive approach has been perfected by Werner, Eriksson, and Westdahl (1985) in the program TREOR. POWDER by Taupin (1968) is another program that uses the exhaustive approach.

Louer and Louer (1964) have developed an approach that is partly exhaustive and partly analytical. Their program DICVOL divides up crystal space into domains and test each domain for the probability that a solution exists in that region. Prospective domains are further subdivided into smaller domains, and successive testing yields a very small volume of crystal space and finally a solution to the indexing question.

Other programs exist, but they are similar in approach to one of the three schemes just described. With the ability to resolve most diffraction peaks even in a complex diffraction pattern and to collect very accurate experimental peak positions, indexing by one or several of the computer programs is usually successful. As more high-quality lattice parameters are determined and archived in databases, the lattice-matching approach to phase identification using the CDIF should become as useful as the Powder Diffraction File.

Cell Refinement

Indexing is usually based on the low-angle diffraction peaks because the first 25 to 40 peaks that have the minimum overlap problems are employed and unobserved peaks are rare. To refine the lattice parameters, it is necessary to use all the experimental data, especially the high-angle peaks whose d-values may be determined with more significance from the experimental 2θ measurements. There are two approaches for the refinement of lattice parameters that have been in use since the 1950s. One method assigns each d-spacing its Miller index, and systematic errors are included in the refinement as functions added to the d-spacing equation that is the basis of the refinement. This method is usually called Cohen's method. The first program using Cohen's method was B-106, written by Mueller and Heaton (1956), and several other programs exist based on this first program.

A well-known refinement/indexing program is generally identified as APPLEMAN (Appleman and Evans, 1973). Although this program does refine the lattice parameters, the program was actually designed to assign indices to unindexed peaks based on a starting cell. Some or all peaks may be preassigned Miller indices, but this procedure is not necessary. The program uses the starting cell to assign a first set of indices for peaks that fit the calculated position within some defined tolerance. Using the positions of the indexed peaks, the cell is refined. The processed is cycled with a reduced tolerance for each cycle. Because the indices are not fixed, it is not possible to include systematic error functions. There are many versions of the APPLEMAN program in use today.

7.4 Structure Refinement

With the availability of good digitized diffraction traces from laboratory equipment, the Rietveld method (Rietveld, 1967, 1969) has become a very important technique for extracting the most material information for a specimen. The many aspects of the technique have been described in detail in Young (1993). This method was initially developed to use the data in a neutron wavelength-dispersive powder diffraction trace to refine the crystal structure of a material; however, it has since been expanded to utilize many types of diffraction. It allows the user to obtain improved crystal structure descriptions even when large single crystals of a material are not available. Structures with over 200 structural parameters have been refined successfully to reveal positional and occupational details previously unavailable for these phases.

The refinement procedure requires careful use of several levels of parameters, including overall pattern parameters (scale and background), experimental factors (zero shift, transparency, displacement, etc.), pattern parameters (profile shape functions, width dependence and asymmetry, etc.), and structure description parameters (unit cell, atom coordinates, occupancy and temperature factors).

7.5 Structure Solution

It has always been the goal of many diffraction experiments to solve the arrangement of atoms in the crystal structure directly from the experimental diffraction data. Initially, simple structures were solved by inspection (trial-and-error) techniques from powder pattern information. As the structures to be solved became more complex, single-crystal methods were developed and employed almost exclusively in the 1950-1970 period. Subsequently, structure solutions were needed on materials for which suitable single crystals were unavailable, and the powder method was revived as a major tool for structure solving. Now there are many approaches to extracting the necessary information from the powder diffraction trace, and many program packages to accomplish the structure solution are available.

The problem with solving structures from powder data for completely unknown structures is more than just the phase problem. Because of the overlap of the diffraction peaks onto one-dimensional linear trace, the individual peaks must be resolved by a decomposition technique to determine an accurate position and intensity for each one. Using the peak positions, the unit cell and symmetry must be identified and all the peaks must be indexed. Then the correct intensity must be assigned to each peak prior to any attempt at solving the structure. If the data are of sufficient accuracy, the indexing problem is essentially solved, and the geometric interpretation is usually correct. The intensity assignment is another matter. Where peaks are partially resolved, good peak decomposition programs do give a reasonable estimate for the proper intensity. When the peaks are fully superposed or where the clusters are composed of too many peaks for reasonable decomposition, the assignment of individual peak intensities may be impossible, and preconceived distribution rules must be followed. Once the set of peak intensities is established; the methods of single-crystal structure determination (direct methods, Patterson synthesis, heavy atom or isomorphous substitution) may then be applied. The reader is referred to the chapter on structure solution (Chapter 42) in this book for detailed information.

It is a common situation that the structure of the material under study is not totally unknown. Once the unit-cell information and symmetry are determined, the user should refer to the appropriate crystallographic database and search for other phases with similar crystallographic information and related chemistry. Probably the most useful of the databases for this first search is the CDIF. Sometimes the structure has already been solved. More commonly, an isostructural phase can be located. If such partial information can be located, the problem becomes more one of improving the structural model than solving a complete unknown.

Where partial information on a structure is already available, such as the framework arrangement of a host zeolite or an isostructural phase, further progress can be accomplished by several approaches. Physical models of packed balls or graphical plots using one of the many modern computer packages usually suggest reasonable arrangements for substitutional or interstitial atoms. One can distribute the known chemistry on the potential interframework atoms and calculate the expected powder patterns from different distributions for comparison with experimental patterns. Once an appropriate model is obtained, further refinement by the Rietveld method, with or without restraints, should lead to an improved structure.

7.6 Crystallite Size and Strain

As indicated in Table 3, the diffraction pattern, in particular, the profile shapes of the diffraction peaks, contains information on the physical state of the crystallites that comprise the sample. Due to incomplete interference of the diffracted X-rays, the sample contribution to the width of the diffraction profile increases as the crystallite size becomes smaller below a value dependent on the instrumental resolution. This broadening is symmetrical and varies as a function of the diffraction angle. Strain, in response to internal or external forces, manifests itself as a change in the crystallite dimensions and appears as shift of the diffraction peak from its unstrained position. The resulting profile is often asymmetrical because the strains are not homogeneous. The powder diffraction technique has been perfected to extract an analytical description of the shape of the diffraction peak and the changes with diffraction angle that allows the user to interpret the profile parameters into measurements of these crystal properties. The Rietveld refinement technique allows different analytical forms of the profile shape and different functions to represent the full-width at half maximum (FWHM), and the interpretation of the selected function can be split into the size and strain components. For example, the Thompson, Cox and Hastings (1987) FWHM function is FWHM $(2\theta) = \sqrt{g} + \sqrt{L}$ where

$$\Gamma_G = (U \tan^2 \theta + V \tan \theta + W + Z/\cos^2 \theta)^{1/2}$$
$$\Gamma_L = (X \tan \theta + Y/\cos \theta). \tag{7}$$

The U, V, W, X, Y and Z are refinable parameters, and $_G$ and $_L$ are interpreted as the FWHM function related to the strain and size, respectively. However, the extraction of physically meaningful size and strain information is much more complex than using a single, set of functions. The reader is directed toward several chapters in this book and the chapter by Delhez et al. (1993) in Young (1993) for more details.

It should be mentioned that an International Conference on Size and Strain was held in Liptovsky Mikulas, Slovakia, in 1995.

The proceedings for this conference should appear shortly as a volume of the International Union of Crystallography Monographs on Crystallography (Snyder, Bunge, and Fiala, 1998).

7.7 Crystalline Defects

Crystalline defects in powder samples affect the diffraction pattern in many ways. The types of possible defects are many, as listed in Table 7. All defects result in localized strain fields that may be spherically symmetric to highly asymmetric. The nearby crystal volume is distorted and causes modifications in the diffracted intensity and in the diffraction angle. Where the defect density is low, the defect distribution may actually be observed directly by diffraction topography. The transmission electron microscope is commonly used to image defects, but methods using X-rays are applicable to large crystals (Tanner, 1976). Point defects appear as cloudy regions; dislocations appear as lines in space; and planar defects appear as phase contrast effects. The theory of imaging is beyond the scope of this section.

Usually the defect density is too high for direct imaging, and the powder diffraction pattern is used to imply the existence and type of defect present. Defects may be introduced to the sample during its formation (growth) or induced by postformation treatments (crushing, grinding). Point and linear defects may distributed uniformly throughout the sample crystal fragments, or they may cluster, yielding subgrain domains where each crystal fragment is composed of many small crystallites isolated by walls of defects. In such situations, the effective crystallite size is that of the small domain. The diffraction profile is a composite of the broadening effect of the small crystallite domain dimensions and the strain effects of the defects on the local lattice and on the domain, where elastic strains may reside as a result of the energy stored in the domain boundaries. This phenomenon is known as polygonization and is common in work-hardened metals. The result is a diffraction pattern with very broad peaks or a pattern of mixed sharp and broadened peaks.

Table 7 Crystal Defects

Point
    Vacancies
    Substitutional atoms
    Interstitial atoms
Linear
    Dislocations
Planar
    Stacking faults
    Antiphase boundaries

Solid solutions are an example of the effect of point defects. The end members of a solid-solution series may show diffraction peaks that approach FWHM values close to the instrument limit, but intermediate members of the series always show profiles considerably broader, due to the random distribution of the substituting elements, especially when they are distinctly different sizes. The substitutions may be as isolated points or clusters, but both will locally distort the crystal lattice and cause peak shifting and broadening. Such point substitutions usually affect all peaks in a similar fashion, but if the elasticity of the crystal lattice is highly anisotropic, then the peak response will also be anisotropic.

Planar defects produce quite different effects compared to

point and linear defects. The planar defect usually leaves one part of the structure in registry while other parts of the structure are displaced. For example, the *001* stacking fault in hexagonal close-packed structures does not affect the 00*1* set of peaks. They also may not affect the *hk*0 peaks, but all *hkl* peaks appear broadened due to the reduction of the coherent crystallite domain due to the fault. As the stacking fault density increases, *hkl* peaks broaden asymmetrically and coalesce, whereas the 00*1* and *hk*0 peaks may broaden very little and such broadening is symmetric. Modern computer programs allow the modeling of stacking fault effects on structures (Treacy, Deem, and Newsam, 1995). Order-disorder structures and structures with antiphase domains also result in mixed peak shapes.

A special type of defect structure is the incommensurate structure, where the structure is composed of two or more parts that do not register exactly in crystal space. Mixed-layer clay minerals, interlayered oxide/sulfide structures, and some magnetic structures are examples. The quasi-crystal is a special case of incommensurate structures where two different unit cells comprise the three-dimensional arrangement of atoms. These structure types result in powder diffraction patterns that are composed of sharp and broad peaks.

7.8 Crystal Orientation and Texture

The microstructure of a sample, the distribution of crystallites and their orientation, has long been of considerable interest in materials applications, but it has been relatively recently that the instrumentation has really allowed full diffraction analysis, Microstructure analysis requires the collection of massive amounts of data to fully describe the texture in terms of a weighted orientation function. The spatial distribution of several selected crystal directions has to be measured over three-dimensional sample space. Initially, with a wavelength-dispersive diffractometer, it was necessary to fix the diffracted beam detector and move the specimen over a programmed path to include as much sample space as physically available. Data collection on a single specimen could take many days. Now the area detector has decreased the data acquisition time to a reasonable value, and the data are easily digitized and converted to orientational distribution functions. The reader is referred to the chapter on texture (Chapter 41) for further details.

8. THE COMPUTER IN DIFFRACTION ANALYSIS

The computer is indispensable in the studies of crystals and crystalline properties measured by diffraction analysis. When computers first became a research tool in the 1950s, crystallographers were among the early scientists to utilize them. Initially, computations were directed at the interpretation of experimental data that was measured visually from films and strip charts. The raw data were mostly in analog form. In the 1960s, a few mechanical instruments were devised to allow the digitization of the raw data, but it was really the 1980s before the computer became an integral part of commercial data-collection diffractometers in industrial laboratories both for instrument control and for data acquisition, storage, and analysis. In the 1990s, computers have become physically smaller, more sophisticated, more powerful and versatile, and faster, and the programs have followed along the same lines. It is now possible to do far more on a small laboratory computer than on the largest available computer of only a few years ago. Smith and Gorter (1995) reviewed the history of the computer impact in powder diffraction over the last 50 years.

In the early 1950s, computer programs were limited in language and the types of computers that existed. The limited size of the computer memory and low-speed input/output (I/O) devices restricted the capability of the calculations, and innovative programming often pushed the capabilities to the ultimate. Computer programs were developed mostly as a necessity within research programs, and the programs were freely distributed. As computers evolved and improved, programs also developed to take advantage of the new innovations. Gorter and Smith (1995) listed over 600 programs devoted specifically to some aspect of powder diffraction analysis, and the number is up considerably in 1998. One major difference in. the computer programs available today is the small fraction of programs that are freely distributed. This trend has developed for several reasons. The personal computer has become the basic computer for most calculations, and users do not have the time or support to develop the programs within the structure of the research projects. Entrepreneurs, who have become programming specialists, develop most of the programs independently of any direct research program, and sales are the only means to support the development efforts. Such commercial software is usually thoroughly tested and well documented, in contrast to freeware today. Freeware is rarely well tested and documented. Well-known freeware exceptions include DBWS and GSAS, the two principal Rietveld refinement programs in use throughout the world.

Diffraction instrument manufacturers (original equipment manufacturers, OEM) now spend considerable effort to supply a complete package of control and analysis programs with the instruments. The OEMs realize that the success of sales is now more dependent on software capabilities than on instrument capabilities because the instruments are really quite similar in quality and functionality. Many of the programs that are incorporated in these packages are modification of programs and algorithms developed in the public domain. Unfortunately, most of these early efforts are not referenced to provide proper credit to the early crystallographers who gave so much of their time to improving the field.

Future computer developments will take advantage of graphical procedures. It is already possible to display a crystal structure and its powder pattern simultaneously on a screen. The user can use a pointer to move an atom in the structure image and watch the powder pattern change in response to the atom shift. Present databases are based on numerical data. Future databases may be composed of data images, and pattern recognition procedures need to be developed for utilization. An example is the PDF-3 database of powder diffraction pattern traces being developed by TODD (Smith, Johnson, and Jenkins, 1995).

9. ADVANTAGES OF DIFFRACTION ANALYSIS

Although the methods of X-ray diffraction (XRD) analysis date back to the 1910s, XRD analysis is still one of the most important characterization and analytical tools available to the scientist even today. This situation is due to two conditions: Crystals and materials in the crystalline state have become extremely important in modern technology, and diffraction methods are sensitive to the crystalline properties. Examples of modern materials that required X-ray diffraction to elucidate the true nature of the material include high-T_c superconductors, fullerenes, and quasi-crystals. It was not until XRD revealed the true structure of these materials that a real understanding was possible.

XRD is phase sensitive, not merely element sensitive. It can measure crystalline characteristics of one phase even in the presence of other phases. It is sensitive to the whole of the crystal structure, not just selected parts, as is true for most other spectroscopic methods, which allows complete interpretation of the experimental data in terms of a single model. An example of phase sensitivity is phase quantification. XRD, for example, is the only method that can be used to analyze for rutile, anatase, and brookite abundance in TiO_2 paint pigments. The principles of XRD phase quantification is sound theoretically; the limitations on accuracy are primarily experimental. Single-peak methods usually result in accuracies that are on the order of ±10%. Careful use of multiple peaks and reproducible procedures may reduce this figure to ±5%. Whole-pattern methods (including Rietveld) seem to produce accuracies that are ±2%. Smith (1993) showed that the limiting factor in XRD is the size of the specimen and that accuracies near ±1% are attainable when the particle size of the specimen is sufficiently small (≤5 μm). Although many users do not feel ±1% is sufficient accuracy for a quantitative analysis technique, there is no other method available that can produce equivalent accuracy for phase abundance.

Powder methods can now substitute for single-crystal methods in characterizing new phases. The crystal structures of many materials that never yielded suitable single crystals for study can now be determined from powder data. Totally unknown structures can be solved by following procedures that extract the decomposed diffraction data from the experimental pattern (pattern-fitting methods), elucidate the crystal parameters (the unit cell dimensions and symmetry), index and assign intensities to each observed peak, and utilize the methods of single-crystal structure analysis (direct methods, etc.).

With the introduction of the synchrotron as a source for experimental X-rays, a whole new realm of measurements is available. High beam intensities allow time-resolved studies and studies on very tiny specimens, and crystallites as small as 1 μm can be examined as single crystals. Most of the experiments on the synchrotron are ones that were not possible in the laboratory because of the relatively low beam intensity of laboratory instruments. However, the synchrotron also allows wavelength tenability, so experimental effects near absorption edges can be utilized to reveal individual element contributions to diffraction intensities (Nichols, Smith, and Johnson, 1982).

10. REFERENCES

Abola, E. E.; F. C. Bernstein; S. H. Bryant; T. F. Koetzfe; and J. Weng (1987). Protein data bank. In *Crystallographic Databases*, F. H. Allen, G. Bergerhoff, and R. Sievers, Eds., pp. 107-132. International Union of Crystallography,

Chester UK.

Alexander, L. E.; and H. P. King (1984). X-ray diffraction analysis of crystalline dusts. *Anal Chem.* 20:886-894.

Allen, F. H.; G. Bergerhoff; and R. Sievers (1987). *Crystallographic Databases*. International Union of Crystallography, Chester, UK.

Allen, F. H.; and O. Kennard (1987). Cambridge Structural Database. In *Crystallographic Databases*, F H. Allen, G. Bergerhoff, and R. Sievers, Eds., pp. 31-76. International Union Crystallography, Chester, UK.

Amoros, J. L.; M. J. Buerger; and M. Canut de Amoros (1975). *The Laue Method*, Academic Press, New York.

Appleman, D. E.; and H. T. Evans Jr. (1973). US. Geological Survey, Computer Contribution 20, U.S. National Technical Information Service, Doc. PB2-16188.

Backhaus, K. O.; H. Krell; and K. Fitchner (1987). Database of OD (Order-Disorder) Structures. In Crystallographic Databases, F. H. Allen, G. Bergerhoff, and R. Sievers, Eds., pp. 178-182. Inter-national Union of Crystallography, Chester, UK.

Bergerhoff, G.; and I. D. Brown (1987). Inorganic Crystal Structure Database. In *Crystallographic Databases*, F H. Allen, G. Bergerhoff, and R. Sievers, Eds., pp. 178-182. International Union of Crystallography, Chester, UK.

Bish, D. L.; and S. A. Howard (1988). Quantitative phase analysis using the Rietveld method. *J. Appl. Cryst.* 21:86-91.

Bish, D. L.; and J. E. Post (1989). *Modern Powder Diffraction.* Reviews in Mineralogy, Vol. 20, Mineralogical Society of America, Washington, D.C.

Buerger, M. J. (1960). *X-Ray Crystallography.* J. Wiley and Sons, New York.

Buerger, M. J. (1964). *The Precession Method.* J. Wiley & Sons, New York. Cambridge University Cambridge, UK.

Chung, F. H. (1974a). Quantitative interpretation of X-ray diffraction patterns, *i*. Matrix-flushing method of quantitative multicomponent analysis. Acta Crystallogl: 7:519-525,

Chung, F. H. (1974b). Quantitative interpretation of X-ray diffraction patterns, II. Adiabatic principle of X-ray diffraction analysis of mixtures. *Acta Crystallogr:* 7:526-531.

Chung, F. H. (1975). Quantitative interpretation of X-ray diffraction pattern, III. Simultaneous determination of a set of reference intensities. *Acta Crystallogr:* 8:17-19.

Davis, B. L. (1992). Quantitative phase analysis with reference intensity ratios. In *Proc. Conf: Accuracy in Powder Diffraction*, C. R. Hubbard and J. K. Stalick, Eds. NIST, Special Publ. 846, Gaithersburg, MD.

Delhez, R.; T. H. de Keijser; J. I. Langford; D. Louer; E. J. Mittemijer; and E. Sonneveld 1 (1993). Crystal imperfections and peak shape in the Rietveld method. In *The Rietveld Method*, R. A. Young, Ed., pp. 132-166. International Union of Crystallography, Chester, UK.

de Wolff, P. M. (1958). Particle statistics in X-ray diffractometry. *Appl. Sci. Res.* 7:102-122.

de Wolff, P. M. (1962). Indexing of powder diffraction patterns. *Adv. X-Ray Anal.* 6:1-17.

de Wolff, P. M.; and J. W. Visser (1964). Absolute Intensities. Report 641.109. Technisch Physische Dienst, Delft, Netherlands. Reprinted (1988) *Powder Diffract.* 3:202-204.

Elton, N. J.; and P. D. Salt (1996). Particle statistics in quantitative X-ray diffractometry. *Powder Diffract.* 11:218-229.

Gilliand, G. L. (1987). NBS Biological Macromolecules Crystallization Database. In *Crystallographic Databases*, F. H. Allen, G. Bergerhoff, and R. Sievers, Eds., pp. 156-157. International Union of Crystallography, Chester, UK.

Glusker, J. P.; and K. Trueblood (1985). *Crystal Structure Analysis: A Primer*. Oxford University Press, New York.

Goebel, J. B.; and A. S. Wilson (1965). Indexing Program for Indexing X-ray Diffraction Powder Patterns. Batelle-Northwest Laboratories, Report BNWL-22.

Gorter, S.; and D. K. Smith (1995). World Directory of Powder Diffraction Programs, Release 2.2, International Union of Crystallography, Chester, UK.

Hanawalt, J. D.; and H. W. Rinn (1936). Identification of crystalline materials. *Ind. Eng. Chem. Anal. Ed.* 8:244-247.

Hanawalt, J. D.; H. W. Rinn; and L. K. Frevel (1938). Chemical analysis by X-ray diffraction-classification and use of X-ray diffraction patterns. *Ind. Eng. Chem. Anal. Ed.* 10:457-512.

Hill, R. J.; and C. J. Howard (1987). Quantitative phase analysis from neutron powder diffraction data using the Rietveld method. *J. Appl. Crystallogr:* 20:467-474.

Hill, R. J.; and C. J. Howard (1990). Australian Atomic Energy Commission report No. M112, p. 15. Lucas Heights research Laboratory, PBM 1, Menai, NSW, Australia.

Himes, V. L.; and A. D. Mighell (1987). NBS crystal data: NBS*SEARCH: A program to search the database. In Crystallographic Databases, F. H. Allen, G. Bergerhoff, and R. Sievers, Eds. International Union of Crystallography, Chester, UK.

Howard, S. A.; and K. D. Preston (1989), Profile fitting of powder diffraction patterns. In *Modern Powder Diffraction*, pp. 217-276. Reviews in Mineralogy, Vol. 20. Mineralogical Society of America, Washington, D.C.

Ito, T. (1949). A general powder X-ray photography. *Nature (London)* 164:755-756.

Jenkins, R.; and D. K. Smith (1987). Powder Diffraction File. In *Crystallographic Databases*, F. H. Allen, G. Bergerhoff, and R. Sievers, Eds., pp. 158-177. International Union of Crystallography, Chester, UK.

Jenkins, R.; and R. L. Snyder (1995). *Introduction to X-ray Powder Diffractometry* J. Wiley & Sons, New York.

Le Bail, A.; H. Duroy; and J. L. Fourquet (1988). The ab-initio structure determination of lithium antimony tungstate (LiSbWO6) by powder X-ray diffraction. Mater Res. Bull. 23:447-455. Leroux J., Lennox, D. H., and Kay, K. (1953). Applications of X-ray diffraction analysis in the environmental field. *Anal. Chem.* 25:740-748.

Louer, D.; and M. Loner (1964). Methode d'essois et erruers pour l'indexatiion automatique des diagrammes successives. *J. Appl. Crystallogr* 15:271-275.

Mighell, A. D.; J. K. Stalick; and V. L. Himes (1987). NBS crystal data: Database description and applications. In *Crystallographic Databases*, F. H. Allen, G. Bergerhoff, and R. Sievers, Eds., pp. 133-143. International Union of Crystallography, Chester, UK.

Mueller, M. H.; and L. Heaton (1956). *Determination of Lattice Parameters with the Aid of a Computer*. ANL-6176. Argonne National Laboratory, Argonne, IL.

Nichols M. C.; D. K. Smith; and Q. C. Johnson (1982). Differential X-ray diffraction by wavelength variation, a preliminary study. *J. Appl. Crystallogr.* 25:301-350.

Nusinovic, J.; and M. J. Winter (1994). DIFFRAC-AT SEARCH: Search/match using full traces as input. *Adv. X-Ray Anal.* 37:59-66.

Parrish, W.; E. A. Hamacher; and K. Lowitzsch (1954). The Norelco X-ray diffractometer. *Philips Tech. Rev.* 16:123-133.

Rietveld, H. M. (1967). Lines profiles of neutron powder diffraction peaks for structure refinement. *Acta Crystallogr.* 22:151-152.

Rietveld, H. M. (1969). A profile refinement procedure for nuclear and magnetic structures. *J. Appl. Crystallogr* 2:65-71.

Rodgers, J. R.; and G. H. Wood (1987). NRCC Metals Crystallographic File (CRYSMET). In *Crystallographic Databases*, F. H. Allan, G. Bergerhoff, and R. Sievers, Eds., pp. 96-106. International Union of Crystallography, Chester, UK.

Runge, C. (1917). Die Bestimmung eines Kriatallsystems durch Roentgenstrahlen. *Phys. Zeit.* 19:509-515, Reprinted (1992) Powder Diffract 7:200-205.

Shirley, R. (1980). Data accuracy for powder indexing. In *Accuracy in Powder Diffraction*, S. Block and C. R. Hubbard, Eds., pp. 361-382. Special Publication No. 567. NBS, Gaithersburg, MD.

Smith, D. K.; G. G. Johnson Jr.; A. Scheible; A. W. Wims; and G. Ullmann (1987). Quantitative X-ray powder diffraction method using the full diffraction pattern, *Powder Diffract.* 2:73-77.

Smith, D. K. (1993). Particle statistics and whole pattern methods in quantitative X-ray powder diffraction analysis. *Adv. X-Ray Anal.* 35:1-15.

Smith, D. K.; S. Q. Hoyle; and G. G. Johnson Jr. (1993). Phase identification using whole pattern matching. *Adv. X-Ray Anal.* 36:287-300.

Smith, D. K.; and S. Gorter (1995). Software development for X-ray diffraction analysis 1950-1995. *Adv. X-Ray Anal.* 39:19-27.

Smith, D. K.; G. G. Johnson Jr.; and R. Jenkins (1995). A full-trace database for the analysis of clay minerals. *Adv. X-Ray*

Anal. 38:117-125.

Smith, D. K.; and G. G. Johnson Jr. (1997). POWD12, VAX Fortran version available from D. K. Smith and G. G. Johnson, Jr., Materials Research Laboratory, Pennsylvania State University, University Park, PA; PC Version available from. Materials Data, Inc., Livermore, CA.

Smith, G. S.; Q. C. Johnson; D. E. Cox; R. L. Snyder; D. K. Smith; R. S. Zhou; and A. Zalkin (1988). The crystal and molecular structure of beryllium hydride. *Solid State Commun.* 67:491-496.

Snyder, R. L.; H. L. Bunge; and J. Fiala, Eds. (1998). *Proc. Size and Strain Conference*, Liptovsky Mikulas. Slovakia, August 1995, International Union of Crystallography, Chester, UK.

Stout, G. H.; and L. H. Jensen (1989). *X-ray Structure Determination*. Wiley Interscience, New York.

Tanner, B. K. (1976). *X-Ray Diffraction Topography* Pergamon Press, Oxford, UK.

Taupin, D. (1968). Une methode generale pour l'indexation des diagrammes de poudres. *J. Appl. Crystallogr.* 1:178-181

Thompson, P.; D. E. Cox; and J. B. Hastings (1987). Rietveld refinement of Debye-Scherrer synchrotron. X-ray data of Al_2O_3 *J. Appl. Crystallogr:* 20:79-83.

Toraya, H. (1986). Whole-powder-pattern fitting without reference to a structural model: Application to X-ray powder diffractometer data. *J. Appl. Crystallogr:* 19:440-447.

Toraya, H. (1993). Position-constrained and unconstrained powder-pattern-decomposition methods. In *The Rietreld Method*, R. A. Young, Ed., pp. 234-275. International Union of Crystallography, Chester, UK.

Treacy, M. M. V.; M. W. Deem; and J. M. Newsam (1995). DIFFaX v1.80, NEC Research Institute, Inc.

Visser, J. W. (1969). A fully automatic program for finding the unit cell from powder data. *J. Appl. Crystallogr* 2:89-95.

Werner, P. E.; L. Eriksson; and M. J. Westdahl (1985). TREOR, a semi-exhaustive trail-and-error powder indexing program for all symmetries. *J. Appl. Cristallogr*: 18:367-370.

Young, R. A. (1993). *The Rietreld Method*. International Union of Crystallography Chester, UK.

Zevin, L. S.; and G. Kimmel (1995). *Quantitative X-Ray Diffractometry*, Springer, New York.

15.

Polymers and Pigments in Paint Industry

> It is better to fail in originality than to succeed in imitation
>
> — Herman Melville

1. ABSTRACT

A paint system is probably one of the most complex mixtures among industrial products. Easily it contains a dozen of components. Each component itself could be a formulated product. Every manufacturer tends to protect its trade secret. As a result, the paint formulators never know really what he is working with. Therefore, paint formulation remains more an art than a science, and paint analysis presents tough challenges. To simplify this situation, the components of a paint are grouped into four components: polymer, pigment, solvent, and additives. Generally, organic components are analyzed by IR, NMR, GC, etc, inorganic components by XRD, XRF, Atomic Absorption, etc. Because XRD & XRF are fast, simple and dependable, both are run first to determine what chemical elements and/or compounds are in a totally unknown sample.

2. INTRODUCTION
2.1 Old Industry
In the prehistoric period, cavemen used paint for decoration and

communication. They used blood, milk, egg, or tree saps (all are natural polymers) as binder, mixed with ashes (charcoal) and minerals (natural metal oxides) as pigments. Some cave paintings survive even today, proof of the remarkable durability of the paints. Noah used pitch to waterproof his ark around 4000 B.C. Egyptians and Greeks painted their statues (hair, eyes, and lips) ca. 3000-600 B.C. The Chinese and the Japanese developed ink and lacquer before 1000 B.C. for painting art and wooden objects. Slowly, the ancients applied paints to their ships, utensils, weapons, mummies, temples, and palaces for decoration and protection. In the mid-18th century, in England, the industrial revolution created an increased demand for paints. It marked the beginning of the modern paint and pigment industry (Morgan, 1990).

After World War I (1914-1918), there was a large excess of guncotton (nitrocellulose) in storage, a dangerous fire/explosion hazard. After a brief period of research, DuPont and Hercules dissolved it in amyl acetate (banana oil), and plasticized it to produce nitrocellulose lacquers for cars and furnitures. Other polymeric binders were subsequently developed, such as alkyds, polyesters, vinyls, acrylics, urethanes, melamines, silicones, and epoxies. During the World War II (1939-1945), due to a shortage of natural rubber latex, synthetic SBR rubber was developed. The war suddenly ended, leaving a large overcapacity for styrene-butadiene latex without a market. By changing the ratio of styrene to butadiene, the now popular latex paint was born. Important developments of modern paints and pigments have taken place in the 20th century, most of which occurred within the last 50 years (Lambourne, 1989).

2.2 Modern Paints and Pigments

The glittering automotive paints, the weather-tough architectural paints, and the heat-resistant aerospace paints represent the state-of-the-art modern paints. Modern paints contain four major components: polymers (binders), pigments (colors), solvents (carriers, evaporated into air during drying), and additives (modifiers to improve performance, such as driers, ultraviolet-absorbers, fungicides, surfactants, antisettling agents, and others). X-ray diffraction (XRD) and X-ray fluorescence (XRF) are quick and easy techniques to identify and confirm the pigment compositions for varied purposes.

In general, inorganic pigments are weak and dull. They provide good covering power at low cost. They have outstanding resistance to the attack by light, heat, and solvents, but often are sensitive to acid and alkali. Organic pigments are almost invariably strong and bright, but their fastness varies widely from poor to outstanding. These two types are often used together to achieve a particular set of properties. Color has been taken as excluding black and white. Most black pigments consist of specially treated carbon. White pigments are invariably of inorganic origin. Titanium dioxide was commercialized in 1918. It replaces white lead for higher hiding power and no toxicity. White pigments with low refractive indices (1.7 or less) are called extenders (fillers). They provide very little hiding, but contribute to other functional qualities such as gloss and texture. The particle-size distribution of extenders has much wider range, 0.01 to 50 µm, than that of pigments, 0.01 to 5 µm (Solomon & Hawthorne, 1983; Carter et al:, 1984). Particle-size distribution influences hiding, viscosity, gloss, and surfactant demand of the paints.

Pigment science could not develop until after the discovery of X-ray diffraction in 1912. Binder science began when nitrocellulose (gun-cotton) was used for lacquers after the end of World War I (1918). The attention to pigment-binder interactions started in the 1950s (Parsons, 1993).

Expanding environmental concerns and shrinking margins led to a series of mergers, acquisitions, and hostile takeovers. The number of paint companies has dropped to less than 1000 in the United States, and about 1500 in western Europe. The leading U.S. paint manufacturers include Sherwin-Williams, PPG, Du Pont, Glidden, Valspar, Benjamin Moore, and RPM. Giant paint companies worldwide include ICI, BASF, Courtaulds, AKZO, Hoechst, Nippon Paints, and others.

2.3 Directions of Modern Paint Research

Since the 1970s, research efforts to the paint industry have been directed toward three issues: energies, materials, and regulations. They are interrelated, and all have great impact on costs. Petroleum is the major source of energy for industry. It is also the source of raw materials for polymers, pigments, solvents, and additives used in paints. Logically, the paint industry must continue to develop low-energy processes for coatings, to search for renewable or alternative sources of raw materials, and to reformulate paints for compliance with U.S. Environmental Protection Agency (EPA) and Occupational Safety and Health Administration (OSHA) regulations yet maintain quality.

Concerted efforts have made great advances in all fronts: Low-energy curing includes ultraviolet (UV) and electron-beam curing, electrocoating, and inmold coating. Low-energy cross-linking includes epoxies, urethanes, and latex systems. Renewable sources of materials include polymers from starch, cellulose, and silicones. Paints of low volatile organic contents (VOC) include water-based paint, high-solids paints and powder coatings. Petroleum solvents have been replaced with water, alcohol, and liquid carbon dioxide. Almost all paints have been reformulated to comply with numerous government regulations limiting VOC, waste disposal, and carcinogens.

Intense analytical supports are needed to carry out these research projects. X-ray diffraction techniques are widely used for pigment identification, polymer crystallinity (spherulite forming), and compliance of government regulations (Chung, 1981a). Generally, XRD, XRF, and atomic absorption (AA) are favored for inorganic/elemental analyses; infrared (IR), gas chromatography/mass spectroscopy (GC/MS), and liquid chromatography (LC) are methods of choice for organic/molecular analysis. Other micro or surface analysis such as ESCA, SIMS, AES, ISS, and MOLE are usually run by outside services, depending upon necessity.

3. X-RAY DIFFRACTION FOR CHEMICAL ANALYSIS
3.1 Analytical Functions in Paint Industry

The analytical department of the paint industry carries out four major functions: R&D support, quality control, problem-solving, and environmental compliance, with occasional outside consultation or services. In order to fulfill its obligations, the analytical lab is usually equipped with an impressive array of expensive instrumentation including:

Elemental analysis: X-ray fluorescence (XRF), atomic absorption (AA), inductively coupled plasma (ICP), wet chemical, etc.

Molecular analysis: X-ray diffraction (XRD), infrared (IR or Fourier-transform infrared, FTIR), nuclear magnetic resonance (NMR), chromatography (GC, LC, GPC), etc.

Property characterization: Particle size distribution, rheological properties, thermal analyses, light scattering, optical microscopy (brightfield), polarizing, phase-contrast, and stereozoom), scanning electron microscopy, and other physical/chemical measurements such as adhesion, tensile (Instron), impact strength, corrosion resistance, etc.

XRD, XRF, IR, and NMR are usually the first and the most frequently used for various projects requesting analytical data.

3.2 Harsh Demands

Chemical analyses in industrial laboratories, particularly in paint industry, are very demanding. Frequently, the samples represent a performance complaint from the field, or a production problem in the factory. The sales contracts, the production runs, or the spraying operations are put "on hold," waiting for the analytical data to derive a solution. The urgency and high profile put a lot of pressure on the analytical chemists, because the sample and the analyses are usually:

> **Non routine:** Except quality control samples, every sample represents a problem looking for solutions.

Totally unknown: No one knows its chemical composition. Only know its source and relevant problem.

Complex mixture: Any paint could have a dozen or more components. Some component itself is a formulated product with trade secrets.

Quantitative: Need numerical data to define the problem or to find the solution.

Multicomponent: Reqesting XRD & XRF analyses for multiple chemical compounds or chemical elements in sample.

ASAP: Need results "as soon as possible".

ACAP: Want cost "as cheap as possible".

In order to cope with these demands, the Sherwin-Williams Company developed and published the matrix-flushing XRD method for quantitative multicomponent analysis (Chung, 1974a) and the thin-film XRF method for quantitative multi-element analysis (Chung, 1976; Chung, Lentz, & Scott 1974). Both methods are designed to get rid of the matrix (absorption) effects, such that intensity is linearly proportional to concentration. Both methods are simple, easy, and fast, generating quantitative data accurate enough to point out the root of a problem or lead to the direction of a solution.

3.3 X-ray Diffraction Analysis
Sample Preparation

Original samples commonly encountered in the paint and pigment industry and their mode of mounting in a scanning diffractometer are cited here. The operator must be alert to avoid the possible preferred orientations of platy, needle, or fibrous components (Jenkins & Snyder, 1996).

Sample form	Mode of mounting
Powder	(1) Free fall into a powder sample holder, or (2) press into a circular cake with a special mold at specific pressure; or (3) sprinkle the powder on a film of petroleum jelly or amorphous glue.
Paint chip	Pulverize, then run as powder or place in a Gandolfi camera.
Paint panel	Cut the panel to fit sample holder.
Wet paint	(1) Make drawdown on Mylar film or aluminium panel, run as a cast film, or (2) separate solids from liquids by centrifuge. The pigment plug is then dried, ground, and run as powder.
Tiny sample	The amount of sample needed for diffractometer scan is about 0.1 to 2 g. For very small samples, 1 to 100 mg, the Debye–Scherrer or Gandolfi camera has to be used. The tiny sample can be loaded into a thin-walled capillary or mixed with glue to form a ball loading on a glass filament. It can also be mounted on single-crystal quartz or silicon.

Qualitative Analysis

Each X-ray lab usually compiles a file of frequently encountered X-ray diffraction patterns as standard references, including 30-40 clean pigments and extenders commonly used in paints, typical raw materials from suppliers, and popular products of the company. Note that preferred orientation is quite common in paint samples due to the paint drying process and the presence of mica, talc, bentonite, kaolinite, zinc phosphate, and metal flakes (Al and Zn).

Commonly, the major components can be quickly identified by the standard procedures using the reference filed compiled in-house. Systematic search/match procedures using the PDF are then used to identify the unassigned peaks (Jenkins & Holomany, 1987). Information from the source of sample, the purpose of analysis, the elemental analysis by XRF, and the functional groups by IR can significantly speed up the identification process.

Quantitative Analysis

Almost all quantitative phase analyses of paint samples, pigment samples, and compliant samples were run by the matrix-flushing XRD method, now known as the reference intensity ratio (RIR) method (Chung, 1974b, 1974c, 1975; Davis, 1986). It consists of three working equations relating intensity (I_i, cps) to concentration (X_i, %weight) through the reference intensity ratios (I_i/I_c), or RIR, which are a set of constants. The powder diffraction file (PDF) defines the RIR (k_i) as the intensity ratio (I_i/I_c) of the most intense lines from a binary mixture made with pure compound and synthetic corundum by one-to-one weight ratio. PDF compiles and updates a comprehensive list of RIR values. The RIR values of many natural or commercial products are source specific or supplier specific and hence have to be determined in the lab. To save time, a set of k_i of interest can be determined simultaneously (Chung, 1975) by using the matrix-flushing principle. The RIR method has been adopted by the American Society of Testing and Materials for the analysis of paints and pigments (ASTM, 1978).

The Chung's Matrix-Flushing theory derives three equations for quantitative X-ray diffraction analysis: (1) Matrix –Flushing (Standardless) equation, (2) External Standard (Reference Standard) equation, and (3) Determinant (Criteria) equation. These three equations are exact deduction, neither assumption nor approximation was made in their derivation. As illustration, a numerical example is presented for each of the three working equations. For fast response instead of best accuracy, all intensities were peak heights, which are the maximum counting rates in counts per second of the peak from the strip charts. For better accuracy, the integrated intensities of the characteristic peaks should be used to accommodate the size/stress broadening, if any. The integrated intensity of the whole pattern would minimize the effect of preferred orientations and extinctions (Smith et al., 1987). Using the peak-height intensity, one may miss the bull's eye, but never miss the bull's head.

(1) Matrix-Flushing Equation: When all components are crystalline and identified, The Chung's Matrix-Flushing Equation (1) is used. It is nicked **Standardless method**, because neither internal standard nor external standard is required. In this case, the percent weights of all components can be obtained from a single scan of the original sample. Its application is illustrated in Table 1.

$$X_i = \left(\frac{k_i}{I_i} \sum_{i=1}^{n} \frac{I_i}{k_i} \right)^{-1} \quad (1)$$

Table 1 Numerical Example of the Matrix-Flushing Equation

Component	Composition (g)	Intensity I_i (cps)	RIR k_i	% Weight Known	Found[a]
ZnO	0.2236	610	4.35	9.87	9.3
NiO	0.5454	1412	3.81	24.26	24.5
CdO	0.6588	3303	7.62	29.07	28.6
KCl	0.8386	2207	3.87	37.00	37.6
Total					100%

$$X_{ZnO} = \frac{1}{1 + \frac{4.35}{610} \left(\frac{1412}{3.81} + \frac{3303}{7.62} + \frac{2207}{3.87} \right)} = 9.26\% \quad (2)$$

(2) External Standard Equation: When some compounents are amorphous or unidentified, the Chung's Matrix-Flushing equation (2) is used. It is nicknamed **External Standard Method**, because an external (reference) standard is required. The slope of the External Standard Equation (2) is a **characteristic constant** free from matrix effect. Each compound or element in any mixture has the same characteristic constant slope,. Note that the slope of the the traditional **Internal Standard method** must be determined by experiments. Because for the same component, its slope changes from sample to sample. Its application is illustrated in Table 2.

$$X_i = \left(\frac{X_c}{k_i}\right)\left(\frac{I_i}{I_c}\right) \tag{2}$$

Table 2 Numerical Example of the External Standard Equation

Component	Composition (g)	Intensity I_i (cps)	RIR k_i	% Weight Known	Found[a]
ZnO	0.6759	2408	4.35	24.38	25.3
TiO$_2$	0.4317	931	2.62	15.57	16.2
CaCO$_3$	1.1309	2558	2.98	40.79	39.2
Al$_2$O$_3$	0.5341	420	1.00	19.26	Int. std.

[a] Use Eq. 2 to calculate all the % weights found. The Al$_2$O$_3$ is an external standard added into sample

$$X_{ZnO} = \frac{19.26}{4.35} \cdot \frac{2408}{420} = 25.3\% \text{ in doped sample}$$

The weight percent of ZnO in the undoped sample is:

$$\%ZnO = \frac{25.3}{100 - 19.26} = 31.3$$

(3) Discriminant Equation: The discriminant equation (3) is used to test the presence or absence of amorphous components in sample. It needs an external standard, but no calibration lines are required. Use the external standard equation to calculate the percent weights of all the crystalline components. Then, use mass balance to calculate the weight fraction of amorphous content. Its application is illustrated in Table 3.

Table 3 Numerical Example of the Discriminant Equation

Component	Composition (g)	Intensity I_i (cps)	RIR k_i	% Weight Known	% Weight Found[a]
ZnO	0.8090	4948	4.35	40.81	38.7
CdO	0.2825	3337	7.62	14.25	14.9
Resin	0.4057	0	---	20.46	21.9
Al_2O_3	0.4854	719	1.00	24.48	Int. std.

[a] Use Eq. 3 to calculate the terms on both sides, compare their values, then decide whether there is amorphous content in sample:

$$\sum_i^n \frac{I_i}{k_i} = \frac{4948}{4.35} + \frac{3337}{7.62} = 1575$$

$$I_c \frac{X_o}{X_c} = \frac{719 \times 75.52}{24.48} = 2218$$

$$\sum_i^n \frac{I_i}{k_i} \overset{\geq}{\underset{<}{=}} I_c \frac{X_o}{X_c} \tag{3}$$

Where

    = indicates that all components are crystalline,

    < indicates the presence of amorphous material.

    > indicates wrong experimental data,

Note that $\Sigma I_i / k_i = 1575$ is less than $I_o X_o / X_c = 2218$. This indicates the presence of amorphous content according to Eq. 3. Mass balance gives 21.9% noncrystalline content (100 - 38.7-14.9-24.5 =21.9). The actual composition of the original sample without the added internal standard can be easily obtained by simple conversion:

    % Amorphous content (resin in undoped sample) = 21.9 / (100 - 24.48) = 29.0

(4) Limitations:

With proper reference standards, which means both the standard and sample have similar purity, crystallinity, and particle size, the precision (S/X) of the matrix-flushing method is about 5% relative or better. The major limitations of quantitative XRD analysis in general, and for paints and pigments in particular, are:

> **Amorphous content:** Some pigments are amorphous, hence escape detection such as carbon black, amorphous silica, and highly processed clays.
> **Particle shape:** Platelet, acicular, or fiberous components such as mica and metal flakes tend to assume strong preferred orientations in sample, which invalidate the linear relationship between intensity and concentration.
> **Different supliers**: The same pigment from different manufacturers may have different degrees of crystallinity; consequently, their RIR could be supplier specific.
> **Source specific:** The natural minerals used as fillers usually have varied impurity, crystallinity, and imperfection. Consequently, their RIR could be source specific (Davis, Smith, & Holomany, 1989).

3.4 Integrated Intensity vs. Peak Height

The integrated intensities can be measured by scanning integration, by step scanning, by peak areas with a planimeter, or by peaks weights of the cut peaks from a strip chart. In order to show what kind of differences can be expected from different mode of intensity data collection, Chung (1982) collected a set of intensity data from a mixture of 53.91% cristobalite (99.6% pure) and 46.09% corundum as listed in Table 4. The flush constants (RIR, k_j), calculated with different versions of intensity data, agree well within experimental errors.

When diffraction peaks are noticeably broad, obviously integrated intensities must be used for most accurate quantitative analysis.

Table 4 Integrated Intensity Versus Peak Height

	Scanning integration counts	Peak area (cm^2)	Peak weight (g)	Peak height (counts/s)
Cristobalite, (101)	325,130	132.85	0.5950	13,800
Corundum, (113)	67,065	26.85	0.1198	2960
RIR, k_i	4.16	4.24	4.26	4.00

Table 5 Polymorphs of Common Pigments

Material	Crystal Polymorph	Property/Use
Phthalocyanine blue	Alpha, stabilized	Reddish blue
	Beta	Greenish blue
	Gamma	Not used as pigments
Quinacridone	Beta	Violet
	Gamma	Red
	Alpha and delta	Not used as pigments
Titanium dioxide	Rutile	Refractive index 2.76, high opacity, resistant to chalking
	Anatase	Refractive index 2.55, pure white, tendency to chalk
	Brookite	Not used as pigments
Lead chromate with lead sulfate	Orthorhombic	Pale primrose
	Monoclinic	Lemon

The choice between peak-height intensity and integrated intensity is really a choice between speed and accuracy. In the paint industry and many other industries, timely response and analytical cost are major concerns. Many cases can be solved by the peak-height intensity instead of integrated intensity as long as the I_i and k_i are treated consistently. The error from using peak-height intensity arises from the differences in particle size distribution, lattice imperfection, and impurity between samples and standards.

3.5 Pigment Polymorphs

General information of pigments can be found in a comprehensive reference work, *Colour Index*, published and updated by the Society of Dyers and Colourists. It lists over 5000 different pigments used in the paint, ink, plastic, rubber, cement, and paper industries (Society of Dyers and Colourists, 1982). Most pigments are crystalline; some have several polymorphs (i.e., the same chemical substance has different crystal structures). Seine of the crystal structures may be more stable than others, and some may not be suitable for use as pigments. The polymorphs of some common pigments are cited in Table 5. X-ray diffraction is the only technique for their identification and analysis (Snider, 1992).

3.6 Pigment Particle Size and Surface Treatment

Particle size distribution of pigments affects free surface area, oil absorption, opacity, color hue, brightness, tinting strength, durability, and gloss of paint films. Particle sizes near the half wavelength of light ($\lambda/2$) provide the best hiding power due to the most effective scattering properties. Particle sizes close to the wavelength of light (2) are least effective scatterers of light and lead to paint transparency and high gloss. Typical ranges of pigment particle sizes are:

Organic pigments	0.01 to 1.00 μm
Inorganic pigments	0.10 to 5.00 μm

Titanium dioxide	0.22 to 0.24 µm
Carbon black	0.01 to 0.08 µm
Extenders	0.01 to 50 µm

Note that titanium dioxide has the best hiding feature for two reasons: high refractive index (2.76) and optimum particle size (0.23 µm, a good match for half the average wave-length of white light in air). The particle surface of titanium dioxide is usually treated with alumina, silica, fatty acids, or amines to improve its dispersibility, durability, opacity, tinting strength, and blocked photoactivity to protect the binder.

The surface of pigment particles is the most important feature. Its polarity governs the affinity for various polymers, hence affects dispersion and stability of the liquid paints. Consequently, pigment particles are deliberately modified by the manufacturers (trade secrets) to improve performance.

When the particle size of a powder is below 1000 Å (0.1 µn), it can be estimated from the width of their diffraction peaks through the Scherrer equation (Klug & Alexander, 1974):

$$L = \frac{K\lambda}{B \cos \theta} \tag{4}$$

where L is the mean crystallite size, λ is the wavelength of X-ray, B is the broadening of the peak, is the Bragg angle of the peak, and K is a physical constant related to the shape of the crystallites and the manner of defining B and L.

Although the average particle size and surface treatment of pigments can be characterized by XRD and XRF, other techniques such as SEM, JLDC and a Coulter counter can give more direct and detailed information including particle size distribution and multi-modal features. However, XRD is the best and probably the only technique for analyzing unconventional particles such as crystalline domains in polymers (100-400 Å), micelles or microvoids in fibers (15-200 Å), grains (1 mm-1 μn), or precipitates (500 Å) in metals, or type, amount, and crystallite size of pretreatment on metals such as the hopeite or scholtzite on galvanized steel (Chung, 1989; Snider, 1992).

3.7 Crystallinity of Polymers

Crystallinity is defined as the weight fraction of the crystalline portion of a polymer. Because the crystallinity of a polymer has profound effects on its performance, five different methods have been developed for determining the crystallinity of polymers: X-ray diffraction, density, infrared, NMR, and heat of fusion. X-ray diffraction method provides a sound physical definition of "order" in molecular packing of the solid: The crystalline fraction diffracts X-rays coherently according to Bragg's law, giving sharp peaks, while the amorphous fraction scatters X-rays incoherently, giving a diffuse halo. The X-ray diffraction pattern of a semicrystalline polymer is the superposition of sharp peaks over a diffuse halo.

There are two major problems involved in the X-ray method: (1) It needs near-perfect crystalline polymer standards, which do not exist, and (2) it needs quantitative separation of intensity contributions from crystalline peaks, amorphous halo, and background. Chung and Scott (1973) reported a technique to solve these two problems. Their technique uses the totally amorphous polymer as standard, which can be made from the partially crystalline sample by the melt-quench-grind procedure, and selects a simple yet effective analytical function (simplest Lorentzian function with only one empirical parameter) to trace the overlapping amorphous halo:

$$y = \frac{a}{x^2 + 1} \tag{5}$$

where: x is the two-theta value, y is the peak height intensity, and a an empirical constant that can be calculated from the X-ray diffraction pattern of the partially crystalline polymer sample.

The partially crystalline polymer is considered as a two-component mixture: the crystalline component (weight% = X_c) and the noncrystalline component (weight% = 1- X_c). The matrix-flushing equation, Eq. 1, degenerate, to the following simple form:

For the unknown:

$$X_c = \frac{1}{1 + k\frac{I_a}{I_c}} \tag{6}$$

$$0 \cdot 80 x_c = \frac{1}{1 + k\frac{I'_a}{I'_c}} \tag{7}$$

Solve for k and x_c

For 80/20 mixture:

The 80/20 mixture is a blend of 80% original unknown sample and 20% amorphous standard. Solve the two equations for the slope k and crystallinity X_c. Twenty numerical examples were cited in the reference paper. Excellent agreement was obtained between results from the XRD method and the density method (ASTM method D153-B). The X-ray diffraction patterns of polyethylene terephthalate and its amorphous standard are shown in Figs. 1 and 2.

For quantitative X-ray diffraction work, the integrated intensity should be preferred to the peak height. However, due to serious overlapping, it is difficult to obtain the integrated intensities with reasonable confidence. The difference between the integrated intensity and peak height will be compensated by the empirical constant k (slope of calibration line), which also takes care of the absorption factor and the Lorentz polarization factor automatically.

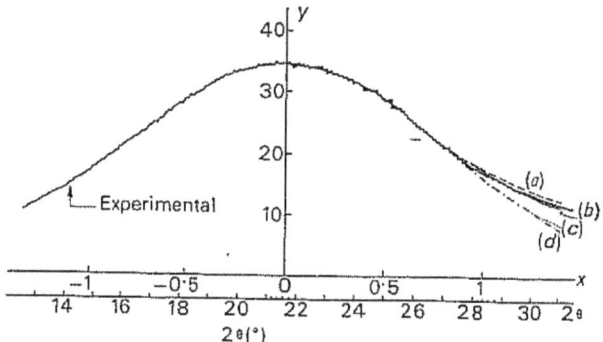

Figure 1 The X-ray diffraction pattern of a completely amorpous PET polymer and its intensity distribution functions (a) $y = a / (x^2 + 1)$; (b) $y = a \, \text{sech} \, x$; (c) $y = ae^{-x^2}$; (d) $a \sin^2 x / x^2$

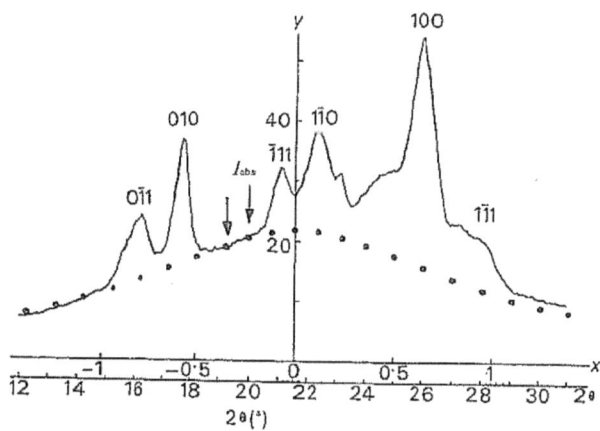

Figure 2 The x-ray diffraction patern of a partially crystalline PET polymer and the calculated amorphous halo. Cu $K\alpha$ radiation: 50 kV, 40 mÅ; full-scale intensity: 3200 c.p.s.; scanning speed: 1° (2θ;) per min; time constant: 2 sec.

The crystallinity of polymer by X-ray diffraction is a measure of the "order" versus "disorder" of molecular packing in the polymer. The Bragg peaks represent "ordered" molecules. The halo represents the "disordered" molecules. The paracrystalline state, the lattice imperfections, the crystalline size effect, the strain/stress effect, the thermal vibrations, and the noncrystalline structure are just different kinds of "disorder."

4. APPLICATIONS
Qualitative XRD Analysis

Standard search/match procedures are used for qualitative analysis. Positive identification of the components in a mixture often leads to the solution of a technical problem, the mechanism of a reaction, or the quality of a product. Some industrial cases are briefly cited next (Chung, 1981b).

- **Color Problem of Paint.** The sheet metal of an appliance was originally coated green. After a brief period of service, its green color changed to blue. X-ray diffraction analysis indicated that the pigment in the original paint was composed of lead chromate and iron blue; the exposed paint contained an extra component, lead chloride. This finding led to the conclusion that the green color is a composite of yellow (lead chromate) and blue (iron blue). Acid fumes (HCl) in that particular environment attacked the lead chromate ($PbCrO_4$, yellow) forming lead chloride ($PbCl_2$, white). The depletion of a yellow component of the green color ended in a blue color (Scott, 1975).

- **False Accusations.** Sherwin-Williams produces a corrosion-inhibiting pigment, Moly-White (patented product), that is quite popular because it is effective, nontoxic, and white. Two separate customers complained that the Moly-White pigment caused seed formation in their paint. X-ray diffraction analysis of the isolated seeds from the seedy paints indicated that the seeds in one paint were bentonite balls, and the seeds in the other paint were spherulite particles of the polymers used in its formulation. Both results were confirmed by the customers, and both complaints were withdrawn.
- **Adhesion Problem.** The sheet metal for "tin" cans has a tin coat by "hot dipping" process. A thin film of tin adheres to the steel surface, thus protecting it from corrosion. An adhesion problem popped up in that certain steel sheets could not be wetted by the molten tin. X-ray diffraction analysis of these faulty steel sheets found a fine deposit of graphite on the steel surface. The presence of graphite prevented the contact between steel and molten tin. The solution is obvious (Chung, 1991).
- **Peeling of New Paint.** A premium paint was used for aluminum windows in a government housing project. The paint peeled before the houses were occupied. It was a crisis situation for the liable paint company. X-ray diffraction and fluorescence analyses indicate that the adhesion failure was caused by the lack of chromate pretreatment of the aluminum windows, a requirement for painting all aluminum metal. The liability was shifted from the paint company to the paint contractor.

- **Grass-killing Fertilizer.** After a proper application of fertilizer, the grass became withered instead of flourishing. The results of elemental analyses of the good and the bad fertilizers were identical. X-ray diffraction analyses indicated the presence of a potassium pyrosulfate in the bad fertilizer, and a potassium sulfate in the good fertilizer. Potassium pyrosulfate is highly acidic in water, hence killing the grass.
- **Optical Crystallography.** Optical microscopes (brightfield, phase-contrast, polarizing, and stereozoom) are frequently used to examine spherulites in resin, painted surfaces, cross sections, crystalline powders, adhesion failures, and various complaint samples. For example, Sherwin-Williams produces isophthalonitrile (IPN), a free-flowing powder. A caking crisis popped up when the 50-pound bags caked to solid blocks. Over a million pounds of IPN piled in the warehouse due to rejection. Huge cash flow was tied up. X-ray diffraction did not show any difference between the good and the bad IPN. Microscopical examination indicates that the IPN crystal habit changed from granular to acicular (needles). Needles have larger contact surface and hence tend to cake. By adjusting the process parameters, the IPN crystal habit returned to granular, and the problem disappeared.

4.2 Quantitative XRD Analysis

Quantitative analyses are frequently requested for quality control, problem solving, monitoring a competitor's new or popular products, and environmental protection. For practical purposes, the results by the matrix-flushing method with peak-height intensities (counts per second) can satisfy most requests. Almost all quantitative XRD work in the paint and pigment industry uses powder diffraction and rarely use single crystals, primarily because people are cost conscious and always want the results as soon as possible (Federal Highway Administration, 1979; Kamarchik, 1980; Curry & Rendle, 1982; Cunningham & Kamarchik, 1984).

- **Anatase in Rutile.** Titanium dioxide is the best white pigment. It replaces the toxic white lead because of its high hiding and nontoxicity. Titanium dioxide has three polymorphs: rutile, anatase, and brookite. Rutile TiO_2 is normally used in paints, because anatase TiO_2 tends to chalk. The specification for rutile TiO_2 limits the anatase content to be below 0.2% by weight. X-ray diffraction is the only technique for its analysis, Quantitative XRD analysis of percent anatase in rutile is routinely run for every batch of titanium dioxide by manufacturers, as well as by users (ASTM, 1978).
- **Chalking Coil Coating.** A chalking problem (white powder migrated to the surface of paint film) of a gray paint created a multi-million-dollar liability issue. Quantitative X-ray diffraction analysis of the paint chip indicates high anatase content (a few percent by weight vs. spec. of <0.2%). The pigment manufacturer admitted that the calcination temperature for making the gray pigment was

lowered to save energy, and consequently the anatase/rutile conversion was incomplete. The liability was shifted from the paint company to the pigment company.

- **Zinc Molybdate Pigment.** The active component of Moly-White pigment (patented product by Sherwin-Williams) is a basic zinc molybdate, a corrosion inhibitor. Its crystal structure was determined by X-ray diffraction. Its reaction kinetics was followed by successive powder diffraction patterns. Its transition point and optimum manufacturing conditions were established. These fundamental quantitative data were used for patent application and factory manufacturing.

- **Silica and Asbestos in Paints.** White pigments constitute by far the largest percentage (some 90%) of all pigments used. White pigments are of two types: (1) hiding pigments, with refractive indices above 1.7, such as TiO_2, ZnO, and white lead; and (2) extender pigments, with refractive indices below 1.7, such as $CaCO_3$, $BaSO_4$, talc, silica, and china clay. Extenders lower the cost and also control gloss, viscosity, texture, suspension, etc. Talc crystals can be platy (micaceous), fibrous (foliated), nodular (steatite), or acicular (tremolite impurity). Silica can be crystalline (quartz) or amorphous (diatomaceous). OSHA regulates asbestos and silica, because they may get into the air by sanding or flaking. X-ray diffraction and optical microscopy are routinely used for the analysis of crystalline silica and asbestos in talcs (Chung, 1978, 1982).

- **Quinacridone Pigments.** Quinacridone pigments have

outstanding light, heat, and chemical fastness. They are nonbleeding, provide a wide gamut of shades ranging from orange to violet, and hence supplement the phthalocyanine blues and greens in extending the availability of high-quality colors. There are four polymorphic forms of linear trans-quinacridones, and some 150 substituted quinacridones have been synthesized, but only a few are useful as pigments. For complete characterization, single crystals of unsubstituted gamma-form quinacridone (red) and beta-form 4,11-dichloroquinacridone (scarlet) were grown by vacuum sublimation, their single crystal structures were determined by X-ray diffraction, and their powder patterns were indexed, together with thermal, particle size, tinting strength, and other analyses (Chung & Scott, 1971; Chung, 1971).

- **Crystallinity of Polymers.** Polyethylene terephthalate (PET) polymer can be amorphous or crystalline, depending on applications. Fibers and plastics can be highly crystalline. Paint films are predominantly amorphous. X-ray diffraction is the most direct technique to identify and quantify its crystallinity. The Chung and Scott procedure for determination of the crystallinity of polymers has been frequently adopted in industrial labs for quality control applications (Chung & Scott, 1973).

5. CONCLUSIONS

A paint system is probably one of the most complex mixtures among industrial products. A paint formula could easily have over a dozen of components; each component itself could be a formulated mixture. Manufacturers of resins, pigments, and special additives are usually reluctant to reveal details of their products (trade secrets). As a result, the paint formulator does not know what precisely he or she is working with. Consequently, paint formulation remains more an art than a science, and paint analysis presents tough challenges. Some minor components or contaminations can never be identified.

X-ray powder diffraction is an essential technique for chemical (phase) analysis of crystalline components in paints, pigments, and related samples. Specimen preparation and search/match procedures are typical and conventional. The matrix-flushing (RIR) method is routinely used for quantitative analyses. It is quick, easy, and nondestructive, yet offers accurate and convincing results. X-ray single-crystal work is rarely done in the paint and pigment industry except for patent claims or basic research.

The major contributions of analytical laboratories to company business are four: (1) supporting research projects, (2) solving problems from customer complaints, (3) trouble-shooting in production processes, and (4)assuring compliance to EPA and OSHA regulations. Unfortunately, these functions are not the main stream of company business. In industry, one gets greater recognition by introducing new products, designing efficient processes, or striking big sales. The array of snappy modern scientific instruments is impressive indeed. However, the analytical chemists are usually unsung heroes. Like FBI agents, their efforts are indispensable yet mostly invisible. Nevertheless, the greatest reward in science is the eternal pleasure of cracking a puzzle or discovering a secret.

6. REFERENCES

American Society for Testing and Materials (1978). Standard Test Method D 3720-78. Philadelphia: ASTM.

Carter, J. R.; et al. (1984). Combined X-ray and IR methods to assure paint pigment value, Research and Development, February, p. 124.

Cluing, F. H.; and R. W. Scott (1971). Vacuum sublimation and crystallography of quinacridones, *J. Appl. Cryst.*, 4, 506.

Chung, F. H. (1971). Crystallography of toluidine red, *J. Appl, Cryst.*, 4, 79.

Chung, F. H.; and R. W. Scott (1973). A new approach to the determination of crystallinity of polymers by X-ray diffraction, *J. Appl. Cryst.*, 6, 225.

Chung F. H.; A. J. Lentz; and R. W. Scott (1974). A versatile thin film method for quantitative emission analysis, X-ray Spectrometry 3, 172.

Chung, F. H. (1974a). A new X-ray diffraction method for quantitative multi-component analysis, Adv. X-Ray Anal. 17, 106.

Chung, F. H. (1974b). Quantitative interpretation of X-ray diffraction patterns of mixtures. *i.* Matrix flushing method, J. Appl. Cryst. 7, 519.

Chung, F. H. (1974c). Quantitative interpretation of X-ray

diffraction patterns of mixtures. II. Adiabatic principle, J. Appl. Cryst. 7, 526.

Chung, F. H. (1975). Quantitative interpretation of X-ray diffraction patterns of mixtures. III. Simultaneous determination of a set of reference intensities, J. Appl. Cryst. 8, 17.

Chung, F. H. (1976). A new approach to quantitative multi-element X-ray fluorescence analysis, Adv. X-Ray Anal., 19, 181.

Chung, F H. (1978). Imaging and analysis of airborne dust for silica, Environ. Sci. Technol., 12, 1208.

Chung, F. H. (1981a). Imaging and analysis of airborne particulates, in *Air/Particulate: Instrumentation and Analysis*, ed. P. N. Cheremisinoff; p. 89. Ann Arbor, MI: Ann Arbor Science.

Chung, F. H. (1981b). X-ray diffraction techniques and instrumentation, in *Analytical Measurements and Instrumentation for Process and Pollution Control*, ed. P. N. Cheremisinoff and H. J. Perlis, p. 151. Ann Arbor, MI: Ann Arbor Science.

Chung, F. H. (1982). Synthesis and analysis of crystalline silica, Environ. Sci. Technol., 16, 796. Chung, F. H. (1989). Industrial applications of X-ray diffraction, American Laboratory, 21(2), 144.

Chung, F. H. (1991). Unified theory and guidelines on adhesion. J. Appl. Polymer Sci., 42, 1319.

Cunningham, G. P.; and P. Kamarchik (1984). Automation and computerization of a coatings research analytical laboratory, Progress in Organic Coatings, 12, 369.

Curry, C. J.; and D. F. Rendle (1982). Pigment analysis by X-ray diffraction, J. Forens. Sci. 22, 173.

Davis, B. L. (1986). *Reference Intensity Method of Quantitative X-ray Diffraction Analysis*, South Dakota School of Mines and Technology Rapid City, SD.

Davis, B. L.; D. K. Smith; and M. A. Holomany (1989). Powder Diffraction, 4, 201.

Federal Highway Administration, U.S. Department of Commerce. (1979). X-ray Diffraction Analysis of Selected Paint Pigments, Sacramento, CA, PB80-13061.

Jenkins, R.; and M. Holomany (1987). Powder Diffraction, 2, 215.

Jenkins, R.; and R. L. Snyder (1996). *Introduction to X-ray Powder Diffractometry*. New York: Wiley and Sons.

Kamarchik, P. (1980). Quantitative crystalline pigment analysis by X-ray diffraction, J. Coating Tech., 52, 79.

Klug, H. P.; and L. E. Alexander (1974). *X-ray Diffraction Procedures*. New York: John Wiley and Sons.

Lambourne, R. (1989). *Paint and Surface Coatings: Theory and Practice*. New York: John Wiley and Sons.

Morgan, W. M. (1990). Outlines of Paint Technology. New York: John Wiley & Sons.

Parsons, P. (1993). *Surface Coatings-Raw Materials and Their Usage*. New York: Chapman & Hall.

Scott, R. W. (1975). X-ray analysis: Diffraction and emission in *Characterization of Coatings: Physical Techniques, Part II*, ed. R. R. Myers. New York: Marcel Dekker.

Smith, D. K., et al. (1987). Quantitative X-ray powder diffraction using the full diffraction pattern, Powder Diffraction, 2, 73.

Snider Jr., A. M. (1992). X-ray techniques for coating analysis, in *Analysis of Paints and Related Materials*, ed. W. C. Golton, ASTM STP 1119, p. 82.

Society of Dyers and Colourists. (1982). *Colour Index: Pigments and Solvent Dyes*, 3rd ed.

Solomon, D. H.; and D. G. Hawthorne (1983). *Chemistry of Pigments and Fillers*. New York: John Wiley & Sons.

16.

Crystallinity of Polymers

> It has been my experience that folks who have no vices have few virtues.
> — Abraham Lincoln

1. ABSTRACT

A new X-ray diffraction method for the determination of crystallinity of polymers is reported. A probability function is used to express the intensity distribution of an amorphous halo. The intensity of the halo buried under any crystalline peak can be calculated by this function. An amorphous-standard addition method was used to determine crystallinity. A linear relationship between intensity and concentration is derived theoretically and applied to polyethylene terephthalate. No previous chemical or structural information about the polymer is necessary for this method. Very good agreement between X-ray data and density measurements were obtained. This method is rapid, practical and suitable for routine analysis.

2. INTRODUCTION

Crystallinity is defined as the weight fraction of the crystalline portion of a polymer. Since the advent of polymer science, two models have been used to correlate the polymer structure and its properties, the earlier fringed-micelle model and the recent folded-chain model. In both models it is assumed that the polymer is composed of crystalline and amorphous regions. The crystallinity of the polymer has a definite effect on its performance and various methods have been used to determine this parameter.

There are five different methods for determining the crystallinity of a polymer: (1) X-ray diffraction, (2) density, (3) infrared, (4) n.m.r. and (5) heat of fusion. The infrared method requires a standard, the crystallinity of which has been determined by other independent methods (Elliott, 1969). Density measurements have been used to give an independent estimate of crystallinity. However, the density method requires the density of the completely crystalline polymer which in turn requires the unit-cell dimensions of the polymer crystals as determined by X-ray diffraction analysis (Meares, 1965). The n.m.r. method is primarily a measurement of *motion*, not *order*. The n.m.r. spectrometer classifies the slower-moving protons as the 'rigid' crystalline fraction, the faster-moving protons as the "mobile" amorphous fraction. The motion of the polymer chains is sensitive to temperature, molecular weight, and crosslinks (Miller, 1966). The calorimetric method has a sound thermodynamic definition of *order*. It requires the heat of fusion of pure-crystalline polymers derived from other thermodynamic measurements by extrapolation (Du Pont, 1968). The X-ray method provides a sound physical definition of *order*. It requires a clear separation of the amorphous halo from the crystalline pattern.

Crystals diffract X-rays coherently according to Bragg's law, giving sharp peaks, while amorphous materials scatter X-rays incoherently giving a diffuse halo. The X-ray diffraction pattern of a semicrystalline polymer is the superposition of sharp peaks over a diffuse halo. There are several different procedures to derive the degree of crystallinity from this diffraction pattern. The currently used crystallinity-index methods (Wakelyn & Young, 1966; Statton, 1963; Bosley, 1964) assign an index to a sample by comparing its pattern with that of the least (index =0) and most (index= 100) crystalline standards obtained by quenching-annealing treatments. The disadvantage of these relative methods is that the indices cannot be compared between different laboratories or for different polymers. The absolute crystallinity methods (Hermans, 1961, 1962; Matthews, Peiser & Richards, 1949; Ruland, 1961) involve many empirical rules, correction factors, and/or abstract functions; hence, they are not convenient for rapid routine analysis.

A new approach to the determination of crystallinity has been applied to polyethylene terephthalate. This approach appears to be sound in theory, and suitable for rapid routine analysis.

3. PROBLEMS AND SOLUTIONS OF THE X-RAY METHOD

There are two major problems involved in all the X-ray methods: (1) the need for quantitative separation of intensity contributions from crystalline peaks, amorphous halo, and background; and (2) the need for near-perfect crystalline polymer standards. Such standards do not exist. These two problems are dealt with in the following manner by this new app

3.1 Intensity Separation

The intensity distribution of the X-ray diffraction pattern is the Fourier transform of the electron-density distribution in the sample. The halo of the diffraction patterns of amorphous polymers is due to the interatomic vectors between adjacent polymer chains (Klug & Alexander, 1954). Hence, the position of this halo is a measure of the interchain separation, and the intensity of this halo is a measure of amorphous material in the sample. For amorphous polymers, the interchain separation is a continuous variable and completely random within certain limits (James, 1965). This suggests that some kind of probability function might be useful for governing this situation. Naturally, the Gaussian distribution function is the first choice. The Fourier transform of a Gaussian function is another Gaussian function. Hence, the amorphous diffraction halo could be represented by a normal distribution function. It is found that amorphous diffraction halo of many polymers (Hermans et al., 1961, 1962; Klug & Alexander, 1954; Barlow & Young, 1970) can be fitted to a Gaussian function. For practical purposes, any of the following functions can be used to fit the intensity distribution of scattered X-rays by choosing proper scales as shown in Fig. 1.

1. Gaussian function

$$y = ae^{-x^2}.$$

2. The Witch of Agnesi

$$y = \frac{a}{x^2 + 1}.$$

3. Hyperbolic function

$$y = a \operatorname{sech} x.$$

4. Trigonometric function

$$y = \frac{a \sin^2 x}{x^2}.$$

The feature of this treatment is that only one observable point of the curve is needed to trace out the whole curve. Of course, more points can be used to improve the precision.

The fitting is slightly off for values of $x > 1$ and therefore, the observable point should be chosen not too far away from the maximum. In very unusual cases where the only observable point is at $x > 1$, then a correction term should be added, such as to obtain a best fit for the whole curve.

$$y = \frac{a}{x^2 + 1} + bx,$$

In the X-ray diffraction pattern of a polymer, Fig. 2, the minimum between well separated peaks, e.g. at $x = 20°$ (2θ value), minus the background, should be solely due to the amorphous portion of the polymer. This experimental point can be used to calculate the constant a, which is the maximum intensity of the amorphous halo, in the intensity distribution functions. Once the scale and the constant are determined, the intensity of the amorphous halo buried under the crystalline peaks can be easily calculated, thus achieving the desired intensity separation.

Because of improved modern instrumentation, the background due to electronic noise and white radiation is suppressed to the minimum. This can be seen from the diffraction patterns of rutile TiO_2 (Swanson & Tatge, 1953) and toluidine red (Chung, 1971), Fig. 3, an inorganic and an organic pigment. The air scattering is significant only below the 15° (2θ) value. Since these crystals have near-perfect lattices, the very low back-ground can be totally attributed to electronic noise and white radiation. Any background higher than this is due to amorphous materials and lattice imperfections. By comparing Figs. 2 and 3, it is apparent that when the observable point is between two peaks separated by about 2° or more, its intensity would not be affected by the tails of the crystalline peaks.

The diffraction patterns were obtained by use of a Norelco X-ray diffractometer equipped with a full-wave rectifier, high-intensity copper tube, curved graphite-crystal monochromator, solid-state scintillation counter and an electronic-circuit panel containing a pulse-height analyzer. All subsequent data were obtained with this diffractometer under the same instrumental conditions. The same background (32 c.p.s.) was also used for all subsequent calculations.

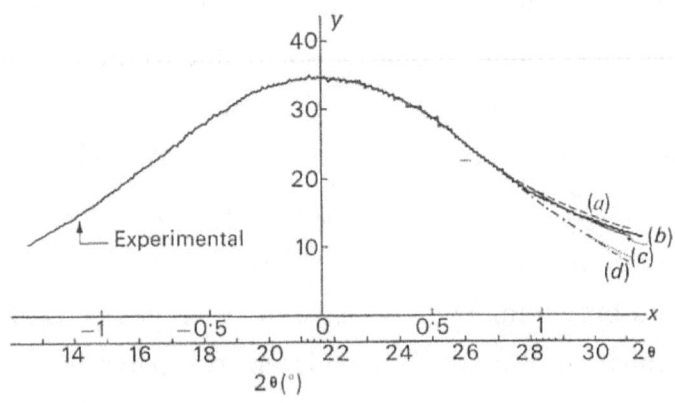

Figure 1 The X-ray diffraction pattern of a completely amorphous PET polymer and its intensity distribution functions. (a) $y = a / (x^2+1)$; (b) $y = a \operatorname{sech} x$; (c) $y = ae^{-x^2}$; (d) $a \sin^2 x / x^2$

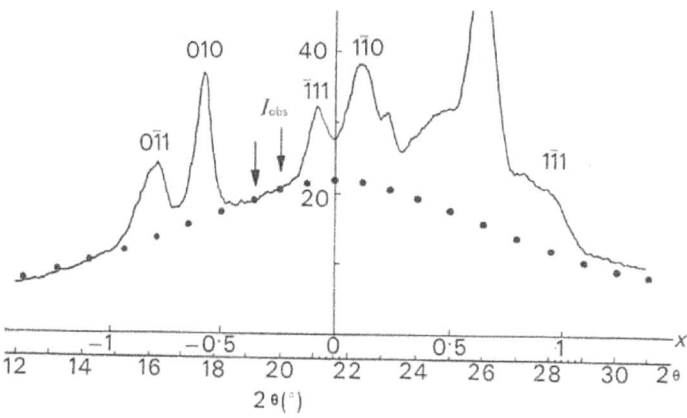

Figure 2 The X-ray diffraction pattern of a partially crystalline PET polymer and the calculated amorphous halo. Cu Kα radiation; 50 kV, 40 mA; Full-scale intensity: 3200 c.p.s.; Scanning speed: 1° (2θ) per min; Time constant: 2 sec.

3.2 Amorphous Standard

Although the perfect crystalline polymer does not exist, a completely amorphous polymer of the same chemical composition is usually available with a few exceptions such as polyethylene and Teflon, *etc*. These completely amorphous polymers can be used as standard to determine the crystallinity.

Consider the polymer as a mixture of two components, the crystalline and the non-crystalline. The crystalline component is defined as the one which diffracts X-rays coherently according to Bragg's law. The non-crystalline component is defined as the one which scatters X-rays incoherently, forming a halo. Thus a polymer is a simple two-phase system. According to a mathematical relationship derived by Klug & Alexander, (1954), the intensity of X-rays diffracted by component a of a mixture is:

$$I_a = \frac{K_a x_a}{\varrho_a \{x_a(\mu_a - \mu_m) + \mu_m\}}$$

where:

K_a = a constant dependent on nature of component a,

x_a = weight fraction of component a,

ϱ_a = density of component a,

μ_a = mass absorption coefficient of component a,

μ_m = mass absorption coefficient of the matrix.

It is obvious that when $\mu_a \neq \mu_m$ the intensity-concentration (I–x) relationship is not linear. However, in the case of a polymer its two components, the crystalline and the noncrystalline, are polymorphic (or allotropic) forms in the sense they are the same compound in two different forms, hence $\mu_a = \mu_m$ and a linear relationship between intensity and concentration should exist.

If subscript a is used for the noncrystalline component, and subscript c is used for the crystalline component, we have

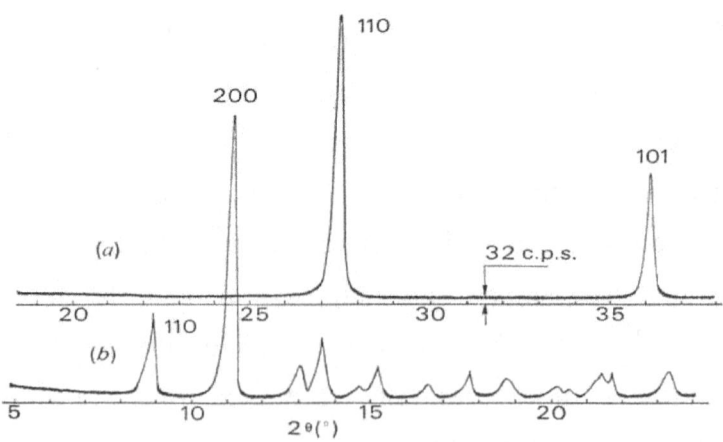

Fig. 3. The X-ray diffraction patterns of (*a*) rutile TiO$_2$ and (*b*) toluidine red.

$$I_a = \left(\frac{K_a}{\varrho_a \mu_a}\right) x_a = k_a x_a$$
$$I_c = \left(\frac{K_c}{\varrho_c \mu_c}\right) x_c = k_c x_c$$
$\bigg\}$ linear relation

$$\frac{x_c}{x_a} = \left(\frac{k_a}{k_c}\right)\frac{I_c}{I_a}$$

$$x_c = \frac{1}{1 + k\frac{I_a}{I_c}} \quad \because x_a + x_c = 1, \; k = \frac{k_c}{k_a}.$$

Note that x_c is the weight fraction of the crystalline phase which is also the crystallinity of the polymer by definition, and k is the slope of the straight line when x_a/x_c is plotted against I_a/I_c. Incidentally, the linear relationship between intensity and concentration was assumed in the method of Hermans et al. (1961, 1962). I_a and I_c are experimental data, k and x_c are two unknowns. One more equation is required to solve for k (slope) and x_c (crystallinity). The completely amorphous polymer rather than the perfectly crystalline polymer (which does not exist) can be used as a standard. If a mixture of resin powders of 80% unknown and 20% amorphous standard is made, then

for unknown:

$$\text{for unknown: } x_c = \frac{1}{1 + k\frac{I_a}{I_c}}$$

$$\text{for 80/20 mixture: } 0.80 \, x_c = \frac{1}{1 + k\frac{I'_a}{I'_c}}$$

$\bigg\}$ solve for k and x_c.

In order to increase the precision, mixtures of different proportions can be made and run. We can have n ($n > 2$) equations for 2 unknowns which can be easily solved by the least-square method to obtain the best possible value of slope (k) and crystallinity (x_c).

This amorphous standard may be (*a*) the quenched melt of the same polymer, (*b*) a polymer made from the same monomer by a different process (Sweeting, 1971), or (*c*) a polymer made from isomeric monomers such as polyethylene terephthalate and polyethylene isophthalate. The scattering factor and absorption coefficient of these standards should be the same.

4. EXPERIMENTAL

A crystalline PET powder SWPET (Sherwin-Williams Company) was used for this study. Three amorphous standards were used: VPE (Goodyear PET), polyethylene isophthalate SWPET (Sherwin-Williams Company), and a quenched melt made from SWPET. Three series of experiments were run by using these three amorphous standards respectively. All the four resin powders were passed through a Fisher 150 mesh sieve. The amorphous-standard powder was added into the unknown sample in various proportions. The mixtures were thoroughly blended by putting them on a roller mill for about two hours. The X-ray diffraction patterns of these mixtures were obtained. The three series of data are listed in Table 1. The intensity data are in chart units. One chart unit is equivalent to 32 c.p.s.

Sample No 1 is the polymer SWPET whose crystallinity is to be determined. Sample No. 10 is the completely amorphous standard. The X-ray diffraction patterns of the three amorphous standards are nearly the same, as shown in Fig. 4. The experimental intensity distribution of the halo is fitted to one of the previously mentioned functions. For the sake of simplicity, the Witch of Agnesi is picked

$$y = \frac{a}{x^2+1},$$

where y is the intensity, x is the 2 value, and a is the maximum intensity. In our case, Fig. 1, $x=0$ at 2 21.6°, $y=a=33\cdot2$ chart units (maximum intensity). When an interval of 6.8° in 2θ is taken as the unit of x, the amorphous halo can be well represented by the above equation.

The 100 crystalline PET peak is at 2θ = 26°, (Fig. 2) which is equivalent to $x = 22/34$. The observable amorphous intensity is chosen at 2θ = 20° which is equivalent to $x = -8/34$. The background is one chart unit (equivalent to 32 c.p.s.) based on the X-ray diffraction patterns of near-perfect crystals under the same instrumental conditions.

The result obtained for sample No. 9 is too far from the average. This sample was prepared with 92·32% of standard and 7·68% of the unknown. Further, the determined crystallinity of this mixture is only 7·68 x 66·0% = 5.0% which is close to the detection limit of the X-ray method.

The average crystallinity of the polymer SWPET is 66·0% (samples Nos. 6 and 9 are excluded). The standard deviation is 0·5%. At the 95% confidence level, according to Student statistics, the crystallinity of this polymer is 66·0 ± 0·3%.

Using the determined crystallinity of polymer SWPET of 66.0%, the crystallinity of all 20 samples was calculated. Their intensity ratios and concentration ratios are listed in Table 2. The data in Tables 1 and 2 are plotted in Fig. 5 and 6. A linear relationship between intensity and concentration does exist as expected from theoretical considerations even with different amorphous standards. Note that in Fig. 5, the I_c line passes through the origin, while the I_a line by extrapolation passes through the point of 100% crystallinity.

Table 1. *Crystallinity of PET from amorphous-standard-addition method*

Sample	% SWPET	I_{obs} at 20°	I_a at 20°	I_{obs} at 26°	I_c at 26°	I_a/I_c	k	Crystallinity %
SWPET + VPE								
1	100	12·6	11·6	75·5	65·9	0·176		
2	88·76	15·0	14·0	68·8	57·4	0·243	2·83	66·8
3	74·74	17·5	16·5	60·0	46·7	0·353	2·88	66·4
4	60·71	21·0	20·0	54·4	38·5	0·519	2·83	66·8
5	50·68	22·8	21·8	49·5	32·3	0·675	2·97	65·7
6	39·08	25·0	24·0	44·5	25·6	0·938	3·19	64·1
7	25·39	28·5	27·5	37·6	16·1	1·708	2·90	66·2
8	19·57	30·0	29·0	35·2	12·6	2·302	2·93	66·0
9	7·68	31·1	30·1	29·5	6·1	4·934	4·55	55·5
10	0	33·2	32·2	24·8	0	∞		
SWPET + SWPEI								
11	83·73	15·6	14·6	65·0	53·2	0·275	2·98	65·6
12	68·16	19·5	18·5	58·8	44·0	0·421	2·87	66·4
13	59·70	20·3	19·3	52·5	37·1	0·520	3·00	65·4
14	49·82	22·8	21·8	49·0	31·8	0·686	3·02	65·3
15	36·15	25·2	24·2	41·8	21·8	1·110	2·85	66·6
SWPET + Quenched melt of SWPET								
16	84·86	15·4	14·4	65·7	54·0	0·267	2·97	65·7
17	76·61	17·0	16·0	61·1	48·2	0·332	2·99	65·5
18	70·83	18·4	17·4	58·4	44·5	0·391	2·90	66·2
19	53·70	21·6	20·6	49·6	33·3	0·619	2·96	65·8
20	40·83	24·3	23·3	43·3	25·0	0·932	2·89	66·3

Table 1.

Table 2. *Intensity ratios and concentration ratios of PET*

Sample	1	2	3	4	5	6	7	8	9	10
I_a/I_c	0·176	0·243	0·353	0·519	0·675	0·938	1·708	2·302	4·934	∞
x_a/x_c	0·515	0·707	1·027	1·496	1·990	2·879	4·967	6·740	18·72	∞

Sample	11	12	13	14	15	16	17	18	19	20
I_a/I_c	0·275	0·421	0·520	0·686	1·110	0·267	0·332	0·391	0·619	0·932
x_a/x_c	0·810	1·223	1·538	2·041	3·191	0·786	0·978	1·139	1·822	2·711

Table 2

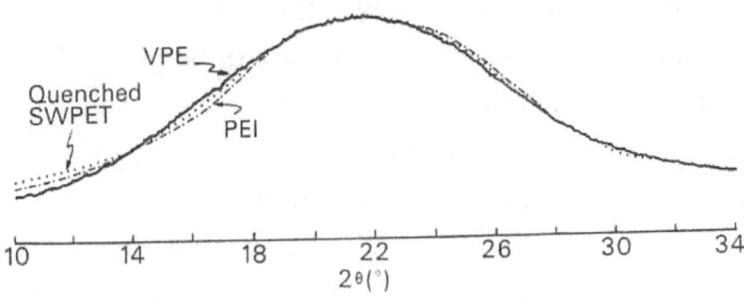

Fig 4 The X-ray diffraction patterns of three amorphous standards: PET, PEI and quenched SWPET.

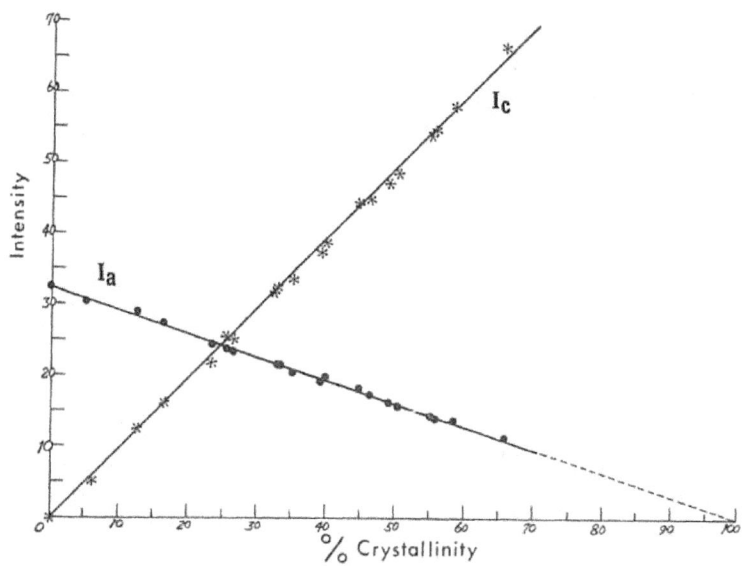

Fig. 5. Linear relationship between intensity and crystallinity.

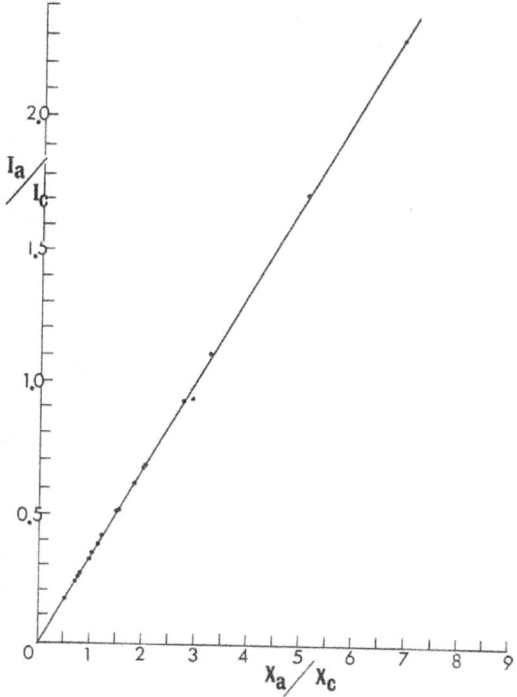

Fig. 6. Linear relationship between intensity ratio and concentration ratio.

5. DISCUSSION

For X-ray diffraction work, the integrated intensity should be preferred rather than simply the peak height. However, because of serious overlapping, it is difficult to obtain the integrated intensities with reasonable confidence. The difference between the integrated intensity and peak height will be regulated by the empirical constant k which also takes care of the absorption factor and the Lorentz-polarization factor automatically.

The diffuse scattering need not come only from amorphous materials in the sample but can also arise from crystal-lattice distortions and crystallite-size effect. This is surmounted by defining the polymer as a system of crystalline and noncrystalline components. The weight fraction of the crystalline component, hence the crystallinity, is a measure of *order*, and the order is represented by that portion of the structure which diffracts X-rays coherently according to Bragg's law. Then the paracrystalline state, the lattice imperfections, the crystallite-size effect, the thermal vibrations, and the noncrystalline structure are merely different kinds of *disorder*.

The crystal structure of PET has been determined by use of X-ray diffraction analysis (Daubeny, Bunn & Brown, 1954) the density of completely crystalline PET is 1·455 g cm^{-3}, and the density of completely amorphous PET is 1·335 g cm^{-3}. The density and the crystallinity are related by the following equation (Alexander, 1969):

where

x = crystallinity

ρ_c = crystalline density

ρ_a = amorphous density

ρ = density of partially crystalline sample.

Given $x = 66.0\%$, the calculated density of SWPET should be $\varrho = 1.412$ g cm^{-3}. The density of the SWPET powder measured by ASTM method D153-B which should give accurate density for powder samples is 1.415 g cm^{-3}. The agreement is remarkably close in this case, even though the agreement between X-ray and density crystallinity is still controversial in the literature (Miller, 1966; Dumbleton & Bowles, 1966).

The features of this new approach are: (*a*) the intensity-separation and standard-addition procedures are not subject to human error. Hence the data should be comparable between different laboratories or for different polymers, (*b*) previous chemical or structural information about the polymer sample is not required for this method, and (*c*) it is simple, rapid and suitable for routine analysis.

6. REFERENCES

Alexander, L. E. (1969). *X-ray Diffraction Methods in Polymer Science*, p. 189. New York: Wiley-Interscience.

Barlow, A.; and M. Young (1970). *J. Appl. Polymer Sci.* 14, 1731-1736.

Bosley, D. E. (1964). *J. Appl. Polymer. Sci.* 8, 1521-1529.

Challa, G.; P. H. Hermans; and A. Weidinger (1961). *Makromol. Chem.* 50, 98-115.

Challa, G.; P. H. Hermans; and A. Weidinger (1962). *Makromol. Chem.* 56, 169-178.

Chung, F. H. (1971). *J. Appl. Cryst.* 4, 79-80.

Daubeny, R. P.; C. W. Bunn; and C. J. Brown (1954). *Proc. Roy. Soc.* A226, 531-542.

Dumbleton, J. H.; and B. B. Bowles (1966). *J. Polymer Sci.* 4, 951-958.

Dupont (1968). DTA Application Brief 12, Jan. 15.

Elliot, A. (1969). *Infrared Spectra and Structure of Organic Long-Chain Polymers*, p. 111. New York: St. Martin's Press.

James, R. W. (1965). *The Optical Principles of the Diffraction of X-rays*, p. 459. Ithaca, New York: Cornell Univ. Press.

Klug, H. P.; and L. E. Alexander (1954). *X-ray Diffraction Procedures*, pp. 413 and 633. New York: John Wiley.

Matthews, J. L.; H. S. Peiser; and R. B. Richards (1949). *Acta Cryst.* 2, 85-90.

Meares, P. (1965). *Polymer Structure and Bulk Properties*, p. 127. London, New York: Van Nostrand.

Miller, R. L. (1966). *Encyclopedia of Polymer Science and Technology*, Vol. 4, pp. 490-492. New York: John Wiley.

Ruland, W. (1961). *Acta Cryst.* 14, 1180-1185.

Statton, W. O. (1963). *J. Appl. Polymer Sci.* 7, 803–815.

Swanson, H. E.; and E. Tatge (1953). *NBS Circular* 539, I, 44.

Sweeting, O. J. (1971). *The Science and Technology of Polymer Films*, p. 620. New York: Wiley-Interscience.

Wakelyn, N. T.; and P. R. Young (1966). *J. Appl. Polymer Sci.* 10, 1421-1438.

17.

Vacuum Sublimation and Crystallography of Quinacridones

> Listen to everyone and learn from everyone, because nobody knows everything, but everyone knows something.
>
> — Dwight Eisenhower

1. ABSTRACT

Quinacridones are pigments of outstanding light, heat and chemical fastness. Single crystals of linear unsubstituted quinacridone and 4,11-dichloroquinacridone were grown by the vacuum sublimation method at about 425°C and 390°C respectively under a vacuum of 0.3 torr. The sublimation products are always λ-form quinacridone and β-form 4,11-dichloroquinacridone no matter what respective polymorphic forms are used as the starting material. Exothermic effects at transition temperatures were observed for both materials on heating by TGA and DSC techniques. The space groups and unit cell dimensions were determined by the precession method. The powder patterns are indexed by single crystal data. The size, shape, and orientation of the molecules as well as the general structure features are discussed.

2. INTRODUCTION

Perhaps the major development in the pigment industry in the last decade has been the commercialization of the quinacridone pigments which are characterized by outstanding light, heat and chemical resistance, non-bleeding, and a wide gamut of shades ranging from orange to violet. They supplement the phthalocyanine blues and greens in extending the availability of high quality pigments of excellent fastness. As a result of continued research, four polymorphic forms of linear trans-quinacridones (I) and some 150 substituted quinacridones have been synthesized (Labana & Labana, 1967; Struve, 1958; *Du Pont Innovation*, 1970) but few find commercial use. Among the polymorphic forms of quinacridones, β and γ are most useful as pigments. There are three polymorphic forms of 4,11- dichloroquinacridone (II) reported in the patent literature (Ehrich, 1964; Wilkinson, 1962). A solid solution of quinacridone with 4,11-dichloroquinacridone yields a light-fast scarlet pigment (Spengeman, 1970).

The starting phases of the components are immaterial to the final result of solid solution product which has a characteristic X-ray pattern different from that of the physical mixtures.

X-ray diffraction patterns of these quinacridones have been obtained on powder samples and are used for identification by Struve (1958), Spengeman (1970) and Schweizer (1968). However, single-crystal structure analyses of these pigments have not been reported. Single crystals of unsubstituted γ-form quinacridone and 4,11-dichloroquinacridone were grown by a vacuum sublimation method, and their crystal structures have been studied by precession techniques in this laboratory in order to obtain a clearer insight into their physical and chemical properties.

3. VACUUM SUBLIMATION

All pigments have the inherent property of insolubility. Unsubstituted and 4,11-dichloroquinacridones are in-soluble and infusible. Single crystals of these pigments were obtained by vacuum sublimation at high temperature.

In the procedure a small amount (about one gram) of crude quinacridone powder (Sherwin-Williams Company) is placed in a specially made test tube of about 1·5 cm in diameter and 30 cm in length. This long test tube is connected to the vacuum pump by a heavy rubber hose. After being evacuated for 5 minutes, this tube is put in a muffle furnace with the rubber hose end sticking out. The door of the furnace is left partially open so as to create a temperature gradient along the test tube. The temperature is increased slowly. The linear unsubstituted quinacridones sublime at about 425 °C, while the 4,11-dichloroquinacridones sublime at about 390 °C under a vacuum of 0·3 torr.

The quinacridones sublime as yellow fumes, which then condense on the wall of the tube as dark red crystals when hot and bright red crystals when cooled to room temperature. After sublimation the tube is taken out of the muffle furnace and allowed to cool to room temperature while continuous vacuum is maintained.

The products of sublimation are highly crystalline and have large crystallite particle sizes and transparent bright red colors.

The linear *trans*-quinacridone single crystals are plate-like, and about 1·0 x 0·8 x 0·1 mm. The dichloroquinacridone single crystals display sphenoid facets and are about 0·8 x 0·6 x 0·2 mm. Their morphologies are shown in Figs. 1 and 2.

The infrared spectra of these two compounds are the same before and after the high temperature vacuum sublimation process, which indicates that the molecular structures are unchanged. However, the X-ray diffraction patterns indicate that the γ and β -forms of quinacridones are converted to the γ-form after vacuum sublimation, while the γ-form remains unchanged, an indication that the γ-quinacridone is the most heat-stable form. Similarly, the X-ray diffraction patterns also indicate that the α - and γ-forms of 4,11-dichloroquinacridone are converted to the β-form after vacuum sublimation while the -form remains unchanged and hence the β-form of 4,11-dichloroquinacridone has the best thermal stability.

It is noteworthy that the vacuum sublimation at high temperature achieves separation, purification, and phase transition at the same time. Impurities are either decomposed or condensed at much lower temperatures. These features could be very useful for pigment manufacturing as well as chemical analysis.

4. THERMAL ANALYSES

While melting point is almost independent of pressure, the sublimation temperature: is greatly influenced by the presence or other gases of vapors. The vapor pressure of any substances is directly proportional to temperature (Clausius-Clapeyron equation). When the vapor pressure of a solid exceeds the pressure of the atmosphere above it, sublimation occurs. Under vacuum, the vapor molecules have the largest mean free path hence the sublimation is achieved at the lowest possible temperature with minimum or no thermal decomposition. Furthermore, the absence of air prevents the oxidation of the subliming molecules.

In order to observe the temperature of sublimation and the accompanying thermal effects, Thermogravimetric Analysis (TGA) and Differential Scanning Calorimetry (DSC) were obtained by use of a DuPont 900 differential thermal analyzer. The results are listed in Table 1. The inflection point on the TGA curve was located by the first derivative method, and the onset point on the DSC curve was determined by extrapolation. A typical set of thermograms are shown in Figure 3.

Figure 1 *Morphology of γ-quinacridone single crystals.*

Figure 2 *Morphology of β-4, 11-dichloroquinacridone single crystals*

The temperatures at the peak maximums for the DSC curves are all higher than were found on the TGA curves. This has been discussed elsewhere (Smothers & Chiang, 1966; Powell, 1957). The onset temperature is dependent on the peak profile which is in turn dependent on the experimental conditions. Judging by the definition of sublimation, the sublimation temperature given by a TGA curve seems to be more reasonably dependable.

The distinguished feature of the DSC thermograms is the sharp exothermic peak at the transition temperatures on dynamic heating. Whether the exothermic effect is due to a crystallographic phase transition or to partial thermal decomposition overriding the sublimation needs further investigation.

Since the area under the DSC peak is directly proportional to the heat of reaction, it is found qualitatively that the heat of transition given off by unsubstituted quinacridone is much more than that by 4,11- dichloroquinacridone.

5. CRYSTAL DATA

Nickel filtered Cu K_α radiation and a Buerger precession camera were used to collect the data. The single crystals studied have the shapes shown in Figure 4.

In γ-quinacridone, the (100) face is prominently developed. The (001) plane is the cleavage plane. No definite end faces were observed; the growth of the crystal generally terminated by fracture across the b axis. In the 4,11-dichloroquinacridone, crystals of perfect end facets were obtained with prominent faces (010) and (100).

Unsubstituted γ-quinacridone (Formula I)

The crystal grown by vacuum sublimation was oriented so that the a^* axis lay on the spindle axis. Two sets of layer photographs, mutually perpendicular, show monoclinic symmetry of Laue group $2/m$. No general extinction was found. The only special extinction is the h0l reflections when l is odd; this indicates a c-glide plane perpendicular to the b axis. However, no 0k0 reflections were observed; hence, whether there are screw axes in this unit cell is left undetermined. The possible space groups that fit these extinction conditions are $P2/c$ and $P2_1/c$.

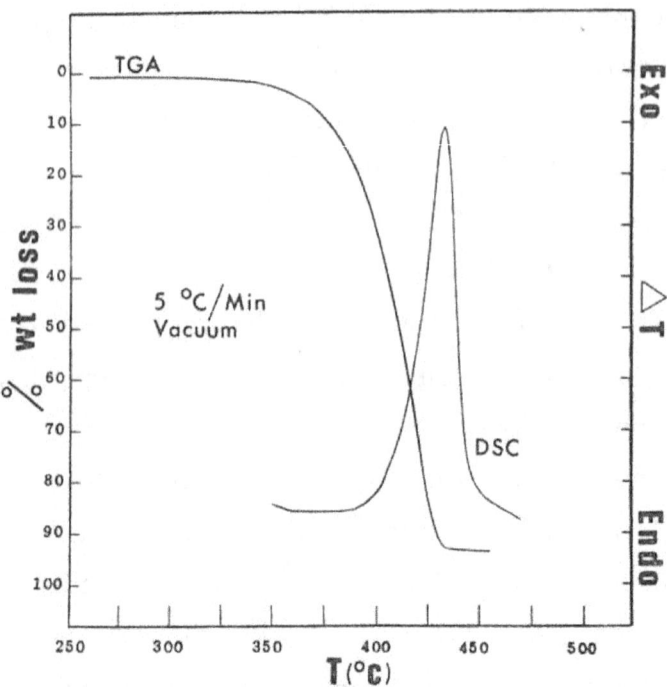

Fig 3 *Thermograms of β-quinacridone at a heating rate of 5°C min⁻¹, under a vaccum of 0-3 torr.*

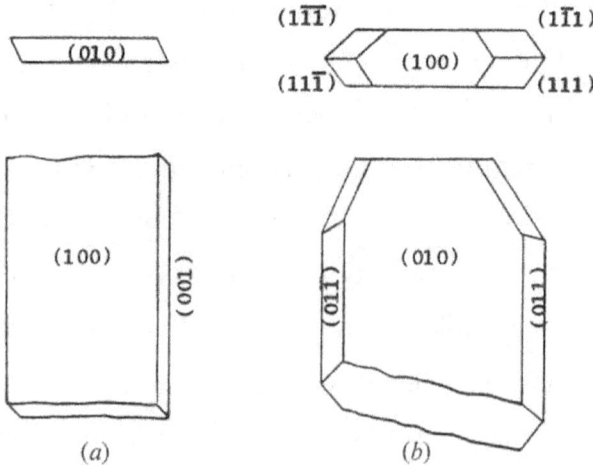

Fig 4 *External symmetry of (a) γ-quinacridone, and (b) β-4, 11-dichloroquinacridone.*

The unit-cell dimensions as determined from pre-cession photographs after film shrinkage corrections are:

$a = 14.116 \pm 0.010$ Å
$b = 3.885 \pm 0.003$
$c = 13.377 \pm 0.010$
$\beta = 107.22 \pm 0.08°$
$V = 700.73$ Å^3
$Z = 2$
$D_x = 1.480$ g.cm^{-3}
$D_m = 1.475 + 0.005$ g.cm^{-3}.

The density was measured by sink-float observations and the conventional pycnometer method. A mixture of ethyl iodide and propanol was used as the immersion fluid.

4,11-Dichloroquinacridone (Formula II)

The crystal was also oriented with the a^* axis parallel to the spindle axis. The precession layer photographs indicate orthorhombic symmetry of Laue group *mmm*. There is no general extinction of hkl reflections which reveals a primitive unit cell. Special extinctions are hk0 reflections when *h* is odd. 0kl reflections when k is odd, and h0l reflections when *l* is odd. Consequently, the extinctions of h00, 0k0, and 00l reflections are observed when h, k, and l are odd respectively. These extinction conditions definitely establish that the space group is *Pbca*.

The unit-cell dimensions as determined from precession photographs after film shrinkage corrections are:

$a = 13.621 \pm 0.010$ Å

$b = 15.580 \pm 0.012$

$c = 7.423 \pm 0.006$

$V = 1575.27$ Å^3

$Z = 4$

$D_x = 1.607$ g.cm^{-3}

$D_m = 1.612 \pm 0.005$ g.cm^{-3}.

The density was measured by sink-float observations of the crystals in mixtures of carbon tetrachloride and iodomethane.

6. POWDER PATTERNS

X-ray powder diffraction patterns were obtained by use of a Norelco diffractometer equipped with a scintillation counter and pulse height analyzer. Graphite crystal monochromatized Cu $K\alpha$ radiation was used. The instrument was calibrated with the 110 line of rutile TiO_2. Intensities were measured as peak heights above back-ground and expressed as a percentage of the strongest peak. The powder patterns were then indexed according to the previously obtained single crystal data. The observed and calculated d-spacings are listed in Tables 2 and 3, together with relative intensities.

Table 2. X-ray powder diffraction data for linear trans-γ-quinacridone

d_{obs}	d_{calc}	I/I_o	hkl
13·56	13·483	100	100
6·75	6·741	24	200
6·58	6·575	21	$\bar{1}02$
6·39	6·388	40	002
5·23	5·207	4	102
4·34	4·327	6	$\bar{3}02$
4·07	4·074	2	202
3·74	3·733	6	110
3·55	3·506	4	111
3·37	3·370, 3·366	23	400, 210
3·34	3·337, 3·344	16	$\bar{1}04$, $\bar{1}12$
3·18	3·178	3	$\bar{2}12$
3·11	3·142	3	211
2·92	2·928, 2·920	4	$\bar{1}13$, 104
2·81	2·811	2	212
2·70	2·696	1	500
2·57	2·571	1	$\bar{4}12$

Table 3. X-ray powder diffraction data for
β-4,11-dichloroquinacridone

d_{obs}	d_{calc}	I/I_o	hkl
7·80	7·790	78	020
6·81	6·810	7·5	200
6·23	6·240	3	210
6·00	6·012	3	111
5·37	5·373	2·5	021
5·12	5·127	10	220
5·00	4·998	15	121
4·07	4·061, 4·129	58·5	131, 230
3·89	3·895	31·5	040
3·45	3·449	100	041
3·345	3·350, 3·343	41·5	022, 141
3·26	3·258	4	202
3·125	3·120	5	420
3·08	3·076	7	241
3·015	3·006	4·5	222
2·88	2·876, 2·873	3	421, 302
2·815	2·815	15	151
2·75	2·746	4	341
2·69	2·686	2	042
2·65	2·647	4	251
2·60	2·596	12	060
2·52	2·523	2	511
2·45	2·451	3	061
2·43	2·427, 2·426	5	351, 260
2·33	2·323	2	123
2·30	2·298	3	451
2·275	2·270	5·5	600

7. STRUCTURE FEATURES

The symmetry of space groups *P2/c* or *P2₁/c* can be attained by arranging four asymmetric units in the unit cell. Since there are only two quinacridone molecules in the unit cell, each molecule must possess twofold symmetry; and since finite molecules, unlike high polymers, can have neither a screw axis nor a glide plane, they must have the other kinds of twofold symmetry-rotation diad or inversion center. The center of symmetry in the quinacridone molecule obviously lies in the middle of the central benzene ring. Aromatic compounds of fused rings are sufficiently rigid, and severe distortions exist only where the rings belong to certain overcrowded molecules. Hence, this symmetry consideration points to trans-planar molecules.

From the unit-cell dimensions, one molecule of γ-quinacridone is contained in a space of ½abc sin β or about 14·1 x 3·9 x 6·4 Å^3. On the basis of the usual interatomic distances (C···C: 1·42; C-N: 1·37; C=O: 1·36; and C-H: 1·10 Å) and hexagonal benzene rings, the dimension of a flat molecule of five linearly fused rings can be calculated to be 14·2 x 5·1 Å^2. Since distances between atoms in neighboring molecules (usually 2·5-4·0 Å) are much greater than those between bonded atoms within the molecule, it is reasonable to have the width of the molecule at 6·4 Å instead of the calculated 5·1 Å. The length of the molecule (14.1 Å) is very close to the estimated 14·2 Å suggesting that the molecule is tilted lengthwise to accommodate longer distances between unbonded atoms. The minimum distance of approach between aromatic rings in the solid state is known to be 3·6-3·7 Å, which may be taken as a measure of the thickness of aromatic rings (*Report of International Conference on Physics*, 1934). γ-Quinacridone molecules of identical orientation recur at intervals of 3·9 Å along the *b* axis which is 0·2-0·3 Å larger than the thickness of the aromatic rings; this also suggests that the flat molecules are not quite perpendicular to the *b* axis (*i.e.*, parallel to the *ac* plane) but rather slightly tilted. Nevertheless, a projection of the structure along the *b* axis on the (010) plane will give a clear resolution of the molecule and its component atoms.

The symmetry of the lattice is derived from the reflections of zero intensity (systematic absences); the orientation of the molecules in the unit cell can be estimated from the reflections of maximum intensity. In the principal zone $hk0$ of γ-quinacridone, the reflection 210 is outstandingly stronger than all the others; hence, there is a considerable concentration of atoms near these planes. However, the rapid decline of the intensity of its higher order 420 indicated that the flat molecules lie nearly but not quite exactly along the 210 planes. The h0l zone, which corresponds to a projection of the structure on the (010) plane, contains the greatest range of reflection planes, among them 002, 200, 102 and 704 having strong intensities. These intensity considerations also support the fact that the planar molecules are nearly parallel to the *ac* plane.

Intermolecular hydrogen bonding of the N-H···O type must have played an important role in the packing of these polar molecules. The obvious (001) cleavage plane and the plate-like shape of the crystals offer good evidence for this conclusion; the former indicates the direction of weak van der Waals forces relative to the stronger hydrogen bonding forces, and the latter indicates the direction of fastest rate of growth.

In 4,11-dichloroquinacridone, the symmetry of space group *Pbca* can be attained by arranging eight asymmetrical units in the unit cell. Since there are only four molecules in the unit cell, each of the dichloro-substituted molecules must also have a twofold symmetry—an inversion center in this case. Hence the *trans*planar configuration of the unsubstituted quinacridone is retained. The substitution of two hydrogen atoms by two chlorine atoms caused both dimensional and symmetry changes of its unit cell. However, each molecule is contained in a space of 15·6 x 3·7 x 6·8 Å^3 which is very close to that in unsubstituted quinacridone. In comparing the bond lengths C-Cl (1·70 Å) with C-H (1·10 Å) the slight increase in size of 4,11-dichloroquinacridone molecule is rather expected. The change in symmetry from a monoclinic unit cell to an orthorhombic one is the natural way of achieving most efficient packing and attaining lowest potential energy. The accurate determination of the positions of individual atoms must rest on the timeconsuming intensity measurements and Fourier synthesis.

8. REFERENCES

Du Pont Innovation (1970). 2, No. 1, 14.
Ehrich, F. F. (1964). U.S. Pat. 3, 160, 510.
Labana, S. S; and L. L. Labana (1967). *Chem. Rev.* 67, 1.
Powell, D. A. (1957) *J. Sci. Instrum.* 34, 225
Report of International Conference on Physics (1934). Part II, p. 46.
Schweizer, H. R. (1968). British Pat. 1, 125, 577.
Smothers, W. J.; and Y. Chiang (1966). *Handbook of Differential Thermal Analysis.* p. 87. Chem. Publishing Co.
Spengeman, W. F. (1970). *Paint and Varnish Production*, 60, 37.
Struve, W. S. (1958). U.S. Pat. 2, 844, 485.
Wilkinson, D. G. (1962). British Pat. 896, 916.

18.

Crystallography of Toluidine Red Pigment

> You try to be greedy when others are fearful.
> And you try to be fearful when others are greedy.
> — Warren Buffet

1. ABSTRACT

Toluidine Red pigment exhibits brilliant red color, excellent resistance to acid and alkali, very good hiding power, and low cost. But it tends to bleeding in paint and fading in sun light. Depending on the intended service conditions, it offers the formulator a choice between the low cost toluidine Red and the expensive quinacridone Red. Its single crystal data on space group, unit cell dimensions, and powder pattern are reported.

2. INTRODUCTION

X-ray diffraction analysis has been a very useful tool in the paint industry for the identification of pigments and extenders in a paint system (Scott, 1969), especially when complemented with infrared spectroscopic methods (McClure, Thomson & Tannahill, 1968). A pigment which can be characterized by both techniques is toluidine red, an azo pigment made by coupling diazotized m-nitro-p-toluidine with β-naphthol. It exhibits a brilliant red color and has very good hiding power. It is excellent in acid and alkali resistance. However, it also has two weak points: insufficient lightfastness and bleeding when incorporated into a paint system. Considerable effort has been directed to improve these properties. During the course of this work, the single-crystal structure of toluidine red was studied by precession techniques. The powder pattern was obtained by use of a Norelco diffractometer.

3. GROWING SINGLE CRYSTALS

Single crystals in the form of needles were grown by the solvent method. Sufficient toluidine red toner (Sherwin Williams Co.) was dissolved in toluene to make a saturated solution at about 85°C. This solution was filtered. The filtrate was put in an oven at 80°C. After 30 minutes, the oven was turned off to let the solution cool very slowly. Crystal needles as large as 2 x 0·1 x 0·1 cm could be obtained without seeding. Infrared spectra of these needles indicated no change in structure from the original compound.

4. CRYSTAL SYMMETRY

Nickel filtered Cu K_α radiation and the Buerger precession camera were used to collect the data. Two sets of layer photographs, parallel and perpendicular to the b axis (needle axis), showed monoclinic symmetry of Lane group $2/m$. The only extinction is the $0k0$ reflection when k is odd ($k \neq 2n$). The possible space groups that show this extinction are either $P2_1$ or $P2_1/m$.

5. UNIT CELL DIMENSIONS

The unit cell dimensions as determined from precession photographs after film shrinkage corrections are:

The density was measured by sink-float observations of crystals in mixtures of cyclohexane and 1-bromonaphthalene.

6. POWDER DATA

The X-ray powder diffraction pattern was obtained with a Norelco diffractometer equipped with a scintillation counter and pulse height analyzer. Graphite crystal monochromatized Cu K radiation was used. The instrument was calibrated using the rutile phase of TiO_2 as a standard. Intensities were measured as peak heights above background and expressed as a percentage of the strongest peak. The observed and calculated d-spacings are listed in Table 1, together with their relative intensities.

Table 1. X-ray powder diffraction data for toluidine red $C_{17}H_{13}N_3O_3$

d_{obs}	d_{calc}	hkl	I/I_0	d_{obs}	d_{calc}	hkl	I/I_0
10·00	10·024	(110)	30	2·56	2·566, 2·559	(132) (530)	0·7
7·97	7·961	(200)	100	2·505	2·506	(440)	2·5
6·80	6·821, 6·819	(001) ($\bar{1}$01)	11·5	2·48	2·470	(521)	2
6·50	6·450	(020)	23	2·46	2·454, 2·454	(250) (620)	0·8
6·02	6·030	(011)	4	2·43	2·431	(322)	0·7
5·82	5·835, 5·832	(101) ($\bar{2}$01)	11·8	2·42	2·414, 2·413	(232) (051)	0·8
5·32	5·316, 5·307	(111) (300)	4·6	2·36	2·353	(402)	0·7
5·00	5·012	(220)	9·9	2·272	2·273, 2·271	(003) (531)	5·3
4·90	4·908	(310)	1·4	2·240	2·239, 2·240	(013) (332)	1
4·72	4·707	(201)	6·7	2·185	2·186	(103)	3
4·43	4·421	(211)	3·6	2·145	2·144, 2·145	(023) (720)	0·8
4·33	4·327	(121)	2·8	2·075	2·075, 2·072	(260) (203)	2
4·17	4·151	(130)	8·5	2·060	2·059, 2·050	(541) ($\bar{1}$61)	4·14
4·10	4·098	(320)	8·9	2·051	2·049	(640)	4·7
4·00	3·980	(400)	0·6	2·015	2·017	($\bar{1}$61)	2·5
3·83	3·812	(301)	9	2·006	2·006, 2·008	(711) (152)	3·6
3·67	3·656	(311)	5	1·965	1·967	(810)	1·2
3·40	3·410, 3·409	(002) ($\bar{2}$02)	42	1·960	1·955	(261)	1·2
3·37	3·370	($\bar{1}$12)	22	1·952	1·949	(133)	1·2
3·30	3·297, 3·296	(012) ($\bar{2}$12)	67	1·945	1·945	(303)	1
3·175	3·174	(231)	3	1·903	1·901, 1·901	(442) (820)	0·8
3·16	3·161, 3·157	(140) (401)	3·2	1·886	1·885	(612)	1·2
3·095	3·091	(510)	13·3	1·858	1·858, 1·857	(043) (551)	1·3
3·00	2·989, 3·015	(240) (022)	0·9	1·835	1·836	(731)	0·8
2·93	2·917, 2·915	(202) (041)	1·8	1·747	1·746, 1·743	(423) (243)	1·2
2·86	2·865, 2·855	(122) (520)	3	1·728	1·729	($\bar{2}$14)	2
2·83	2·822, 2·835	(141) (421)	1·6	1·716	1·716	(271)	2
2·76	2·756	(340)	0·4	1·684	1·687	(503)	2
2·67	2·672	(032)	0·4	1·631	1·632	(523)	0·7
2·625	2·625	(302)	1	1·375	1·375	($\bar{1}$15)	0·5
2·615	2·619	(511)	1·2	1·358	1·359	(254)	0·4
2·575	2·572	(312)	0·7	1·357	1·356	(015)	0·5

7. REFERENCES

McClure, A.; J. Thomson; and J. Tannahill (1968). *J. Oil Col. Chem. Ass.* 51, 580.

Scott, R. W. (1969). *J. Paint Tech.* 41, 422.

19.

Synthesis and Analysis of Quartz, Cristobalite, and Tridymite

> Motivation is the art of getting people to do what they want to do, because they are happy to do it.
>
> — Dwight Eisenhower

1. ABSTRACT

Inadequate interlaboratory precision of silica analysis is shown in the round robin studies sponsored by NIOSH and AIHA. The inconsistency in analytical results is caused by loose analytical procedures and lack of primary standards. Primary standards of cristobalite and tridymite were synthesized from high-purity quartz to cope with this situation. A matrix-flushing X-ray diffraction procedure is described for silica analysis. The integrated intensities, flush constants, and detection limits for the three forms of crystalline silica are presented.

2. INTRODUCTION

Chronic exposure to crystalline silica may cause silicosis or fibrosis in the pulmonary system. Hence the amount of airborne silica in work places is regulated and monitored by the Occupational

Safety and Health Administration (OSHA). Recent eruptions of the Mt. St. Helens volcano brought up a controversy between the environmentalists and the geologists over the amount of crystalline silica (quartz, criatobalite, and tridymite) in the volcanic ash (*1*). The environmentalists at the Washington State Department of Labor and Industries, the University of Washington, and the National Institute for Occupational Safety and Health (NIOSH) found typically 5-10% cristobalite, 1-2% quartz, and <1% tridymite, while the geologists at the U.S. Geological Survey, the Washington State University, and the Battelle Pacific Northwest Laboratories found little or no free silica (*2*), although both sides used X-ray diffraction (XRD) as the analytical tool.

Apparently the controversy has stemmed from three factors: different sampling practice, lack of primary standards, and varied analytical procedures. The first factor should be easy to settle. For health effects, the respirable portion (<15 μm size) of the airborne dust should be collected; for geological studies, a composite sample might be more representative. The other two factors, however, need some deliberation and collaboration.

3. ROUND ROBIN DATA

Since 1975, the National Institute for Occupational Safety and Health (NIOSH) has sponsored a Proficiency Analytical Testing (PAT) program to monitor the performance of various analytical laboratories in their ability to analyze for various air pollutants. The data of a NIOSH silica study (*3*) in 1980 are cited in Table I in order to show the extent of agreement. The samples are quartz dust deposited on PVC membrane filters (0.5- μm pore size, 37 mm diameter, MSA) that are deliberately contaminated with silicates. Note that a total of 61 laboratories did the analysis of four different silica samples; among them 28 laboratories used X-ray diffraction, 21 laboratories used a colorimetric (Talvitie) method, and 12 laboratories used an infrared method. Details of the methods are found in the NIOSH Manual of Analytical Methods (4), although some laboratories may make minor modifications to these methods.

Only the quartz form of silica was included in the PAT program. Naturally, the analysis of other forms of silica should be explored. The American Industrial Hygiene Association (AIHA) sponsored two round robins for cristobalite analysis, the first round in 1977-1978 and the second round in 1979-1980. A "secondary standard" was sent to participating laboratories for determination of its cristobalite content. Each laboratory was to choose its own analytical procedure and find its own primary standard although there is no known source of certified cristobalite. The "secondary standard" was a homogenized Celite distributed by George Swallow at Johns-Manville. It contains cristobalite as the major component with a very small amount of α-quartz in an amorphous matrix. Seven laboratories participated in the first round robin (5) and twelve laboratories participated in the second round robin (6). The X-ray diffraction technique was used by all participating laboratories. The results of the two round robins are listed in Table II, where X stands for arithmetic mean and S for standard deviation.

Table II. AIHA Cristobalite Study

	cristobalite found, %		$\overline{X}$	S	$S/\overline{X}$, %
	J-M	other labs			
1st round	50.9	49.5, 51.8, 65.0, 68.5, 77.1, 86.9	64.2	14.4	22.4
2nd round	57.3	49.0, 53.0, 53.9, 56.7, 58.2, 59.0, 59.1, 64.0, 75.0, 83.0, 100	64.0	14.8	23.1

Table II.

With the X-ray diffraction techniques of analysis, Klug and Alexander (7) reported a precision (S/X) of ± 5%, and Chung (8) demonstrated a precision of ± 8% or better. Evidently the round robin data indicate that the interlaboratory precision for silica analysis is inadequate whether the sample is quartz on membrane filter (PAT data) or cristobalite in bulk powder (AIHA data).

4. EXPERIMENTAL

X-ray diffraction is about the only technique that can differentiate quantitatively among the three forms of crystalline silica. However, it needs validated analytical procedures to attain precision; it requires certified primary standards to achieve accuracy. The wide spread of silica data is likely due to the lack of pure standards. Efforts were made to synthesize pure Cristobalite and Tridymite from pure Quarts. Some results are reported below:

4.1 Synthesis of Standards. For quantitative X-ray diffraction analysis, very few certified primary standards are available. Each laboratory has to seek its own primary standards. When the compound sought is not readily available in pure state such as cristobalite and tridymite, problems crop up.

During the course of carrying out the analysis for the AIHA cristobalite round robin, different experiments were tried in this laboratory to synthesize primary standards starting from pure quartz. The primary standards for the three forms of crystalline silica were obtained as follows:

(1) α-Quartz. High-purity α-quartz under the trade name Min-U-Sil in 5-, 10-, 15-, and 30-μm size is available from Pennsylvania Glass Sand (PGS) Corp., Pittsburgh, PA 15235. The Chemical Reference Laboratory of NIOSH in Cincinnati, OH, recommended the 5-μm Min-U-Sil as primary standard for quartz analysis.

(2) Cristobalite. The primary standard for cristobalite was prepared by calcination of straight α-quartz as follows (9): Put a few grams of 5-μm Min-U-Sil in a platinum crucible. Heat the crucible up to 1450 °C in an electric furnace (Deltec 30T Hi Temp oven) and maintain that temperature for 48 h. Turn off the furnace, and let the crucible cool to room temperature.

X-ray diffraction analysis of the product by the matrix-flushing method indicates a clean cristobalite with 0.38 ± 0.02% α-quartz. Hence 99.6% conversion was attained. When the calcination was run for 2 h at 1100 °C in a Thermolyne muffle furnace, a binary mixture of 61.8% cristobalite and 38.2% quartz was produced. Consistent results were obtained by using either of these two preparations as primary standards for cristobalite.

(3) Tridymite. The primary standard for tridymite was made by fusion of α-quartz in molten salt, described as follows: Mix 1 part of 5-μm Min-U-Sil and 5 parts of sodium chloride (mp 801 °C, bp 1413 °C). Grind the mixture for 30 min in a Fisher auto grinder. Transfer the mixture into a platinum crucible and heat to 1100 °C for 72 h in a muffle furnace. Take out the crucible and let it cool to room temperature. Grind the fused pellet to a powder. Drop the powder into hot water to dissolve the sodium chloride. Repeat the hot-water washing until the water is free from salt. Filter out the silica and dry at 105 °C for an hour.

X-ray diffraction analysis of the product thus obtained indicates a pure tridymite. The role of molten salt is to isolate the silica from the atmosphere and to change the activation energy of phase transition. When the fusion time was shortened to 3 h at 1100 °C, a binary mixture of 62.6% cristobalite and 37.4% tridymite resulted. Either the pure tridymite or the binary mixture can be used as primary standard, and both give consistent results of analysis. The relevant data of the primary standards are summarized in Table III.

Table III. Primary Standards

	quartz	cristo-balite	trid-ymite
Min-U-Sil, PGS	~100%	0	0
calcination of Min-U-Sil			
2 h, 1100 °C	38.2%	61.8%	0
48 h, 1450 °C	0.38%	99.6%	0
fusion of Min-U-Sil/NaCl			
3 h, 1100 °C	0	62.6%	37.4%
72 h, 1100 °C	0	0	~100%
k_i (integrated intensity)	3.80	4.24	0.836
k_i (peak height)	3.80	4.00	0.740
sensitivity, counts/(s %)	254	256	57
detection limit, %[a]	0.1	0.1	0.3

[a] Silica in light matrix without interference.

4.2. Analysis by X-ray Diffraction.

Anderson (*10*) reviewed the analytical methods for free silica, including colorimetric, X-ray diffraction, infrared, thermal, dye adsorption, and microscopic techniques. To date, only X-ray diffraction (XRD) is practical to assay all forms of crystalline silica. The simplest approach for quantitative XRD analysis is the matrix-flushing procedure (*11*). It involves the determination of a flush constant k_i, $k_i = I_i/I_c$, which is the intensity ratio for a 50/50 mixture of primary standard (i) and corundum (c). The flush constant k_i is independent of matrix effect. Once k_i is known, the weight fraction X_i of compound i in any unknown samples can be obtained with the following intensity-concentration equation (*12*):

$$\alpha_i = (X_c / k_i)(I_i / I_c)$$

Where X is the weight fraction and I is the XRD intensity of the strongest resolved reflections of component i and corundum, c. The routine for collecting intensity data is presented below (8, 13). More details can be found in the cited references.

Add 0.8-1.2 g of Al_2O_3 (Linde A, 1 μm) to a primary standard or 0.4-0.6 grams of Al_2O_3 to an unknown sample to make a mixture of about 2 g. Note that the amounts of Al_2O_3 added to standards and unknowns to obtain optimum intensity ratios, are different. A noninterfering compound can be added to reduce the intensities if necessary. Grind the mixture for 20 min to ensure optimum particle size and sample homogeneity. Load the mixture into sample holder by the free-falling method (14) recommended by NBS to avoid preferred orientation and density gradient, if any. Measure the integrated intensity of the strongest resolved reflection of Al_2O_3, and silica, preferably the quartz peak at $d = 3.34$ Å, the cristobalite peak at $d = 4.05$ Å, the tridymite peak at $d = 4.33$ Å, and the corundum peak at $d = 2.085$ Å.

The integrated intensities can be measured by scanning integration (or step scanning if possible), peak area, or peak weight. The scanning integration uses the accumulated counts in the scaler while the counter scans through the peak of reflection. The peak area is the area in cm^2 under the peak on a strip chart (scan speed, 0.25°/ min; chart speed, 0.5 in./min) measured with a planimeter. The peak weight is the weight in grams of the peak cut from the strip chart. For semiquantitative work, peak height can be used which is the maximum counting rate in counts/s of the peak. A set of intensity data collected from a mixture of 53.91% cristobalite (99.6% pure) and 46.09% Al_2O_3 is listed in Table IV. The flush constant k_i calculated from these different versions of intensity data agree well within experimental errors. The coefficient of variation (S/X) of the matrix-flushing procedure was statistically evaluated to be 8% or better.

Table IV. Integrated Intensity

	scanning integration, counts	peak area, cm^2	peak weight, g	peak height, counts/s
cristobalite				
(101)	325 130	26.57 × 5	0.1190 × 5	13800
(200)	55 200	23.06	0.1011	1980
corundum				
(113)	67 065	26.85	0.1198	2960
(104)	58 150	23.52	0.1040	2690
(110)	24 810	10.20	0.0446	1150
k_i	4.16	4.24	4.26	4.00

A regular XRD scan of the "secondary standard" from Johns-Manville for the AIHA round robin indicates that the line profile of the natural cristobalite (Celite) is much broader than that of the synthetic cristobalite due to different lattice imperfections. The use of integrated intensities for its analysis can compensate this difference. Peak height data can be used for quartz analysis because there is no noticeable difference in line profiles between quartz samples from various sources.

Statistical considerations show that an intensity ratio is determined with the greatest precision in the shortest time when the numerator and denominator are equal. In practice it is advantageous to make the intensity ratio (I_i/I_c) close to unity by adding the proper amount of Al_2O_3 or choosing the right reflections, although this is not always possible. A set of intensity data for the second round-robin sample from AIHA is presented in Table V. Note that the second strongest peaks of cristobalite (200) and corundum (104) were measured, besides the statistical considerations, because they are better resolved and more convenient. The corresponding flush constant for this pair of reflections is $k_i = 0.832$, calculated from the data in Table IV. The amount of cristobalite in celite was found to be 58.2% which was reported to AIHA and checks well with the results from Johns-Manville.

Table V. Cristobalite in Celite

	I		II	
	Celite (200)	Al_2O_3 (104)	Celite (200)	Al_2O_3 (104)
% wt	36.53	63.47	52.15	47.85
peak wt, g	0.0422	0.1502	0.0565	0.1078
% cristobalite in Celite	58.7		57.8	
av, %			58.2	
Johns-Manville, %			57.3	

5. CONCLUSIONS

X-ray diffraction is the only technique feasible for assaying the three forms of crystalline silica. However, inadequate interlaboratory precision data created confusion. Therefore, concerted efforts are needed to develop practical analytical procedures and certified primary standards.

Depending on the purpose of analysis, silica samples may come as a bulk powder or as dust-laden filters. For silica on membrane filters, a multiple-exposure XRD method with imbedded standards was reported by Chung (*15*). For silica in bulk powder, a matrix-flushing (*8*) approach is presented here. The primary standards of cristobalite and tridymite can be synthesized from high-purity quartz. It is preferable to have these primary standards made in larger quantity and distributed by a reputable source to avoid ambiguity.

6. REFERENCES

Anal. Chem. 1980, 52, 1136A.

Anal. Chem. 1980, 52, 1272A.

Anderson, P. L. *Am. Ind. Hyg. Assoc. J.* 1975, 36, 767.

Bhargara, O. P.; A. S. Alexiou; H. Meilach; and W. G. Hines. *Am. Lab.* 1979, Sept, 27.

Chung, F. H. *Adv. X-ray Anal.* 1974, 17, 106.

Chung, F. H. *Environ. Sci. Technol.* 1978, 12, 1208.

Chung, F. H. *J. Appl. Crystallogr.* 1974, 7, 519.

Chung, F. H. *J. Appl. Crystallogr.* 1974, 7, 526.

Chung, F. H. *J. Appl. Crystallogr.* 1975, 8, 17.

Groff, J. H.. Memorandum to PAT participants, May 7, 1980, National Institute for Occupational Safety and Health,

Cincinnati, OH.

Mug, H. P.; and L. E. Alexander. "X-Ray Diffraction Procedures," 2nd ed.; Wiley-Interscience: New York, 1974; p 524.

National Bureau of Standards, Monograph 25, "Standard X-Ray Diffraction Powder Patterns"; Government Printing Office: Washington, D.C., 1971; p 3.

Swallow, G. L. Annual AIHA Meeting, Los Angeles, 1978, and AIHA Accreditation News & Notes, Aug 1978, American Industrial Hygiene Assoc., Akron, OH.

Swallow, G. L. Laboratory Directors' Meeting, Houston, 1980, and AIHA Accreditation News & Notes, Aug 1980, American Industrial Hygiene Assoc., Akron, OH.

U.S. Dept. of Health, Education and Welfare, "NIOSH Manual of Analytical Methods"; Government Printing Office: Washington, D.C., 1977.

20.

Imaging and Analysis of Airborne Dust for Silica

> A man must be big enough to admit his mistakes, smart enough to profit from them, and strong enough to correct them.
> — John C. Maxwell

1. ABSTRACT

A technique to develop the image of dust collected on membrane filters is reported. The developed images reveal that the dust particles from either aerosol or suspensoid are not uniformly distributed over the filter. This uneven distribution affects the precision of in situ x-ray diffraction analysis. To cope with this situation, a multiple-exposure x-ray diffraction method is developed and applied to free silica analysis. This method achieves substantial timesaving (15 min instead of 3 h/sample) and delivers reliable data.

2. INTRODUCTION

Chronic exposure to high concentrations of silica would cause lung damage (silicosis). Therefore, the amount of air-borne silica in factories is regulated and monitored by OSHA. Among the six techniques available for the analysis of crystalline silica in airborne dust collected on membrane filters (1), chemical separation (colorimetric) and x-ray diffraction are most widely used. The advantages of the x-ray method over the chemical method are simplicity and rapidity. However, these advantages are lost in NIOSH P&CAM Method No. 109 (X-ray diffraction) where an internal standard and a silver membrane are utilized to increase precision (2, 3). An in situ x-ray technique is more attractive when a large number of samples are to be screened and particularly when many of the samples contain little or no free silica. A crucial factor causing the spread of results of an in situ x-ray method is the distribution of dust particles on the filter. However, the dust distribution is generally invisible and hence overlooked.

3. IMAGING THE DUST PARTICLES

Membrane filters used for airborne dust sampling are made of either polyvinyl chloride (PVC) or mixed esters of cellulose (nitrocellulose/cellulose acetate). PVC filters are recommended for free silica monitoring, while cellulose filters are routinely used for asbestos and heavy metals.

A simple technique has been found to develop the image of dust particles collected on these membrane filters as described below. A developer is made of one part methylene chloride and five parts chloroform by volume. A few drops of this developer are spread on an aluminum disk about 45 mm in diameter (or square). Surface tension prevents the developer from spilling over the edge of the disk. The PVC filter is then carefully placed on the developer. The filter is dissolved immediately by the developer, forming a transparent thin film. The dust particles are insoluble and thus are imbedded into the PVC film leaving a sharp image. The thin film dries in about a minute, giving a permanent sample or standard. Other than optimum solubility and fast drying, the developer system is chosen so that the filter retains its shape and the dust particles are not disturbed in the process of image development. Some typical examples are shown in Figure 1.

A study of these developed images indicates that the distribution of dust particles on filters tends to be uneven when the dust level in a factory fluctuates, or when a short sampling period is necessary due to high dust levels. Uneven distributions are also likely to be found in standards made by using a wet suspension/filtration procedure. Generally speaking, neither aerosol nor suspensoid collected on a filter would produce a statistically uniform distribution of particles over the filter. Improperly collected, overloaded, or locally compacted samples may cause larger errors in subsequent analysis. The imaging technique can reveal the presence of these problems and warrant precautions or extra care. Besides image development, another application of this technique is the fixation and preservation of standards or samples in their undisturbed condition for shipping, stock, rescanning, or reference.

By the same principle, for filters made of mixed esters of cellulose such as MF-Millipore, the image of dust particles can be developed with straight ethyl acetate or acetone in a similar manner.

4. IN SITU X-RAY DIFFRACTION ANALYSIS FOR SILICA

The quickest way to determine the quantity of free silica in airborne dust collected on a membrane filter is to measure the x-ray intensity of the characteristic silica peak from the filter as received without tedious sample preparation. The silica concentration is then read from a calibration curve previously prepared by use of external standards. Two conditions must be met to make this simplest scheme feasible: First, the matrix effect must be negligible; second, the sample composition must be uniform. In the case of a few milligrams of dust collected on a PVC filter, the matrix effect is indeed negligible (*4*, *5*), but the dust particle distribution is far from uniform. Note that even though the dust distribution is uniform, the silica distribution might still not be uniform, especially when the type and level of dust in air fluctuate. Therefore, some sort of remedy is needed to obtain a meaningful average X-ray intensity. Naturally, spinning the sample during exposure to the X-ray beam would probably be the first choice (*6*). However, the x-ray intensity obtained from a spinning sample grants too much weight to dust particles deposited near the center of the filter as explained below. The ideal condition of spinning would be that the x-ray beam coincides with the full radius or diameter of the spinning filter (Figure 2a) so that all dust particles are exposed to the primary x-rays. Even under this ideal condition, the exposure of the central area of $0.1r$ is 19 times longer than that of the annulus of the outermost $0.1r$ (Figure 2b) because:

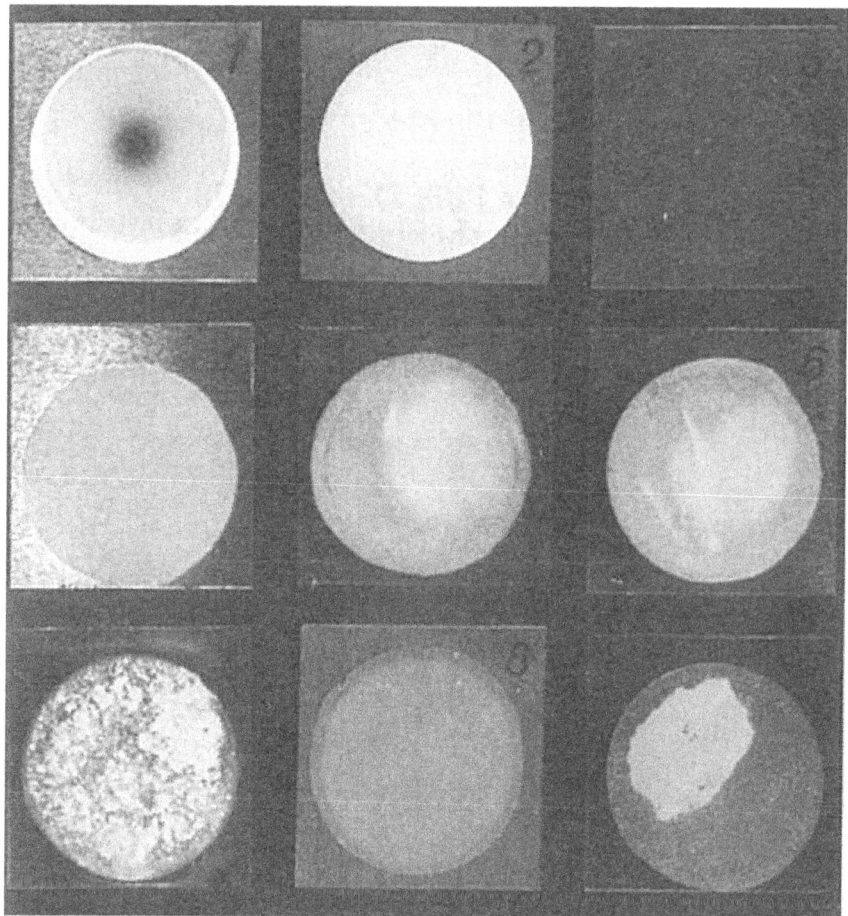

Figure 1. Image of dust on PVC filters

Under any other conditions (Figure 2c) the difference in exposure is worse. A multiple-exposure approach (Figure 2d) has been adopted in our laboratory to cope with this situation. The calibration curve of external standards was prepared according to a wet filtration procedure (CRL-001) obtained from the Chemical Reference Lab of NIOSH (Robert A. Taft Laboratories, Cincinnati, Ohio). To attain stability for frequent manipulation, the external standards were in the form of imbedded films. For the sake of simplicity, however, routine filter samples can be analyzed directly as received except those samples of legal or technical importance that should be imbedded for preservation. The multiple-exposure procedure was applied to both standards and samples to cope with possibly uneven dust or silica distribution on filters.

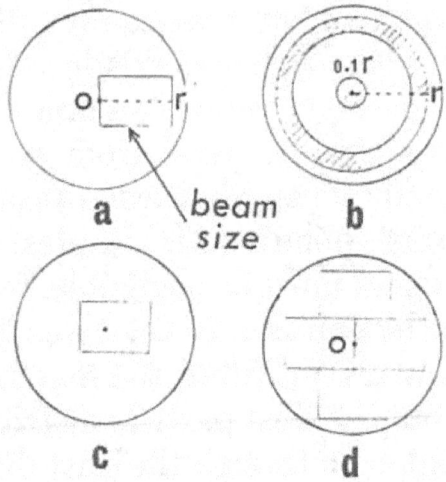

Figure 2. Spinning-sample vs. multiple-exposure

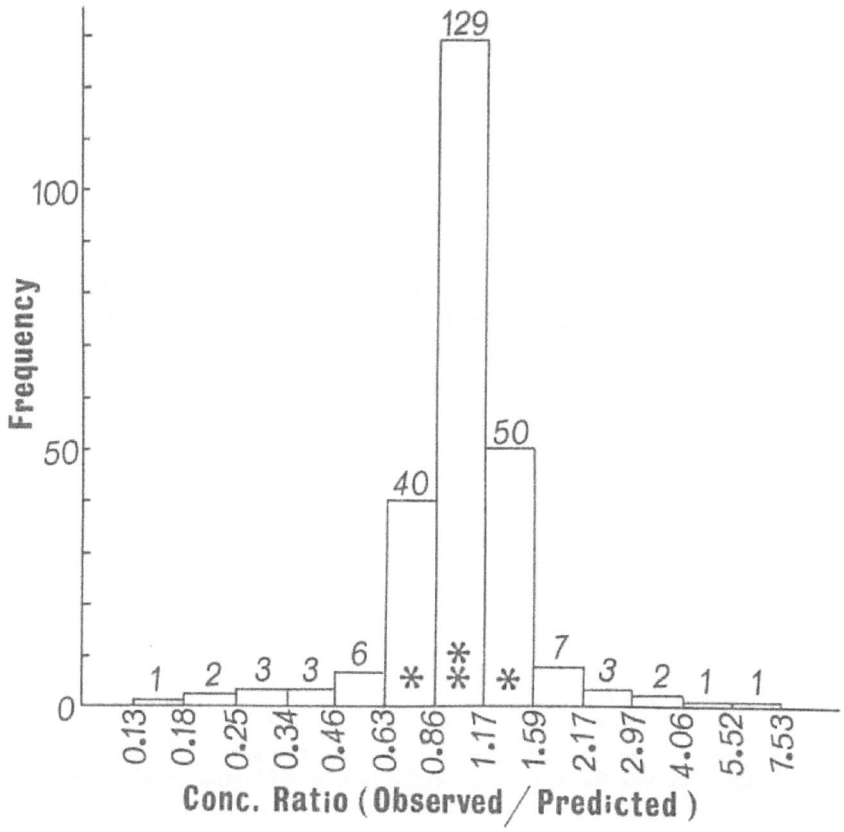

Figure 3. Results of Proficiency Analytical Testing, Round no. 41

We have participated in NIOSH's Proficiency Analytical Testing (PAT) program since February 1976. The multiple-exposure x-ray diffraction method for silica analysis has been used for all PAT round robin samples as well as routine airborne dust samples. In a most recent series of PAT samples (Round 41), 62 laboratories turned in results of silica analysis for a set of four controlled unknowns made of silica and silicates. The frequency histogram of the 248 data field depicts a log-normal distribution as shown in Figure 3 where all the results are expressed as error ratios, i.e., observed concentration to predicted concentration of silica. A semilogarithmic scale was used for plotting. The abscissa of the histogram has a constant interval size if labeled in log of error ratios. However, for easy reference, the histogram intervals are labeled directly in terms of error ratios.

The four stars in Figure 3 indicate the location of results obtained in our laboratory. To illustrate the scatter of x-ray intensities from multiple exposure of different areas of the same filter, the experimental data for the two samples which exhibited the best agreement are presented in Table I.

Table I. Scatter of X-ray Intensities and Silica Concentration

	PAT round robin sample no.	
	S41-2	S41-4
x-ray intensity, c/s (multiple exposures of the strongest reflection)	270	190
	280	260
	185	305
	190	150
av intensity, c/s	231	226
SiO_2 found, mg/filter	0.093	0.092
PAT arithmetic mean, mg/filter	0.097	0.097
PAT geometric mean, mg/filter	0.077	0.064

There were no outliers in this round of silica analysis according to the statistics used for the PAT program. Therefore, the arithmetic and geometric means were calculated from the results of all 62 laboratories submitting data. The wide scatter of x-ray intensities from the same membrane filter confirms the conclusion from image observation that the distribution of dust particles over the filter is not uniform. The excellent agreement of the amount of silica found demonstrates that the multiple-exposure approach is effective and realistic.

6. CONCLUSIONS

For a maximum of 5 mg total dust collected on a membrane filter (overloaded if higher than 5 mg), the effect of particle distribution overrides the matrix effect in X-ray diffraction analysis for silica by an external standard method. Through an image development technique, it has been shown that the dust distribution on the filter is not uniform. To cope with this non-uniform distribution and to preserve the simplicity of analysis, a multiple-exposure X-ray diffraction method for silica analysis has been developed for routine and PAT samples. The results obtained are in good agreement with those of other laboratories participating in the PAT program. This in situ X-ray diffraction method achieves substantial timesaving, delivers reliable data, and may be used to substitute the more demanding NIOSH method for fast screening analysis of silica in airborne dust.

6. REFERENCES

Allen, G. C.; B. Samimi; M. Ziskind; and H. Weill. *Am. Ind. Hyg. Assoc.* J., 35 (11), 711-7 (1974).

Anderson, P. L. Am. Ind. Hyg. Assoc. J., 36 (10), 767-78 (1975).

Chung, F. H. *Adv. X-ray Anal.*, 19,181-90 (1976).

Chung, F. H. *J. Appl. Crystallogr.*, 8, 17-19 (1975).

Chung, F. H.; A. J. Lentz; and R. W. Scott. *X-Ray Spectrum*, 3, 172-5 (1974).

U.S. Dept. of Health, Education and Welfare, "NIOSH Manual of Analytical Methods," GPO, 1974.

21.

Characterization of Airborne Particulates

> A person needs just three things to be really happy in the world: Someone to love, something to do, and something to hope for.
> — Tom Bodett

1. ABSTRACT

In modern society, the automobiles, power plants, incinerators, and space heaters created the air pollution problem, a serious health hazard. The U.S. Environmental Protection Agency (EPA) promulgated numerous regulations to curb air pollution. Consequently, a variety of techniques and instrumentation for imaging and analysis have been developed or perfected. The air we breathe is an aerosol, a mixture of gases and particles. The chemical compositions of airborne particles have important bearings on human health. Trace amount of contaminants in complex mixture is a unique feature of air pollutants. The analytical techniques are pressed for sensitivity and selectivity. The many analytical techniques are divided into three categories: inorganic/elemental, organic/molecular and micro/surface. Each technique is very briefly described here. A treatise on each technique is given in references. In practice, XRD, XRF, and AA are favored for inorganic/elemental analysis; IR, NMR, GC/MS, LC are method of choice for organic/molecular analysis; ESCA, SIMS, and MOLE are preferred for micro/surface analysis.

2. INTRODUCTION

Prometheus stole fire from Olympus and brought it to mankind. He was rightly punished, not for his defiance of Zeus, but because the fire started air pollution. Over 80% of the total U.S. air pollution originates from combustion in power plants, automobiles, incinerators and space heaters. Fostered by combustion, three events marked the expanded scope of air pollution. In the 1760s the industrial revolution initiated the spreading of chemicals into the air. In the 1940s nuclear explosions began to load the troposphere with radioactive dust. In the 1970s supersonic transport set out to eject pollutants directly into the stratosphere.

With the increasing concern about the health hazards of air pollution, the U.S. Congress passed the Clean Air Act of 1970, which created the U.S. Environmental Protection Agency (EPA). EPA promulgated numerous regulations to curb air pollution. To prescribe and enforce these regulations, the sources and behaviors of pollutants have to be learned, and their monitoring and abatement must be implemented. Consequently, a variety of techniques and instrumentation for imaging and analysis have been developed or perfected, and a wealth of information concerning our environment has been generated.

The air we breathe is an aerosol, a mixture of gases and particles. Depending on locale and time, the atmospheric aerosol contains hundreds of contaminants that may be harmful to human health. In 1971 EPA published the National Ambient Air Quality Standards (NAAQS) for six contaminants: SO_2, NO_2, CO, oxidants, hydrocarbons and particulate matter. The primary standard for particulate matter is 75 $\mu g/m^3$ annual geometric mean, and 260 $\mu g/m^3$ in 24 hours, not to be exceeded more than once per year. An air quality standard for airborne particles based on gross weight is convenient for control purposes but is inadequate in terms of health. The penetration and retention of particles in the respiratory tract are functions of particle size. The physiological effects of particles in the human body are dependent on the chemical composition of particles. Therefore, the size distribution and chemical composition of airborne particles have important bearings on human health.

To study these properties of the atmospheric aerosol and to define their health effects, extensive research has been in progress for some eight years. Our current understanding of airborne particles and their imaging and analytical techniques are summarized here.

3. NATURE OF AIRBORNE PARTICLES

In an unpolluted countryside, airborne particles may amount to 50 $\mu g/m^3$. At the height of pollution episodes, airborne particles may reach 100 times that amount. The health effects of acute exposure to high concentrations of toxic pollutants, such as episodes in Donora (Pennsylvania), London (England), Meuse Valley (Belgium) and Poza Rica (Mexico), are well recognized and documented [1]. But the health effects of chronic exposure to low concentrations of toxic pollutants are still uncertain. Some generalizations of the nature of airborne particles are discussed below.

3.1 Size Distributions

The size of airborne particles ranges from a few tens of Angstroms to a few hundred of microns. Most fine particles (diameter $\phi < 2$ μm) are formed by two condensation processes: (1) condensation of hot supersaturated vapors produced by combustion, such as metallic vapors and high boiling oils; and (2) condensation of cold supersaturated vapors produced by in situ reaction, such as the photochemical smog and sulfate aerosols. Most coarse particles ($\phi < 2$ μm) are generated from two mechanical processes: (I) natural turbulence such as gusts, windstorms or volcano eruption; and (2) industrial operations such as coal-grinding and cement-making.

The size distribution of airborne particles tends to be relatively constant. Typical size distribution of urban aerosol consists of three modes [2]: Aitken nuclei (≤ 0.1 μm), accumulation (0.1 ~ 2.0 pm) and coarse particles (≥ 2.0 μm). Nuclei and accumulation modes make up the fine particles. Coagulation of nuclei, growth of nuclei and the smallest industrial dusts all tend to accumulate in the 0.1 to 1.0 μm size range, peaking at about 0.2 μm and called the accumulation mode [3,4]. A large number of nuclei are transferred to the accumulation mode, but little mass is transferred from the accumulation mode to the coarse mode. Particulates of 0.1 to 1.0 μm size range scatter and absorb light most efficiently. They are largely responsible for haze and turbidity. Particulates larger than 1μm have little effect on visibility.

Whitby [5] observed that most particles are approximately 0.01 μm in diameter. Most of the surface area is provided by particles averaging 0.2 μm in size. The mass (volume) distribution is bimodal: 0.3 μm and 10 μm in diameter. The mass of fine particles (ϕ < 2 μm) is almost equal to the mass of coarse particles (ϕ > 2 μm).

Coarse particles such as soil and fly ash have appreciable settling velocity. They will be removed from the aerosol before they travel very far. Coarse particles larger than 25 μm can be observed only near their source. Fine particles are not much affected by the sedimentation process. The dispersion of fine particles is governed by eddy diffusion and advection. Accumulation mode aerosols may be blown several thousands of miles away from their Sources.

The airborne particles lodge in the human body through three mechanisms [6]: (1) impaction (for particles > 1.5 μm), (2) sedimentation (for particles 0.5 ~ 1.5 μm), and (3) diffusion (for particles < 0.5 μm). Both sedimentation and diffusion produce rather uniform surface deposition in small bronchioles, alveolar ducts and alveolar sacs. During the past several decades, it was believed that the particulates of health consequence were those less than 3 μm in size, which could penetrate deep into the lung. Recent evidence indicates that larger particles of 3 ~ 15 μm size deposited in the upper respiratory system (nasopharynx and conducting airways) also can be associated with health problems [7]. This assessment led to the designation of this size range ($\leq$ 15 μm) as Inhaled Particulates. Particles greater than 15 μm are either rejected by the nose or restricted to the nasopharyngeal region (the head area). Dichotomous samplers are chosen by EPA to separate the inhaled particulates into two fractions: fine (ϕ < 2.5 μm) and coarse (ϕ > 2.5 μm). The fine particulates penetrate to the gas exchange region of the respiratory tract. Their size distribution and chemical composition are important parameters for epidemiological studies. Note that the Total Suspended Particulates (TSP) were defined as particulates of approximately 100 μm or less, and were collected by Hi-Vol samplers [6].

The major chemical species in atmospheric aerosols may fall into different size groups. Elements in the soil such as Si, Ca, Fe and Al are mostly in the coarse fraction. Elements produced by condensation such as Pb, As, sulfate (S) and ammonium (N) are mainly in the fine fraction. The trends of other elements in airborne particulates are listed in Table I [8,9].

3.2 Chemical Compositions

It has been estimated that some 300 new chemicals are released into the environment annually. These chemicals, old and new, traverse from atmosphere to hydrosphere to lithosphere to biosphere. Various chemical changes take place during the long journey. Some of these chemicals (natural, synthetic, or transformed) eventually take residence in the human body. The health effects of variable amounts of diverse chemicals over long periods of time are mostly uncertain or controversial. It has been postulated that a mutation occurring in somatic cells may cause cancer [10]; if it occurs in the gonads, the mutation is heritable. Increasing amounts of data suggest that most human cancers (60% or more) are environmentally derived [11], and 25% of our health burden is of genetic origin [12]. That is to say, mutagens may promote cancer today that may perpetuate in future generations. Certainly this legacy deserves serious concern.

The chemical composition of airborne particles varies widely, depending on site, activities, season and time. Nevertheless, a time-average composition serves as a frame of reference for research and control. The chemical composition of particulates in air is measured in terms of elemental, molecular or ionic species. Perhaps the toxic effects of particulates are mostly due to molecular rather than elemental or ionic character because it is the molecular form of pollutants that shows solubility, reactivity and toxicity.

Table I. Size Groups of Elements

Fine (<2.5 μm)	Coarse (>2.5 μm)	Found in Both Fractions	Variable
H^+	PO_4^{3-}	NO_3^-	Zn
NH_4^+	Fe	Cl^-	Cu
$SO_4^=$	Ca		Ni
Pb	Ti		Mn
As	Mg		Sn
Se	K		Cd
C (Soot)	Si		V
Organics	Al		Sb
	C		
	Pollen		
	Spores		

It is convenient to divide the particulate pollutants into inorganic/elemental pollutants and organic/molecular pollutants. This division usually suggests the origin of particles, their biological effects, the techniques for their analysis, and the rationale and policy of effective control. The inorganic and organic pollutants found in airborne particulates are compiled in Tables II and III, respectively. When national average (urban) data are not available, local research data are entered to indicate likely levels. Note that many low-boiling organics such as solvents, gasoline, other hydrocarbons and derivatives are present in the air as vapors. These vapors could toxicate cilia and mucus of the respiratory system, thus breaking down its defense mechanism.

Table II. Inorganic Pollutants in Airborne Particles

Pollutant	$\mu g/m^3$	Site	Pollutant	$\mu g/m^3$	Site
TSP	105	Nat. Avg. [13]	Br	0.61	Pasadena
$SO_4^=$	10.6	Nat. Avg.	K	0.32	Pasadena
NO_3^-	2.6	Nat. Avg.	Mn	0.10	Nat. Avg.
F^-	< 0.05	Nat. Avg.	Cu	0.09	Nat. Avg.
NH_4^+	1.3	Nat. Avg.	Cl	0.07	Pasadena
$(NH_4)_2SO_4$	3.4	Seattle [14]	V	0.05	Nat. Avg.
NH_4HSO_4	1.0	Seattle	Ba	0.04	Pasadena
NH_4Cl	0.6	Seattle	Ti	0.04	Nat. Avg.
NH_4NO_3	0.5	Seattle	Ni	0.03	Nat. Avg.
$(NH_4)_2SO_3$	0.16	Seattle	As	0.02	Nat. Avg.
$NaNO_3$	2.0	Seattle	Sn	0.02	Nat. Avg.
SeO_2	0.28	Seattle	Sr	0.02	Tucson [15]
H_2SO_4	1.1	Seattle	Cr	0.015	Nat. Avg.
PbBrCl	0.5	Lancaster [16]	Zr	0.015	Toronto [17]
$PbSO_4 \cdot (NH_4)_2SO_4$	0.2	Lancaster	Rb	0.01	Tucson
$PbSO_4$	0.1	Lancaster	Se	0.008	St. Louis [8]
$PbCO_3$	0.1	Lancaster	I	0.006	Pasadena
S	4.01	Los Angeles [18]	Mo	<0.005	Nat. Avg.
Si	2.42	Los Angeles	Co	<0.005	Nat. Avg.
Fe	1.58	Nat. Avg.	Be	<0.005	Nat. Avg.
Mg	1.11	Pasadena [19]	Bi	<0.005	Nat. Avg.
Ca	1.00	Pasadena	Ag	0.004	Chicago [20]
Na	1.00	Pasadena	Li	0.004	Tucson
Al	0.81	Pasadena	La	0.003	Toronto
Pb	0.79	Nat. Avg.	Cd	0.002	Nat. Avg.
Zn	0.67	Nat. Avg.	Sb	0.001	Nat. Avg.

Gross Beta Radioactivity: 0.8 pCi/m^3 (Max. 12.4 pCi/m^3)

Table III. Organic Pollutants in Airborne Particles

Pollutant		ng/m^3	Site
	TSP	105,000	Nat. Avg. 1965 [13]
	Benzene-soluble organics	6,800	Nat. Avg.
C2	Dimethylnitrosamine	100	Baltimore, 1975 [21]
C3	Malonic Acid	80	Riverside, 1976 [22]
C4	Methyl Malonic Acid	20	Riverside
	Succinic Acid	500	Riverside
C5	Methyl Succinic Acid	240	Riverside
	Glutaric Acid	460	Riverside
	Glutaraldehyde	30-300	Pasadena, 1972 [14]
	Glutaraldehydate	400-1400	Pasadena
	Gluteraldehydic Acid Nitrite	70-1000	Pasadena
	Glutaric Acid	170-1350	Pasadena
	5-Hydroxypentanal	100-300	Pasadena
	5-Hydroxypentanoate	70-2000	Pasadena
C6	Methyl Glutaric Acid	160	Riverside
	Adipic Acid	230	Riverside
	Adipaldehyde	200	Pasadena
	Adipaldehydate	300-3000	Pasadena
	Adipaldehydic Acid Nitrate	40-160	Pasadena
	Adipic Acid	100-800	Pasadena
	1,4-Benzoquinone	15-80	Seattle, 1974
	1,2-Dihydroxybenzene	15-65	Seattle
	1,4-Dihydroxybenzene	15-125	Seattle
	6-Hydroxyhexanal	30-400	Pasadena
	6-Hydroxyhexanoate	400-3500	Pasadena
C7	Dimethyl Glutaric Acid	12	Riverside
	Methyl Adipic Acid	38	Riverside
	Pimelic Acid	48	Riverside
	Benzoate	90-380	Pasadena
	7-Hydroxyheptanal	90-180	Pasadena
	7-Hydroxyheptanoate	170-650	Pasadena
	Pimelaldehyde	70-550	Pasadena
	Pimeladehydate	50-800	Pasadena
C8	Suberic Acid	56	Riverside
	Phenyl Acetic Acid	400	Pasadena
C9	Azelaic Acid	120	Riverside
	3-Phenylpropionic Acid	60-500	Pasadena
	2-Methylindole	2	Rome, 1972
	Quinoline	0.6	Rome
C10	Sebacic Acid	30	Riverside
C11	2,6-Dimethylquinoline	0.3	Rome

3.3 Legal Aspects

The Clean Air Act of 1970 created the U.S. Environmental Protection Agency. EPA has published numerous rules and standards concerning emission of pollutants into the air. The regulated pollutants fall into three categories [23-26]:

(1) Criteria pollutants: In 1971 EPA published NAAQS for six pollutants: SO_2, NO_2, CO, oxidants, hydrocarbon and particulate matter. The standards are of two types: primary standards, designed to protect human health, and secondary standards, designed to protect property and aesthetics.

(2) Hazardous pollutants: The toxic pollutants belonging to this category may cause or contribute to an increase in mortality or an increase in serious irreversible or incapacitating reversible illness. EPA has established emission standards for hazardous pollutants including mercury, lead, beryllium, asbestos, benzene and vinyl chloride. New compounds will be added to this list when evidence warrants that action.

(3) Source-specific pollutants: There are four major sources of air pollutants: automotive, combustion (non-automotive), industry and incinerators. EPA has set maximums on emission of specified pollutants for each category of sources, such as SO_x, NO_x and particulate matter emission from power plants; NO_x, CO, hydrocarbon and lead emission from automobiles; SO_x, CO, H_2S, particulate matter, methyl mercaptan, dimethyl sulfide and disulfide from the Kraft pulp industry, etc.

4. IMAGING OF AIRBORNE PARTICLES

The size of the smallest perceptible particle with the unaided human eye is about 0.1 mm (100 μm). Airborne particles are generally much smaller than this. Both the morphological and compositional features of airborne particles can be made visible through various imaging techniques.

The basic principle of all the imaging techniques is that the specimen is made to interact with a beam of particles (photons, electrons or ions). After the interaction, the primary particles are scattered by the specimen and secondary particles are generated from the specimen. The scattered or generated particles are then gathered by optical or electronic devices and focused into images characteristic or diagnostic of the specimen. The imaging techniques can be grouped into three categories: mass images, morphological images and chemical images. The modern trend is to combine the imaging and analytical techniques into a major instrument to enhance its capabilities for specimen characterization (vide infra).

4.1 Mass Images

The airborne particles are usually collected on membrane filters, which are then analyzed for their size distribution and chemical composition. For certain techniques of analysis, only a portion of the filter is sampled. Therefore, the distribution of dust on the filter would affect the results of analysis. The dust distribution on the filter is normally invisible and readily overlooked. Currently, polyvinyl chloride (PVC) filters are recommended for free silica monitoring, and cellulose-ester filters are used routinely for asbestos and heavy metals. Other membrane-filter materials used for various purposes include nylon, Teflon, silver, Nucleopore, acrylonitrile, polysulfone and polyamide.

A simple technique to develop the image of dust particles on these membrane filters was reported by Chung [27]. The imaging procedure using PVC filters is described below.

A developer is made of one part methylene chloride and five parts chloroform by volume. A few drops of this developer are spread on an aluminum disk (for X-ray diffraction), glass plate or Mylar film (for microscopy) about 45 mm in diameter (or square). Surface tension prevents the developer from spilling over the edge of the disk. Then the PVC filter is placed carefully on the developer. The filter is dissolved immediately by the developer, forming a transparent thin film on drying. The dust particles are insoluble and thus are embedded into the now transparent PVC film, leaving a sharp image. The thin film dries in about one minute giving a permanent sample or standard. Other than solubility and fast drying, the developer system is chosen so that the filter retains its shape and the dust particles are not disturbed in dip process of image development. Some typical examples are shown in Figure 1.

This procedure can be applied to filters made of other materials by formulating a developer that dissolves the filter yet does not upset the particles. The information on proper solvents for a specific filter is usually available from the filter supplier.

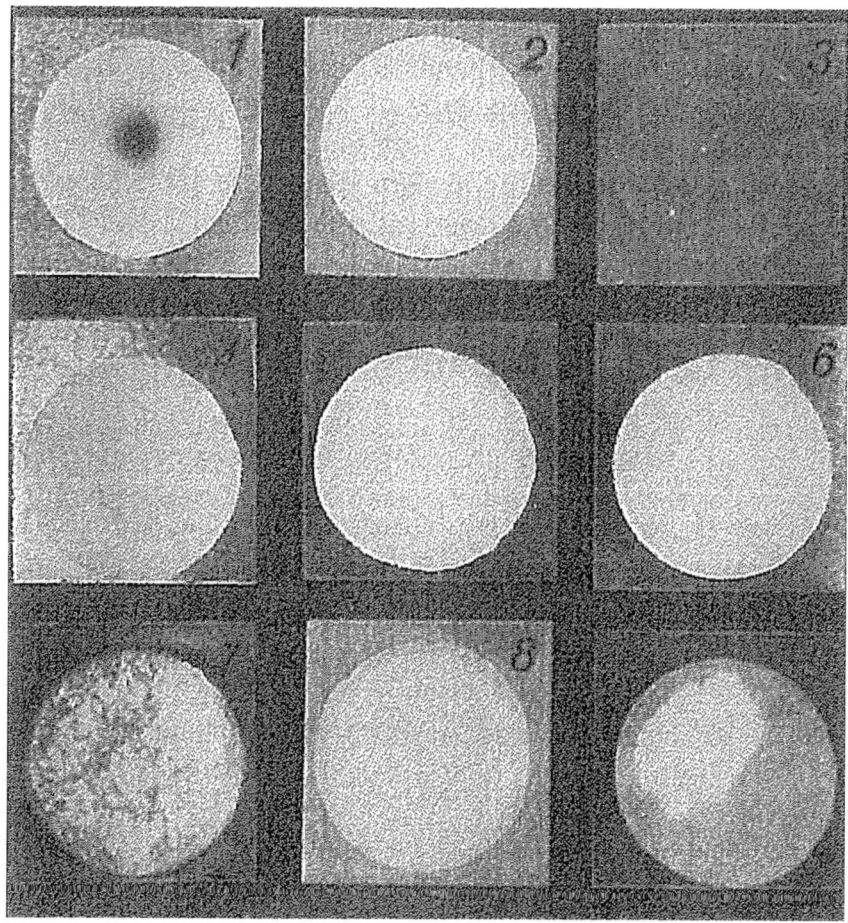

Figure 1. *Image of dust on PVC filters. Except for blank and three PAT samples, all samples were taken from paint factories where various resins, pigments and extenders were loaded into mills and processed. (1) As received, result of improper sampling. (2) As received, dust laden, dust particles invisible. (3) After imaging, clean blank filter. (4) After imaging, PAT sample S41-B, no free silica. (5) After imaging, PAT sample S41-1, 0.105 mg SiO_2 filter. (6) After imaging, PAT sample S41-2, 0.077mg SiO_2 / filter. (7) After imaging, overloaded sample. (8) After imaging, even distribution of dust particles. (9) After imaging, uneven distribution of dust particles. Reprinted with permission from Environ. Sci. Technol. [27]*

A study of these developed images indicated that the distribution of dust particles on filters tends to be uneven when the dust level fluctuates or when a short sampling period is necessary due to high dust level. Uneven distributions are also likely to be found in standards made by using wet suspension/ filtration procedure. Improperly collected, overloaded or locally compacted samples may cause larger errors in subsequent analysis. The imaging technique can reveal the presence of these problems and warrant precautions or extra care. Besides image development, another application of this technique is the fixation and preservation of standards or samples in their undisturbed condition for shipping, stock, rescanning, reference or legal use.

4.2 Morphological Images

For centuries, optical microscopes have been used for morphological examination of micro samples. The maximum useful magnification of optical microscopes is about 1000X. The advent of electron microscopes extended the maximum useful magnification over another 1000X. The resolving power of the human eye is about 0.1 mm (10^6 Å); that of optical microscopes is about 0.2 μm (2000 Å), which is close to the theoretical limit, i.e., half the wavelength of the illuminating radiation. Note that the wavelength of the visible spectrum lies between 0.3 and 0.7 μm. The resolving power of the electron microscope has reached 2 Å, which is yet far from its theoretical limit. Note that the wavelength of 60 keV electron is 0.05 Å.

(1) Optical Microscopes [28,29]

When visible light interacts with an object it is transmitted, reflected, refracted or diffracted. Any of these scattered fights can be gathered to form a magnified image. Depending on the object type and viewing interest, one kind of microscope may work better than another. There are four types of objects:

- **Amplitude object** changes amplitude (intensity) of incident light.
- **Phase object** changes the phase (alignment of wave maximum) of incident light.
- **Birefringent object** splits the incident light into two parts, each of which follows different laws.
- **Fluorescent object** absorbs incident ultraviolet (UV) and emits light of characteristic color.

The operating principle and unique merit of various microscopes are concisely given below. More details can be found in corresponding reference works.

>**Bright-Field Microscope:** It is the conventional ordinary microscope that gathers the transmitted light forming an amplitude image due to varying density (hence different absorption) in the object. Certain objects do not form sharp images in bright-field microscopes. Therefore, various physical principles have been used in special designs to enhance or introduce contrast. These special designs include the following varieties.
>
>**Dark-Field Microscope:** It uses oblique illumination. The transmitted light falls out of the aperture of the ob-

jective lens. Only the refracted light from the specimen is focused into a bright image against a dark background. It is most useful for viewing specimens of high scattering power, such as suspensions of small particles.

Polarizing Microscope: The incident light is plane polarized (vibration in a single direction) before entering the specimen. The transmitted light is analyzed for polarization inflected by the specimen. Birefringent (optically anisotropic) substances show extinctions, optical orientations and interference figures under a polarizing microscope. It is widely used for optical crystallography and chemical microscopy.

Phase-Contrast Microscope: Most biological specimens are phase objects, as opposed to more common amplitude objects. Phase objects such as living cells consist of thin transparent structures of the same optical density, hence give images of minimal contrast. However, phase objects have different optical paths (thickness x refractive index), which cause phase delays on the incident waves. Zernike designed a phase plate converting such phase changes (invisible) to intensity changes (visible). The phase plate produces a quarter wavelength ($\lambda/4$) shift in the direct beam. The direct beam that does not pass through the specimen is $\lambda/4$ out of phase with the diffracted beam that passes though the specimen. The interference between the direct and the diffracted beams imparts contrast in the image.

Interference Microscope: The discovery of the phase-contrast principle promoted the development of other interference microscopes for observing phase objects. The plane polarized light is split into two beams of perpendicular vibrations. The measuring beam passes through the central third of the field of view. The reference beam passes through another third of the field of view. The specimen will retard the measuring beam according to its structure. The two beams are then recombined and interference takes place to impart contrast in the image.

Fluorescence Microscope: Many chemical compounds absorb ultraviolet radiation and emit light of longer wavelength (primary fluorescence). Others do not fluoresce naturally but can be induced to fluoresce after staining with fluorescing dyes (secondary fluorescence). The specimen is excited with a near ultraviolet (365 nm) beam. A filter subsequently removes this UV radiation. The fluorescent structure appears as colored images against a dark background.

Ultraviolet Microscope: Ultraviolet microscopy is frequently confused with fluorescence microscopy. Since the resolving power depends on the wavelength of the illuminating beam, far ultraviolet (254 or 275 nm) is used to improve the resolving power rather than enhance contrast. But it happens that nucleoproteins such as chromosomes strongly absorb UV, giving good contrast. The image is formed through absorption differences within the specimen. The UV image is invisible, however, hence must be recorded on photographic film or on a fluorescent screen. Note that glass systems are adequate for near UV in fluorescence microscopes, but quartz optics are necessary for far UV in ultraviolet microscopes.

X-ray Microscope: X-rays combine the properties of short wavelength and deep penetration. They have been used to produce images in contact microradiography and Point projection microscopy. At present, the resolving power of X-ray microscopes only approaches that of light microscopes due to technical difficulties. One feature of X-ray microscopy is that it deals with specimens too thick or too opaque for other types of microscopy.

Stereozoom Microscope: The stereo microscope is a low-power (~ 100X) microscope that gives three-dimensional images. The impression of depth (relief) arises because the two eyes receive slightly different images of the same object. The stereozoom microscope is a stereo microscope with a zoom system that provides continuously variable magnifications. The variable magnifications are achieved by adjusting the space between lenses, hence changing the focal length, yet the image stays in the same focal plane. It is a product of computer optimization and modern optical materials.

(2) Electron Microscopes [30,31]

When a beam of electrons strikes a specimen, three phenomena occur: firstly, the incident electrons can be transmitted or backscattered; secondly, orbital electrons of atoms in the specimen may be knocked out (secondary electrons, including photoelectrons and Auger electrons); and thirdly, photons may be emitted from atoms in the specimen (visible light and X-rays). All these phenomena have been used to form images of the specimen.

Transmission Electron Microscope (TEM). This is the exact electron counterpart of the transmission light microscope (TLM). Image formation in TLM is due to absorption, but that in TEM is due to scattering. A beam of high-energy electrons intercept a relatively large area (10-10^6, μm^2) of the specimen, which must be thin enough to transmit over 50% of the incident electrons. The emerging electrons are refracted by electromagnetic lenses to form a magnified real image of the specimen on a fluorescent screen or a photographic emulsion. Its maximum useful magnification is about 10^6X, with a resolving power of about 2 Å. The mass thickness of specimen should lie between 2 and 10 $\mu m/cm^2$ (20-100 urn thick if density = 1). A massive specimen must be sliced with an ultramicrotome using a glass or diamond knife. Airborne particles may be placed on a supporting film for direct observation. All TEM specimens must be supported on a very thin film (< 20 nm, Formvar or evaporated carbon), which is, in turn, supported on a thin grid (100 μm thick, 3 mm in diameter) made of copper, gold or platinum. The grid is then mounted on the stage for examination. Specimen preparation for TEM is an art demanding experience and imagination. With an energy-dispersive X-ray attachment, the EPA Cincinnati Lab has used a TEM/SAED/EDX procedure for asbestos analysis.

Scanning Electron Microscope (SEM). A beam of focused electrons of under 50 urn (500 Å) size sweeps over the specimen in rectangular raster synchronous with the luminous spot on a cathode ray tube (CRT). The brightness of the spot on CRT is modulated by signals received from a detector. The detector selectively collects any of the radiation produced by interaction between the incident electrons and the specimen, including secondary electrons, backscattered electrons, Auger electrons, cathodoluminescence, absorbed or induced current, etc. The magnification of SEM can vary from 15X to 100,000X, with resolution about 100-200 Å. Due to its great depth of field and optimum contrast mechanism, the SEM image appears three-dimensional and life-like. Figure 2 contains scanning electron micrographs of fly ash showing the surface structure and epitaxial growth [32]. Sample preparation for SEM examination is minimal. The scanning electron microscope is usually coupled with an energy-dispersive X-ray analyzer to identify chemical elements, and with a transmission stage to produce electron diffraction patterns. Note that the raster scanning mode of operation in "surface" SEM can be applied to TEM, forming a "transmission" SEM or so-called STEM. In general, a surface SEM has a long focal-length lens (1 ~ 2 cm) for imaging with secondary electrons, giving a resolution of about 100 Å; a STEM has a short focal-length lens (typically 0.1.cm) for imaging with transmitted electrons, giving a resolution of about 5 Å or better. The differences between stereoscope, TLM, TEM and SEM are listed in Table IV [33].

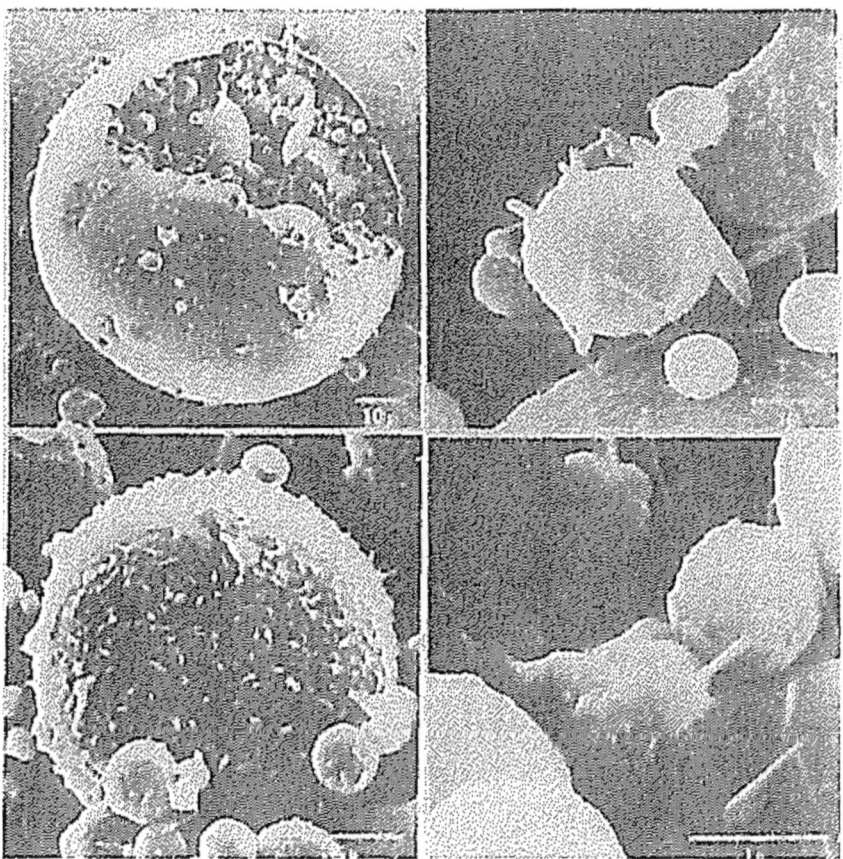

Figure 2 *Scanning electron micrographs of fly ash collected from electrostatic precipitator hopper indicating (A) typical plerosphere and (B-D) variety of observed crystal habits on fine fly ash particles. Reprinted with permission from Environ, Sci. Technol. [32].*

Table IV. Features of Microscopes

	Stereo Light Microscope	Compound Light Microscope	Transmission Electron Microscope	Scanning Electron Microscope
Magnification	8–212X	8–2500X	200–1,000,000X	15–130,000X
Resolution	20,000 Å	2500 Å	2 Å	200 Å
Depth of Field at				
50X	200 μ	20 μ	–	10,000 μ (= 1 cm)
500X	–	2 μ	800 μ	1000 μ
5000X	–	–	80 μ	100 μ
50,000X	–	–	8 μ	10 μ
Working Distance at				
50X	40 mm	16 mm	–	20 mm
500X	–	0.4 mm	2 mm	20 mm
5000X	–	–	2 mm	20 mm
50,000X	–	–	2 mm	20 mm
Field of View at				
50X	4000 μ	2100 μ	–	2000 μ
500X	–	210 μ	200 μ	200 μ
5000X	–	–	20 μ	20 μ
50,000X	–	–	2 μ	2 μ

4.3 Chemical Images

When particles (photons, electrons, ions, etc.) interact with matter, the properties of the incident particles are modified and additional particles may be emitted. Certain physical or chemical information of the matter is impressed on these modulated particles emerging from the interaction. Separating these modulated particles according to their energy and intensity, the information they carry can be sorted out to produce images characteristic of the matter. Morphological images show the physical or topographical features of the matter. Chemical images show the elemental or molecular structure of the matter. Depending on the depth of penetration of the incident-probing particles, the chemical images may represent the surface or bulk composition of the matter (see Table V).

Elemental Images

The electron microprobe [34] is an instrument for X-ray spectrochemical analysis of specimen surfaces as small as 1 μm in diameter. When atoms in a specimen are excited by an electron beam, characteristic X-rays are emitted. The wavelength and intensity of the X-rays of a given element can be mapped on a cathode ray tube forming X-ray images that reveal the distribution of this element on the surface of the specimen. Other instruments capable of producing elemental images include ion microprobe, secondary ion mass spectrometer (SIMS), Auger electron spectrometer and ionoluminescence.

Molecular Images

The Laser Raman microprobe is an instrument for Raman spectral analysis of a specimen surface. It uses a laser beam to excite the molecules in the sample, thus inducing the emission of Raman spectra. Besides identification, the Raman lines have been gathered to produce molecular images. One type of laser Raman microprobe is dubbed MOLE for molecular optics laser examiner. By tuning the instrument to the frequency of a Raman line characteristic of a component, it allows for "mapping" the distribution of the component in a heterogeneous specimen. Other instruments having the potential of producing molecular images include laser microprobe mass spectrometer, secondary ion mass spectrometer and cathodoluminescence.

5. ANALYSIS OF AIRBORNE PARTICLES

An important aspect of environmental science is the chemical analysis of pollutants. Trace amount of contaminants in complex mixtures is a unique feature of pollutant analysis. "Trace amount" demands sensitivity; "complex mixture" suggests likely interference. Therefore, the analytical techniques are hard pressed for better sensitivity/detection limit and specificity/selectivity.

About a dozen reference methods have been published by EPA for the analysis of regulated pollutants, mostly for gases [24,35,36]. As to particulate matter in ambient air, EPA only specifies the high-volume sampler for gross weight of TSP. Nearly 400 procedures for air sampling and analysis are validated and published in the *NIOSH Manual of Analytical Methods* [37]. These EPA and NIOSH methods are used routinely for compliance or control purposes. Numerous other analytical techniques have been applied to the characterization of airborne particles by various research groups. Apparently the trend favors physical (instrumental) methods over the conventional wet-chemical methods, due to speed and sensitivity. These physical methods for chemical analysis can be divided into two groups: inorganic/elemental and organic/molecular. This division is by no means mutually exclusive. In fact, several techniques such as X-ray diffraction, infrared and mass spectroscopy are quite versatile and can be efficient for both. No one can master all these techniques. However, a glimpse of the whole field is always refreshing. The concept and operation of these physical methods are described very briefly below. A treatise on each technique is given in the reference, but for most of us, a clear concept on each technique is probably all we need.

5.1 Techniques for Inorganic/Elemental Pollutants

Colorimetry

The sample is brought into solution through conventional wet chemistry. The solution is treated with a special reagent and thereupon a color (complex, lake or dye) is usually developed. The intensity of the color is measured at a specific wavelength with a spectrophotometer. Beer's law relates the intensity data (absorbance or transmittance) to the concentration of the substance of interest. The colorimetric method is very sensitive. It has been applied to inorganic, organic and biological analyses [38].

Emission Spectrometry

Historically, optical emission spectrometry has been one of the most powerful tools for trace metal analysis. The sample is placed in a flame, electric arc, high-voltage spark, induced plasma, or laser beam. The atomic emission spectrum is recorded. It consists of sharp well-defined lines. The wavelengths of the spectral lines identify the elements; the intensities of the spectral lines relate to their concentrations [39].

X-ray Fluorescence Spectrometry (XRF)

The sample is bombarded with primary X-rays from X-ray tubes (usually with a W or Cr target). All elements in the sample become excited and emit secondary X-rays characteristic to each element. The wavelength and intensity of the X-ray spectra identify and quantify the fluorescing elements in the sample. Elements of atomic number 12 (Mg) and higher can be analyzed by this technique [40,41].

Atomic Absorption Spectrometry (AA)

The sample is brought into solution through wet chemistry. The solution is converted into atomic vapor by an atomizer/burner system. The atomic vapor absorbs the characteristic spectrum from a hollow cathode tube. The absorbance is related to concentration through calibration. A different hollow cathode tube must be used for each different element sought. The atomic absorption method is highly sensitive and usually free from matrix interference. Perfection of flash vaporization directly from solid samples by a laser beam or a plasma jet may expand the scope of this technique [42]

Neutron Activation Analysis (NAA)

The sample is bombarded with a flux of neutrons. If a thermal neutron is captured by a nucleus, the nucleus gains a large excess of energy. This excess energy is promptly emitted as gamma rays and a radioactive isotope of the parent nucleus is produced. Each radioisotope emits several γ-rays of various energies and intensities. Each radioisotope has its own unique decay characteristics. The X-ray spectra or decay constants can be measured to determine the composition of the sample. Sometimes, post-irradiation radiochemical separation is necessary to isolate the isotope of interest from interfering activities. Thermal neutrons are generally used in NAA because their capture favors the most important (n,γ) reaction [43].

Electron Microprobe (EMP)

The sample is bombarded with a focused electron beam (~1 μm diameter). The atoms in the sample are excited and emit their characteristic X-rays. The wavelength and intensity of the characteristic X-rays identify and quantify the elemental composition just like the X-ray fluorescence technique. With the aid of scanning coils, the focused electron beam can be made to sweep across the specimen surface in a raster pattern. Both the electron image and the X-ray image of the specimen surface can be displayed on a cathode ray tube (CRT) and photographed with an oscilloscope camera [44].

Ion Microprobe (IMP)

The sample is bombarded with a focused beam (2 ~ 300 μm in diameter) of ions (Ar^+, F^-, O^-, etc.) generated in a duoplasmatron. The atoms at the surface are ejected by a sputtering process. A small fraction, typically 0.01-1%, of the sputtered atoms is charged, the so-called secondary ions. The secondary ions are attracted into a mass spectrometer and dispersed according to their mass charge ratios. The mass spectrum is then analyzed for elements and isotopes. The early ion microprobes were designed for ion beam machining, scanning ion microscopy or ion implantation without secondary ion analysis. The first instrument designed for analytical applications is called an ion microprobe mass analyzer (IMMA). It is a combination of an ion microprobe with a secondary ion mass spectrometer (SIMS). IMMA detects all elements including hydrogen. It extends the capability of the electron microprobe and overcomes most of its shortcomings [45,46]. A combined ion and electron microprobe analyzer UMPA (Universal microprobe analyzer) was designed by H. Liebl.

Photoelectron Spectrometry (PES, ESCA)

When an atom is irradiated with X-ray photons, typically AlK_α or MgK_α, electrons may be ejected from its inner electron shells. The kinetic energy of the ejected electrons (photoelectrons) can be measured precisely and the binding energy of the electrons calculated accurately. The binding energy of an inner core electron can identify the atom from which the electron was ejected. A small change in the binding energy of a given electron is called "chemical shift" of binding energy, which indicates the oxidation state of the element. This technique has been dubbed as Electron Spectroscopy for Chemical Analysis (ESCA) by its developer K. Siegbahn. When ultraviolet radiation instead of X-rays is used to excite the atom, outer-shell valence electrons are ejected. The ejected photoelectrons are then energy analyzed. The binding energies of valence electrons provide theoretical chemists with a wealth of data on the electronic structure of molecules to verify their molecular orbital theories [47, 48].

Auger Electron Spectroscopy (AES)

When an atom is bombarded with electrons or X-rays, it may eject an inner-shell electron (photoelectron). The remaining electrons of the excited atom undergo rearrangement. During this process, a photon may be emitted (X-ray fluorescence) or, instead, an additional electron may be emitted (Auger electron). An Auger electron actually emerges from a doubly ionized atom. The kinetic energy of the Auger electron is characteristic of the emitting atom independent of the excitation methods. AES is very sensitive to all elements except H and He, both of which do not have core electrons [49]. The Scanning Auger Microprobe (SAM) performs elemental analysis at the surface or in depth. It provides electron micrographs, Auger images and Auger spectra.

Ion Scattering Spectrometry (ISS)

The sample is bombarded with a monoenergetic primary beam of low energy (0.5 ~ 3.0 keV) noble-gas ions ($^3He^+$ $^4He^+$ or $^{20}Ne^+$). ISS is based on the physical principle of elastic binary collisions that every element of mass greater than that of the primary beam causes scattering, giving a peak at a specific energy. The scattered ions are extracted and their energies measured. By contrast to the experiments of higher energy ions, multiply charged ions are not observed in the secondary ion distribution. The energy of the scattered ions identifies the elements on the outermost atomic layer of the sample. Their relative abundance is determined by use of standards [46].

Secondary Ion Mass Spectrometry (SIMS)

The sample is bombarded with a primary beam of high energy ions (1~40 keV, typically and $^{16}O^-$, $^{16}O^+$ and $^{40}Ar^+$). Neutral atoms and secondary ions are sputtered off the specimen surface. The secondary ions are withdrawn and focused into a mass spectrometer then dispersed according to their mass-to-charge ratios. Monoatomic ions identify all elements (including hydrogen) present in the surface of the sample. Polyatomic ions, such as CH_3^+, NH_4^+ and SOH^+, convey information pertaining to the molecular structure of the surface [48].

Both ESCA and AES probe roughly 20 Å or less of the top surface. Both SIMS and ISS probe only the top atomic layer. These four techniques are used primarily for surface analysis [50].

5.2 Techniques for Organic/Molecular Pollutants
Gas Chromatography (GC and GC/MS)

Chromatography is a technique for separating mixtures into their components. The principle of separation is that different molecules have different solubility, adsorption or volatility. There are various types of chromatographic techniques among which gas-liquid chromatography (GC) is most widely used [51].

For GC analysis, the sample (liquid, solution or extract) is injected into a stream of carrier gas (the mobile phase), typically He, Ar or N_2 which carries the sample onto a column. The column contains a thin film of liquid on a powder support (the stationary phase). The components of the mixture flow through the column at varying rates depending on the partition coefficient of each component. Each component is detected as it emerges from the column. Frequently used detectors are flame ionization (FID), electron capture (ECD), nitrogen phosphorus (NPD) and Hall or Coulson detector (HD). The retention time of a compound often permits qualitative identification in simple cases. This qualitative identification usually requires confirmation by IR, NMR or MS techniques. In complex cases, a mass spectrometer is coupled to the gas chromatograph as a detector (GC/MS). The mass spectra are processed and analyzed by a minicomputer. For successful GC separation, the components must have an appreciable vapor pressure at temperatures below 400°C.

Liquid Chromatography (LC)

About 85% of all known chemicals are not sufficiently volatile or stable to be separated by gas chromatography (GC).

Obviously, liquid chromatography is a ready alternative where the mobile phase is a solvent. The mobile solvent provides the analyst with an additional powerful parameter to work with [43].

The sample is injected into a stream of solvent flowing through a column. The column is packed with a polar absorbent (for absorption), an organic liquid bonded chemically to a porous support (for partition), an exchange resin (for ion exchange), or a porous, rigid gel (for size exclusion). The components with affinity for the stationary phase (the packing materials in the column) will move through the column much more slowly than those favoring the mobile phase. Thus, the components emerge from the column in order of their affinity for the stationary phase. Ultraviolet, fluorescence or refractive index detectors are widely used for their detection. The eluting component can be collected for further identification/confirmation by IR, NMR or MS. Thin-layer chromatography (TLC) and gel permeation chromatography (GPC) are versions of liquid chromatography.

Infrared Spectrometry (IR, FT/IR and ATR)

Infrared spectroscopy is unsurpassed in its application to a wide variety of problems and materials involving organic and most inorganic molecules. Solids, liquids and gases can be analyzed.

Insoluble materials can be analyzed by using the KBr or Nujol mull techniques. For such materials the sample is made into a pellet after dispersion in KBr or a film after dispersion in Nujol (mineral oil). The pellet or film is irradiated with infrared light.

Certain frequencies are preferentially absorbed corresponding to the modes of vibration of the molecules in the sample. Each pure compound has a characteristic infrared spectrum. Many functional groups in organic compounds such as carbonyl and hydroxyl absorb at approximately the same frequencies regardless of the molecular structure to which they are attached. These are called group frequencies, which make the infrared technique extremely useful [52,53].

Besides the conventional dispersive spectrometer, infrared spectra also can be obtained with a Fourier Transform spectrometer (FT/IR), which features a rapid scan (a few seconds). For materials difficult to sample, such as rubber, plastic and biological tissues, infrared spectra can be obtained by using the attenuated total reflectance (ATR) technique.

Nuclear Magnetic Resonance Spectrometry (NMR)
The sample is dissolved in a proper solvent. The solution is placed in a magnetic field and irradiated with radiofrequency (rf) power. The nuclei of atoms in a molecule absorb rf radiation and thus change their directions of spin orientation. This interaction between nuclei and rf radiation is called nuclear magnetic resonance. The resonance absorption spectrum is used for structure analysis or compound identification. Hydrogen is the nucleus most often observed. The chemical shift indicates what types of hydrogen are present (methyl, methylene, aromatic, etc.), the spin-spin splitting indicates what groups are next to

each other in the molecule. The relative area of the absorption peaks in the spectrum indicates how many hydrogen atoms are in each group [54].

Besides hydrogen, carbon is the other atom of greatest interest in organic chemistry. However, the signal of the carbon-13 isotope is much weaker than that of the proton. Fourier transform techniques and coherent signal averaging are used to enhance the ^{13}C resonance signal. Recent advances in Fourier Transform ^{13}C NMR (FT/NMR) greatly expanded the scope of this analytical technique.

Mass Spectrometry (MS)

In its most common configuration (electron ionization), the MS sample is bombarded with electrons and breaks into ionized fragments. The ions are accelerated in an electric field, where the ions of different mass attain different velocity. The ions are directed through a magnetic field, where the fast-moving ions take circular paths due to the effect of magnetic force. Under constant electric and magnetic fields, the radius of the circular path of an ion depends on its mass:charge ratio. This is the basis of separation according to mass. Under the same instrumental conditions, a given molecule always breaks into the same fragments. The number and size of these fragments is called the "cracking pattern" of the molecule. It is characteristic of each compound. Mass spectra are usually complicated. A digital computer is normally coupled to handle the data processing [55].

Mass spectrometry has been used for the analysis of elements as well as molecules. Due to its sensitivity (ppb range) and speed (a few seconds), MS is also used as an online detector with other analytical techniques such as gas chromatography, electron diffraction, Auger spectrometry, ion probe and laser probe.

Laser Raman Spectrometry

Since the advent of laser sources, there has been a resurgence of interest in Raman spectrometry. Raman and infrared are two different techniques to obtain the same type of data on the modes of molecular vibrations. Both techniques are complementary to each other. Symmetrical modes of vibration are Raman active. Asymmetrical modes of vibration are infrared active. Many fundamental modes of vibration are active in both but with wide disparity in intensities.

The sample is irradiated with a laser beam (argon, krypton or dye laser). A shift in frequency (Raman effect) of the scattered photons is detected. The wavelength and intensity of the scattered Raman lines are used to identify the components in the sample. The interpretation of Raman spectra is very similar to that of IR spectra. Both make use of the group frequency concept and fingerprint technique. To date, Raman spectrometry suffers from the paucity of a spectral file of known compounds.

Two types of Laser Raman instruments are currently used for micro-analysis: the French Raman Microprobe/Microscope, dubbed MOLE (molecular optical laser examiner) and the NBS Raman Microprobe. Both have been applied to the analysis of pollutants. When the Raman Microprobe is tuned to a Raman line characteristic of a component, an image mapping the distribution of that component in a heterogeneous sample can be displayed [56].

X-ray and Electron Diffraction (XRD and SAED)

There are three diffraction techniques for the analysis of crystalline solids:

(1) **X-ray diffraction,** where the X-rays interact with the atomic electrons, and the coherently scattered X-rays disclose the electron density distribution within the crystal;
(2) **Electron diffraction,** where the electrons interact with the Coulomb field in the atom, and the elastically scattered electrons reveal, the potential distribution within the crystal; and
(3) **Neutron diffraction,** where the neutrons interact primarily with the nuclei of atoms, thus the diffracted neutrons elucidate the positions of atomic nuclei within the crystal.

Electron diffraction should not be used if the analysis can be done by X-ray diffraction, more so for neutron diffraction.

The sample is bombarded with a monochromatic beam of X-rays (electrons or neutrons). The elastically scattered X-rays (electrons or neutrons) produce a pattern (direction and intensity of diffraction lines) that identifies its components [57,58].

The X-ray diffraction technique (XRD) has been indispensable for the analysis of powder samples such as the molecular form of lead in airborne dust. Selected area electron diffraction

(SAED) has been used for the identification of asbestos in airborne dust.

5.3 Techniques for Micro/Surface Analyses

Due to the impact of the drive for environmental control, a host of new analytical techniques have appeared within the last decade. Most of these newer techniques can be applied to trace/micro/surface analyses and many with imaging/mapping capability. Trace analysis deals with concentrations of less than 0.1%, usually in the ppm range. Microanalysis deals with samples of microscopic size on the order of 1 μm^3 or 10^{-12} g. Surface analysis deals with the outermost atomic layer(s) of samples in the order of 1 nm thick. Surface analysis is of particular interest because a number of toxic trace metal and organic species are highly enriched at the surface of airborne particles, and the surface of airborne particles is in direct contact with body fluids following inhalation or ingestion. Two generalizations can be made that most trace elements are all concentrated on the surface of airborne particles and most surface analyses are done with the techniques of micro-analysis. Techniques for micro/surface analyses are summarized in Table V [59]. By ion etching of the surface, a depth-profile analysis can indicate the variation of chemical composition with depth below the original surface. Both surface analysis and depth profiling may be run on a single particle or on a field of particles.

Table V. Micro/Surface Analytical Techniques

Technique	Primary Radiation	Secondary Radiation	Constituents Measured	Lateral Resolution	Depth Resolution	Detection Limit	Typical Accuracy	Image Mapping
1) Electron Microprobe	Electron	X-ray	Elements $Z \geq 11$ (EDS) $Z \geq 4$ (WDS)	1 μm	1 μm	750 ppm (EDS) 100 ppm (WDS)	1-5%	Yes
2) Photoelectron Spectrometry (ESCA)	X-ray or UV	Electron	Elements $Z \geq 3$	10 μm	1 nm	0.1%	5%	No
3) Auger Electron Spectrometry (AES)	Electron	Electron	Elements $Z \geq 3$	0.1-1 μm	1 nm	0.1%	10-25%	Yes
4) Secondary Ion Mass Spectrometry (SIMS)	Ion	Ion	All elements Molecules	1 μm	1 nm	1 ppm	20%	Yes
5) Ion Scattering Spectrometry (ISS)	Ion	Ion	Elements $Z \geq 3$	100 μm	1 atom layer	100 ppm- 0.1%	20%	Yes
6) X-Ray Fluorescence (XRF)	X-ray	X-ray	Elements $Z \geq 11$ (EDS)	1 mm-1 cm	100 μm	1-10 ppm	5%	No
7) Cathodoluminescence	Electron	Photon	Molecules, other	1 μm	1 μm	1-1000 ppm, varies	10%	Yes
8) Laser Raman Microprobe (MOLE)	Photon	Photon	Molecules	1 μm	1 μm	1%	—	Yes
9) Laser Microprobe Mass Analyzer	Photon	Ion	All elements, molecules	1 μm	1-10 μm	1 ppm	5%	No

Unified Theory and Practice | 483

6. CONCLUSIONS

The national average of airborne particles in urban areas is about 100 $\mu g / m^3$. This level may be highly unpleasant in Los Angeles because 95% of the airborne particles there are of respirable size. It may also be offensive in New York because the dust particles there have a high sulfate content (mainly ammonium sulfate). However, this level is innocuous in Denver because 95% of the particles in the Rocky Mountain area are larger than the respirable size and the suspended sulfate is mainly gypsum from soils (calcium sulfate). Obviously, both the size distribution and chemical composition of the airborne particles are intimately related to their health effects.

Airborne particles of less than 2 μm in size are most likely to enter and lodge deep in the lung. They are also largely responsible for making the atmosphere hazy and soupy. Those larger than 2 μm in size have little effect on health and weather since they neither penetrate beyond the nose nor scatter light efficiently.

As to chemical constituents in airborne particles, Hg, Se, Pb, V, As, Cr, Cd and Ni are potent toxic elements; asbestos, nitrosamines and polycyclic aromatic hydrocarbons (PAH) are potentially carcinogenic molecules. These harmful elements and molecules are highly enriched at the surface of fly ash or mineral dust. The major source of these mainly submicron particles and their toxic constituents is combustion in power plants and automobiles. These facts deserve deliberation in forming policy and planning control.

Responding to the demand for faster and more sensitive techniques for pollutant analysis, an array of new techniques have appeared and conventional techniques have been refined within the last decade. The rapid evolution in analytical techniques is still going strong. Although each technique has it unique capabilities, generally XRD, XRF and AA are favored for inorganic/elemental analysis; GC/MS and LC are methods of choice for organic/molecular analysis; and ESCA, SIMS and MOLE are preferred for micro/surface analysis. For pollutant characterization, imaging usually precedes analysis. Polarizing/phase contrast microscopes and scanning electron microscopes are popular for morphological identification [60,61].

Although better techniques are desirable, currently available instruments are adequate for most environmental measurements. To maximize the benefits to health and environmental science, it is critical to design intelligent programs to make efficient use of these imaging and analytical techniques.

7. REFERENCES

ACS Committee on Environmental Improvement. "Clean Our Environment-A Chemical Perspective" (1978).

ASTM. "Methods for Emission Spectrochemical Analysis," Philadelphia, PA (1964).

ASTM. "Quantitative Surface Analysis of Materials," *Spec. Tech. Publ.* 643, Philadelphia, PA (1978).

"ASTM Symposium on X-Ray and Electron-Probe Analysis," *Spec, Tech. Publ.* 349 (1964).

Becker, E. D. *High Resolution NMR* (New York: Academic Press, Inc., 1969).

Benarde, M. A. *Our Precarious Habits* (New York: W. W. Norton Co., 1973).

Berlin, E. P. *Principles and Practice of X-Ray Spectrometric Analysis* (New York: Plenum Publishing Corp., 1970).

Biggins, P. D. E.; and R. M. Harrison. *Environ. Sci. Technol* 13:558 (1979).Dzubay, T. G., and R. K. Stevens. *Environ. Sci. Technol.* 9:663 (1975).

Birk, L. S. *Electron Probe Microanalysis* (New York: Wiley-Interscience, 1963).

Brar, S. S., et al. *J. Geophys. Res.* 71:2939 (1970).

Chang, C. C. *Surface Sci.* 25:53 (1971).

Chung, F. H. *Adv. X-Ray Anal.* 19:181 (1976).

Chung, F. H. *Environ. Sci. Technol.* 12:1208 (1978).

Chung, F. H. *J. Appl. Cryst.* 8:17 (1975).

Colthup, N. B.; L. H. Daly; and S. E. Wiberley. *Introduction to Infrared and Raman Spectroscopy* (New York: Academic Press, Inc., 1964).

Cosslett, V. E. *Modern Microscopy* (London: Bell, 1966).

Dhamelincourt, P., et al. *Anal. Chem.* 51:414A (1979).

Dunlap, R. W.; and B. J. Goldsmith. *Environ. Sci. Technol.* 13:173 (1979).

EPA Health Effect Res. Lab. "Health Effect Considerations for Establishing a Standard for Inhalable Particulate" (July 1978).

Federal Register 36:15486 (August 14, 1971).

Federal Register 40:7042 (February 18, 1975).

Fine, D. H., et al. *Environ. Sci. Technol.* 11:581 (1977).

Fisher, G. L., et al. *Environ. Sci. Technol.* 12:447 (1978).

Fox, D. L.; and H. E. Jeffries. *Anal. Chem.* 51:23R (1979).

Gray, P., Ed. *The Encyclopedia of Microscopy and Microtechnique* (New York: Van Nostrand Reinhold Co., 1973).

Grosjean, D., et al. *Environ. Sci. Technol.* 12:313 (1978).

Hay, W. H.; and P. A. Sandberg. *Micropal.* 13:407 (1967).

Junge, C. E. *J. Geophys. Res.* 77:5183 (1972).

Heinrich, K. F. J.; and D. E. Newbury, Eds. "Secondary Ion Mass Spectrometry," *NBS Special Publ.* 427 (1975).

Hoffman, A. J., et al. *Science* 190:243 (1975).

Klug, H. P.; and L. E. Alexander. *X-Ray Diffraction Procedures for Polycrystalline and Amorphous Materials*, 2nd ed. (New York: Wiley-Interscience, 1974).

Lederberg, J. "Foreword" in *The Mutagenicity of Pesticides*, S. S. Epstein and M. S. Legator, Eds. (Cambridge, MA: The M.I.T. Press, 1971).

Liebel, H. *Anal. Chem.* 46:22A (1974).

Malissa, H., Ed. *Analysis of Airborne Particles by Physical Methods* (Cleveland, OH: CRC Press, 1979).

Martin, L. C. *The Theory of the Microscope* (New York: Elsevier North-Holland, Inc., 1966).

McCrone, W. C.; and J. G. Deily. *The Particle Atlas* (Ann Arbor, MI: Ann Arbor Science Publishers, Inc., 1973).

McLafferty, F. W. *Interpretation of Mass Spectra* (Menlo Park, CA: W. A. Benjamin, Inc., 1966).

McNair, H. M.; and E. J. Bonelli. *Basic Gas Chromatography* (Berkeley, CA: Consolidated Printer, 1969).

Miller, J. A. *Cancer Res.* 30:559 (1970).

Miller, M. S.; S. K. Friedlander; and G. M. Hidy. In: *A Chemical Element Balance for the Pasadena Aerosol, Proceedings of the American Chemical Society*, G. M. Hidy, Ed., Los Angeles, CA (1971).

Miller, S. S. *Environ. Sci. Technol.* 12:1355 (1978).

Moyers, J. L.; L. E. Ranweiler; S. B. Hopf; and N. E. Korte. *Environ. Sci. Technol.* 11:789 (1977).

Newburg, D. E. "Microanalysis in the SEM-Progress and Prospects in Scanning Electron Microscopy" (1979).

NIOSH. NIOSH *Manual of Analytical Methods*, Vol. 1-4, U.S. Government Printing Office, Washington, D.C. (1 977, 1978).

NRC Subcommittee on Airborne Particles. *Airborne Particles* (Baltimore: University Park Press, 1979).

Oatley, C. W. *The Scanning Electron Microscope* (London: Cambridge University Press, 1972).

Paciga, J. J.; and R. E. Jervis. *Environ. Sci. Technol.* 10:1124 (1976).

Plocchini, R. G., et al. *Environ. Sci. Technol.* 10:76 (1976).

Richardson, J. H.; and R. V. Peterson. *Systematic Materials Analysis* (New York: Academic Press, Inc., 1974).

Robinson, J. W. *Atomic Absorption Spectrometry*, 2nd ed. (New York: Marcel Dekker, Inc., 1975).

Sandell, E. B. Calorimetric Detection of Trace Metals (New

York: Wiley-Interscience, 1959).

Sawiki, E. *Proc. 2nd Hanover Int. Carcinogenesis Meeting, Air Pollution and Cancer in Man*, Lyon, France (1977).

Siegbahn, K., et al. *Electron Spectroscopy for Chemical Analysis* (Uppsala, Sweden: Almquist and Wiksells, 1967).

Steele, D. *The Interpretation of Vibrational Spectra* (New York: Barnes and Noble, Inc., 1971).

Train, R. E. *Science* 195:443 (1977).

U.S. Dept. of Health, Education and Welfare. "Air Quality Criteria for Particulate Matter" (1969), p. 16.

Wagner, C. D. *Electron Spectroscopy* (Amsterdam: North Holland Publ., 1972).

Whitby, K. T.; and B. Cantrell. "Atmospheric Aerosols-Characteristics and Measurements," in *Proc. Int. Conf. on Environmental Sensing and Assessment*, Las Vegas, NV, 1975 (New York: Institute of Electrical and Electronics Engineers, 1976).

Whitby, K. T.; R. B. Husar; and B. Y. H. Liu. *J. Colloid Interface Sci.* 39:177 (1972).

Willeke, K.; and K. T. Whitby. *J. Air Poll. Control Assoc.* 25:529 (1975).

APPENDICES

Appendix 1. Chapter Titles vs. Published Articles

Each chapter of this book has been peer reviewed and published in well-known international journals or books. The titles of the original articles may be slightly revised to fit the style and format of books. For easy crosscheck, the sequence of the following twenty original articles matches that of the twenty chapters in this book. Each chapter/article has its own references.

1. Chung, F. H. *Unified Theory and Guidelines on Polymer Adhesion.* Journal of Applied Polymer Science, Vol. 42, pp. 1319 - 1331 (1991).
2. Chung, F. H. "Unified Theory and Practice of Quantitative X-ray Diffraction and X-ray Fluorescence," an Invited Speech for the 2018 Denver X-ray Conference. Full article will be published in the Proceedings of "Advances in X-ray Analysis," Vol. 61.
3. Chung, F. H. *Quantitative Interpretation of X-ray Diffraction Patterns of Mixtures. Part i. Matrix Flushing Techniques.* Journal of Applied Crystallography, Vol. 7, pp. 519 - 525 (1974).
4. Chung, F. H. *Quantitative Interpretation of X-ray Diffraction Patterns of Mixtures. Part II. Adiabatic Principle.* Journal of Applied Crystallography, Vol. 7, pp. 526 - 531 (1974).
5. Chung, F. H. *Quantitative Interpretation of X-ray Diffraction*

Patterns of Mixtures. Part Simultaneous Determination of Reference Intensities. Journal of Applied Crystallography, Vol. 8, pp. 17 - 19 (1975).

6. Chung, F. H. *Unified Theory for Decoding the Signals from X-ray Fluorescence of Mixtures.* Applied Spectroscopy, Vol. 71 (5), pp. 1060 - 1068 (2017).

7. Chung, F. H.; A. J. Lentz; and R. W. Scott. *A Versatile Thin Film Method for Quantitative X-ray Emission Analysis.* X-ray Spectrometry, Vol. 3, pp. 172 - 175 (1974).

8. Chung, F. H. *A New X-ray Diffraction Method for Quantitative Multi-component Analysis.* Advances in X-ray Analysis, Vol. 17, pp. 106 - 115 (1974).

9. Chung, F. H. *Industrial Applications of X-ray Diffraction.* American Laboratory, February, Vol. 21, pp. 144 -156 (1989).

10. Chung, F. H. "X-ray Diffraction Techniques and Instrumentation," a chapter in *Analytical Measurements and Instrumentation for Process and Pollution Control."* Edited by P. N. Cheremisnoff and H. J. Perlis. Ann Arbor Science, pp.151 - 173 (1981).

11. Chung, F. H.; and D. K. Smith. "Progress and Potential of X-ray Diffraction," a chapter in *Industrial Applications of X-ray Diffraction.* New York: Marcel Dekker, pp. 37 - 54 (2000).

12. Chung, F. H.; and D. K. Smith. "The Principle of X-ray Diffraction Analysis," a chapter in *Industrial Applications of X-ray Diffraction."* New York: Marcel Dekker, pp. 3 - 11

(2000).

13. Chung, F. H.; and D. K. Smith. "The Practice of X-ray Diffraction Analysis," a chapter in *Industrial Applications of X-ray Diffraction."* New York: Marcel Dekker, pp. 13 - 35 (2000).

14. Chung, F. H. "Polymers and Pigments in Paint Industry," a chapter in *Industrial Applications of X-ray Diffraction."* New York: Marcel Dekker, pp. 511 - 525 (2000).

15. Chung, F. H.; and R. W. Scott. *A New Approach to the Determination of Crystallinity of Polymers.* Journal of Applied Crystallography, Vol. 6, pp. 225 - 230 (1973).

16. Chung, F. H. *Vacuum Sublimation and Crystallography of Quinacridone Pigments.* Journal of Crystallography, Vol. 4, pp. 506 - 511 (1971).

17. Chung, F. H. *Crystallography of Toluidine Red Pigment.* Journal of Applied Crystallography, Vol. 4, pp. 79 - 80 (1971).

18. Chung, F. H. *Synthesis and Analysis of Crystalline Silica: Quartz, Cristobalite and Tridymite.* Environmental Science and Technology, Vol. 16, No. 11, pp. 796 - 799 (1982).

19. Chung, F. H. *Imaging and Analysis of Airborne Dust for Silica.* Environmental Science and Technology, Vol. 12, pp. 1208 - 1210 (1978).

20. Chung, F. H. "Imaging and Analysis of Airborne Particulates," a chapter in *Air/ Particulate Instrumentation and Analysis."* Edited by P. N. Cheremisinoff. Ann Arbor Science, pp. 89 –117 (1981).

Appendix 2.

Professional Activities

As the better half of Frank Chung, I play multiple roles. One role is to coordinate and remind his professional activities. Over the years, I accumulated a lot of memos and letters. I found som of these documents are worthy to save as milestones of his professional life. A short list of these documents is compiled here as souvenirs of his career.

Doris Chung

1. MIT textbooks on XRD include the book on Industrial Applications by Chung & Smith.
2. Industrial Applications of X-ray Diffraction, Book by Chung & Smith (2000).
3. Photo of Chung, Smith, Patricia and the lovely collies, near Chung's home in Chicago.
4. Matrix Flushing Theory in Contents of "Quantitative XRD Analysis" by Prof. B. Davis.
5. ICDD Education, XRD & XRF Technical Programs (2009), teaches Matrix Flushing.
6. LABCON Exposition and Symposium, Technical programs.
7. Letter from Prof. Henry Chessin, State University of New

York at Albany.

8. Letter from Prof. G. W. Brindley, Pennsylvania State University.
9. Letter from E. L. Bahn, Jr., Southeast Missouri State University.
10. Letter from Prof. Guy Mattson, University of Central Florida.
11. Letter from Prof. Lewis Cohen, University of California, Riverside.
12. Letter from Prof. Choh Hao Li, University of California, San Francisco.
13. Letter from Prof. Claude A. Lucchesi, Chairman, Northwestern University, Chicago.
14. Letter from Prof. Briant L. Davis, Head, Institute of Atmospheric Science, S. Dakota.
15. Letter from Dr. R. D. Heidenreich, Bell Laboratories, Murray Hill, New Jersey.
16. Letter from Dr. H. K. Herglotz, Du Pont, Wilmington, Delaware.
17. Letter from Dr. S. Y. Hobbs, General Electric Company, New York.
18. Letter from Dr. Michel Sotton, Institute Textile de France, Paris, France.
19. Letter from Dr. B. A. Bellamy, Harwell, Oxford shire, England.
20. Letter from Lisa Honski, Editor, Marcel Dekker, Inc. New York, NY.
21. Letter from Susan G. Farmer, Editor, Technomic Pub-

lishing Co., Lancaster, PA.
22. Letter from Int'l Union of Crystallography, Satellite Meeting, Australia (1987)
23. Letter from Int'l Union of Crystallography, Congress and General Assembly, China (1993).
24. Letter from United Nations, TOKTEN Consultancy to Beijing, China (1985).
25. Denver X-ray Conference, Invited Speaker, Denver, Colorado, U.S.A. (2018).

www.ingramcontent.com/pod-product-compliance
Lightning Source LLC
Chambersburg PA
CBHW061502180526
45171CB00001B/3